DYNAMICS IN ENGINEERING

Classical yet Modern, General yet Comprehensive, Original in Approach

ELLIS HORWOOD SERIES IN MECHANICAL ENGINEERING

Series Editor: J. M. ALEXANDER, formerly Professor of Applied Mechanics, Imperial College of Science and Technology, University of London, *and* Stocker Visiting Professor of Engineering and Technology, Ohio University, Athens, USA

The series has two objectives: of satisfying the requirements of postgraduate and mid-career engineers, and of providing clear and modern texts for more basic undergraduate topics. It is also the intention to include English translations of outstanding texts from other languages, introducing works of international merit. Ideas for enlarging the series are always welcomed.

Alexander, J.M.	**Strength of Materials: Vol. 1: Fundamentals; Vol. 2: Applications**
Alexander, J.M., Brewer, R.C. & Rowe, G.	**Manufacturing Technology Volume 1: Engineering Materials**
Alexander, J.M., Brewer, R.C. & Rowe, G.	**Manufacturing Technology Volume 2: Engineering Processes**
Atkins A.G. & Mai, Y.W.	**Elastic and Plastic Fracture**
Beards, C.	**Vibration Analysis and Control System Dynamics**
Beards, C.	**Structural Vibration Analysis**
Beards, C.	**Noise Control**
Besant, C.B. & C.W.K. Lui	**Computer-aided Design and Manufacture, 3rd Edition**
Borkowski, J. and Szymanski, A.	**Technology of Abrasives and Abrasive Tools**
Borkowski, J. and Szymanski, A.	**Uses of Abrasives and Abrasive Tools**
Brook, R. and Howard, I.C.	**Introductory Fracture Mechanics**
Cameron, A.	**Basic Lubrication Theory, 3rd Edition**
Collar, A.R. & Simpson, A.	**Matrices and Engineering Dynamics**
Cookson, R.A. & El-Zafrany, A.	**Finite Element Techniques for Engineering Analysis**
Cookson, R.A. & El-Zafrany, A.	**Techniques of the Boundary Element Method**
Ding, Q.L. & Davies, B.J.	**Surface Engineering Geometry for Computer-aided Design and Manufacture**
Dowling, A.P. & Ffowcs-Williams, J. E.	**Sound and Sources of Sound**
Edmunds, H.G.	**Mechanical Foundations of Engineering Science**
Fenner, D.N.	**Engineering Stress Analysis**
Fenner, R.T.	**Engineering Elasticity**
Ford, Sir Hugh, FRS, & Alexander, J.M.	**Advanced Mechanics of Materials, 2nd Edition**
Gallagher, C.C. & Knight, W.A.	**Group Technology Production Methods in Manufacture**
Gohar, R.	**Elastohydrodynamics**
Gosman, B.E., Launder, A.D. & Reece, G.	**Computer-aided Engineering: Heat Transfer and Fluid Flow**
Haddad, S.D. & Watson, N.	**Principles and Performance in Diesel Engineering**
Haddad, S.D. & Watson, N.	**Design and Applications in Diesel Engineering**
Haddad, S.D.	**Advanced Operational Diesel Engineering**
Hunt, S.E.	**Nuclear Physics for Engineers and Scientists**
Irons, B.M. & Ahmad, S.	**Techniques of Finite Elements**
Irons, B.M. & Shrive, N.	**Finite Element Primer**
Johnson, W. & Mellor, P.B.	**Engineering Plasticity**
Juhasz, A.Z. and Opoczky, L.	**Mechanical Activation of Silicates by Grinding**
Kleiber, M.	**Incremental Finite Element Modelling in Non-linear Solid Mechanics**
Kleiber, M. & Breitkopf, P.	**Finite Element Methods in Structural Engineering: Turbo Pascal Programs for Microcomputers**
Leech, D.J. & Turner, B.T.	**Engineering Design for Profit**
Lewins, J.D.	**Engineering Thermodynamics**
Malkin, S.	**Materials Grinding: Theory and Applications**
McCloy, D. & Martin, H.R.	**Control of Fluid Power: Analysis and Design, 2nd (Revised) Edition**
Osyczka, A.	**Multicriterion Optimisation in Engineering**
Oxley, P.	**The Mechanics of Machining**
Piszcek, K. and Niziol, J.	**Random Vibration of Mechanical Systems**
Polanski, S.	**Bulk Containers: Design and Engineering of Surfaces and Shapes**
Prentis, J.M.	**Dynamics of Mechanical Systems, 2nd Edition**
Renton, J.D.	**Applied Elasticity**
Richards, T.H.	**Energy Methods in Vibration Analysis**
Ross, C.T.F.	**Computational Methods in Structural and Continuum Mechanics**
Ross, C.T.F.	**Finite Element Programs for Axisymmetric Problems in Engineering**
Ross, C.T.F.	**Finite Element Methods in Structural Mechanics**
Ross, C.T.F.	**Applied Stress Analysis**
Ross, C.T.F.	**Advanced Applied Stress Analysis**
Roznowski, T.	**Moving Heat Sources in Thermoelasticity**
Sawczuk, A.	**Mechanics and Plasticity of Structures**
Sherwin, K.	**Engineering Design for Performance**
Szczepinski, W. & Szlagowski, J.	**Plastic Design of Complex Shape Structured Elements**
Thring, M.W.	**Robots and Telechirs**
Walshaw, A.C.	**Mechanical Vibrations with Applications**
Williams, J.G.	**Fracture Mechanics of Polymers**
Williams, J.G.	**Stress Analysis of Polymers 2nd (Revised) Edition**

DYNAMICS IN ENGINEERING

Classical yet Modern
General yet Comprehensive,
Original in Approach

J. C. MALTBAEK, M.Sc., Ph.D.
Senior Lecturer, School of Engineering
University of Exeter

ELLIS HORWOOD LIMITED
Publishers · Chichester

Halsted Press: a division of
JOHN WILEY & SONS
New York · Chichester · Brisbane · Toronto

First published in 1988 by
ELLIS HORWOOD LIMITED
Market Cross House, Cooper Street,
Chichester, West Sussex, PO19 1EB, England
The publisher's colophon is reproduced from James Gillison's drawing of the ancient Market Cross, Chichester.

Distributors:
Australia and New Zealand:
JACARANDA WILEY LIMITED
GPO Box 859, Brisbane, Queensland 4001, Australia
Canada:
JOHN WILEY & SONS CANADA LIMITED
22 Worcester Road, Rexdale, Ontario, Canada
Europe and Africa:
JOHN WILEY & SONS LIMITED
Baffins Lane, Chichester, West Sussex, England
North and South America and the rest of the world:
Halsted Press: a division of
JOHN WILEY & SONS
605 Third Avenue, New York, NY 10158, USA
South-East Asia
JOHN WILEY & SONS (SEA) PTE LIMITED
37 Jalan Pemimpin # 05–04
Block B, Union Industrial Building, Singapore 2057
Indian Subcontinent
WILEY EASTERN LIMITED
4835/24 Ansari Road
Daryaganj, New Delhi 110002, India

British Library Cataloguing in Publication Data
Maltbaek, J.C.
Dynamics in engineering
1. Engineering systems. Dynamics. Analysis
I. Title
620.7′2

Library of Congress Card No. 88–8865

ISBN 0–7458–0435–7 (Ellis Horwood Limited — Library Edn.)
ISBN 0–7458–0548–5 (Ellis Horwood Limited — Student Edn.)
ISBN 0–470–21125–3 (Halsted Press)

Phototypeset in Times by Ellis Horwood Limited
Printed in Great Britain by Hartnolls, Bodmin

Contents

To my Parents

Preface

The basic subject of dynamics is taught in many different types of courses of varying length. A textbook for one particular course would, therefore, not be very useful for other courses. The present book has been written to give the general background material for a two-year course in classical dynamics for engineers. I have included what I consider to be essential material that should form the basic knowledge of the subject for any graduate engineer in mechanical, civil, or electrical engineering, or engineering science.

The first seven chapters deal with the main parts of the subject, including sufficient material in kinematics for a first course in that subject.

Vector methods have been used wherever their application offers an advantage. In general I have used the most useful mathematics in each particular situation.

Chapters 1–8 assume a knowledge of elementary statics including force systems and resultant forces and moments, equilibrium conditions, etc. The analytical background is the usual differential and integral calculus up to and including the solution of second-order differential equations.

Vector algebra, differentiation and integration and the vector dot and cross-products are used throughout the first eight chapters.

Chapter 3 deals with Lagrange's equations. The time is long overdue for the inclusion of this great method in any university course on dynamics. A knowledge of functions of several variables and partial differentiation is necessary for this chapter. For students on more elementary courses, where Lagrange's equations are not considered, Chapter 3 may be omitted without disturbance to the remainder of the book.

Because of the importance of a special class of dynamics, that is mechanical vibrations, a chapter has been included in this field.

I hope that the book will be found useful for students on different courses at universities, at polytechnics, and — in parts — at technical colleges, both for the actual courses and for home study.

I would like to express my appreciation and thanks to my wife Kathleen for typing the original manuscript.

Exeter, January 1988 J. C. Maltbaek

1

Kinematics of a point

The most elementary type of motion is rest. The science of bodies at rest, their equilibrium and forces acting between them is called statics. Statics is the oldest of the engineering sciences; important contributions were made by Archimedes (287–212 BC), who discovered the principle of the lever and scale balance; the modern science of statics started, perhaps, with the discovery of the law of the parallelogram of forces by Stevin in 1586.

Dynamics is the science of motion of bodies and the forces acting during the motion. This science started much later than statics, with the work by Galileo (1564–1642) on falling bodies. The late start of dynamics was due to the fact that accurate measurements of time are involved in dynamics experiments. In Archimedes' time, the necessary measurements were sometimes attempted by measuring a person's heartbeat, which was, of course, highly unreliable. A step forward was taken by Leonardo da Vinci (1452–1519), who used a leaking water tap as a time-measuring instrument, the time interval between the falling drops being considered constant. Real precision time-measuring devices did not appear until the first pendulum clock invented by Huyghens in 1657, and the balance-wheel watch by Hooke somewhat later.

The scientific basis of dynamics is provided by Newton's laws, published in 1687. The first uses of dynamics were in astronomy, and the new science was hardly used by engineers until the last decades of the last century, the machinery before then being so slow moving that it could be designed by using the principles of statics. With the invention of the steam turbine, the internal combustion engine and the electric motor, a great increase in speed took place, which created design problems that could only be solved by the application of the principles of dynamics.

In some problems only the motion itself, or the geometry of the motion without consideration of the forces acting, is of importance; this branch of dynamics is called *kinematics*. The branch of dynamics dealing with the forces acting during the motion is called *kinetics*. It is necessary to study kinematics in some detail before any useful work can be done in dynamics, because kinematical concepts appear in the basic laws of dynamics.

The only units involved in kinematics are the two basic units of length (metre) and time (seconds); these units are defined in the Appendix.

1.1 RECTILINEAR MOTION

One of the most useful and important types of motion of a point is motion in a straight line, and since this is also the most elementary type of motion, we will consider the kinematics of rectilinear motion first.

1.1.1 Displacement, velocity and acceleration

To describe the rectilinear motion of a point P, we take the line of motion, the *path*, as the x-axis (Fig. 1.1). If the distance x of P from a fixed point O on the x-axis is given

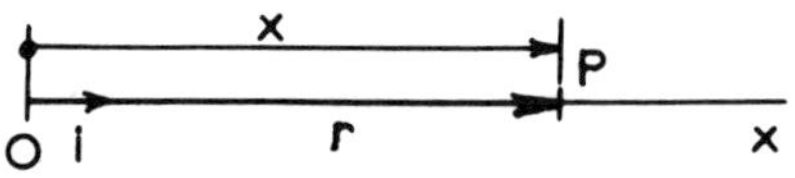

Fig. 1.1.

as a function of time t by $x = f(t)$, the position of P may be determined at any instant of time.

The distance x is measured in metres (m), and the time t in seconds (s).

The distance x is called the *displacement* of P at the time t, and $x = f(t)$ is called the displacement–time equation.

The displacement, velocity and acceleration are all vector quantities, but in rectilinear motion they are all directed along the line of motion and we need to consider only the scalar magnitudes. During a time element Δt from the time t to $t + \Delta t$, the point moves through a distance Δx; we define now the *average speed* during the time Δt by the ratio $\Delta x/\Delta t$. Letting $\Delta t \to 0$, we obtain the *instantaneous speed* of P as

$$V = \lim_{\Delta t \to 0} \frac{\Delta x}{\Delta t} = \frac{dx}{dt} = \dot{x}$$

The speed is considered positive if x is increasing with time and negative if x diminishes with time. The unit for speed is the ratio of the units for displacement and time, that is m/s.

If the point P moves along the x-axis with a variable speed $V = F(t)$, the speed is changing with time, and we say that P has an *acceleration*. In the time element Δt the speed has changed by an amount ΔV; the *average* acceleration in the time interval is defined by the ratio $\Delta V/\Delta t$; the *instantaneous* acceleration at time t is defined by

$$a = \lim_{\Delta t \to 0} \frac{\Delta V}{\Delta t} = \frac{dV}{dt} = \dot{V} = \frac{d^2x}{dt^2} = \ddot{x}$$

The acceleration is taken as positive if ΔV is positive in the interval Δt, and negative if ΔV is negative in this interval. A negative acceleration is sometimes called a *deceleration*.

The units for acceleration are metres per second per second, or m/s^2.

Example 1.1

A point P is moving in rectilinear motion with *constant acceleration a*; if the displacement is x_0 and the velocity V_0 when $t = 0$, determine the velocity–time and displacement–time functions. Determine also the displacement as a function of the velocity and time, and the velocity as a function of the displacement.

Solution

Taking the x-axis in the direction of motion, we have

$$\ddot{x} = a(\text{constant})$$

Integrating gives

$$\dot{x} = V = \int a \, dt + C_1 = at + V_0$$

using the condition that $V = V_0$ when $t = 0$.

$$x = \tfrac{1}{2}at^2 + V_0 t + x_0$$

since $x = x_0$ when $t = 0$.

The last two equations may be combined to give

$$x_2 - x_1 = \tfrac{1}{2}(V_2 + V_1)\,(t_2 - t_1)$$

so that the distance moved in a certain time is the *average* velocity during that time multiplied by the length of time.

We find also that

$$V_2^2 - V_1^2 = (V_2 + V_1)\,(V_2 - V_1) = 2\frac{(V_2 + V_1)}{2}\,a(t_2 - t_1) = 2a(x_2 - x_1)$$

valid for constant acceleration a.

1.2 ROTATION OF A RADIAL LINE IN A PLANE

If a radial line (Fig. 1.2) is rotating about a point O in the plane, the position of the

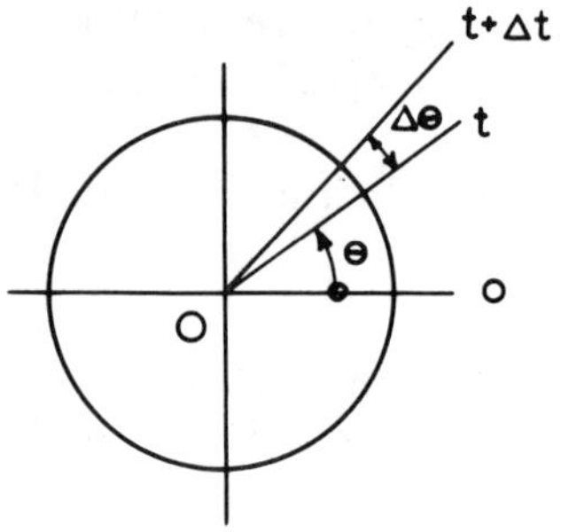

Fig. 1.2.

line at any time t may be given by the angle $\theta = f(t)$ between the rotating line and a line with a fixed direction in the plane. The position angle θ is measured in radians (rad).

If in a time element Δt the position changes by $\Delta\theta$, the average *angular velocity* of the line in the time Δt is defined as the ratio $\Delta\theta/\Delta t$; the *instantaneous angular velocity* is defined as

$$\omega = \lim_{\Delta t \to 0} \frac{\Delta\theta}{\Delta t} = \frac{d\theta}{dt} = \dot{\theta}$$

The angular velocity is measured in radians per second (rad/s). If, in the time element Δt, the angular velocity changes by an amount $\Delta\omega$, we say that the line has an average *angular acceleration* $\Delta\omega/\Delta t$; the *instantaneous angular acceleration* is defined by

$$\lim_{\Delta t \to 0} \frac{\Delta\omega}{\Delta t} = \frac{d\omega}{dt} = \dot{\omega} = \ddot{\theta}$$

The units for angular acceleration are radians per second per second (rad/s^2). If the angular velocity ω is *constant*, then $\dot{\omega} = 0$.

Example 1.2

A radial line is rotating in a plane about a point O with *constant angular acceleration* $\ddot{\theta} = \dot{\omega}$. Determine the displacement–time and angular velocity–time functions, given $\theta = \theta_0$ and $\omega = \omega_0$ when $t = 0$. Determine also θ as a function of ω and t, and ω as a function of θ.

Solution

We have

$$\ddot{\theta} = \dot{\omega} = \text{constant}$$

Integrating gives

$$\omega = \dot{\theta} = \dot{\omega}t + \omega_0$$

Integrating again gives

$$\theta = \tfrac{1}{2}\dot{\omega}t^2 + \omega_0 t + \theta_0$$

The last two equations may be combined to give

$$\theta_2 - \theta_1 = \tfrac{1}{2}(\omega_2 + \omega_1)(t_2 - t_1)$$

The rotational displacement in a time $t_2 - t_1$ may then be found by using the *average* angular velocity in the time interval.

We may also state that

$$\omega_2^2 - \omega_1^2 = (\omega_2 + \omega_1)(\omega_2 - \omega_1) = \frac{2(\omega_2 + \omega_1)}{2}\dot{\omega}(t_2 - t_1) = 2\dot{\omega}(\theta_2 - \theta_1)$$

for a constant angular acceleration $\dot{\omega}$.

Example 1.3

Fig. 1.3 shows a slider–crank mechanism, widely used in internal combustion

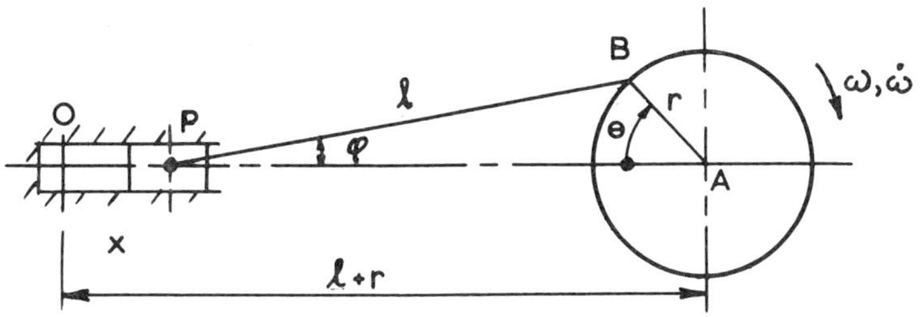

Fig. 1.3.

engines. The radial arm AB, the crank of length r, rotates in the plane about the fixed point A, and the angular position of the crank is given by the angle $\theta = f(t)$ as shown. The crank is pin-connected to a rod BP which is called the connecting rod; BP is of length l. The rod BP is pin-connected at P to a piston, which slides in a fixed cylinder; the motion of the piston is then rectilinear.

Determine the exact expressions for the piston displacement, velocity and acceleration.

Solution

From the geometry of the slider–crank mechanism in Fig. 1.3, the displacement of the piston is determined by $x = r\cos\theta + l\cos\phi$. From the relationship $r^2\sin^2\theta + l^2\cos^2\phi = l^2$ we find $\cos^2\phi = 1 - (r/l)^2\sin^2\theta$; introducing the ratio $n = r/l$ we have $\cos\phi = (1 - n^2\sin^2\theta)^{\frac{1}{2}}$, and introducing *the auxiliary function*

$$\beta = (1 - n^2\sin^2\theta)^{\frac{1}{2}}$$

we find $x = r(\cos\theta + l\beta/r)$, or *the piston displacement*

$$x = r\left(\cos\theta + \frac{\beta}{n}\right)$$

Differentiating the functions β and x with respect to time and introducing the instantaneous angular velocity $\omega = \dot{\theta}$ of the crank arm, there results

$$\dot{\beta} = -\frac{n^2}{2}\beta^{-1}\omega\sin 2\theta$$

and *the piston velocity*

$$\dot{x} = -r\omega(\sin\theta + \frac{n}{2}\beta^{-1}\sin 2\theta)$$

Differentiating $\dot{x}$ with respect to time and introducing the angular acceleration $\dot{\omega} = \ddot{\theta}$ of the crank arm, *the piston acceleration is*

$$\ddot{x} = -r\omega^2\left[\cos\theta + n\beta^{-1}\cos 2\theta + \frac{n^3}{4}\beta^{-3}\sin^2 2\theta\right]$$

$$-r\dot{\omega}\left[\sin\theta + \frac{n}{2}\beta^{-1}\sin 2\theta\right]$$

1.3 GRAPHICAL KINEMATICS. MOTION CURVES

It is sometimes convenient and illustrative to construct motion curves for rectilinear motion, especially in experiments where a series of values of displacements, velocities or accelerations is obtained.

Fig. 1.4 shows a displacement–time curve $x = f(t)$. The slope of the tangent is $dx/dt = V$; the speed at any particular position or time may then be obtained from the curve by graphical differentiation.

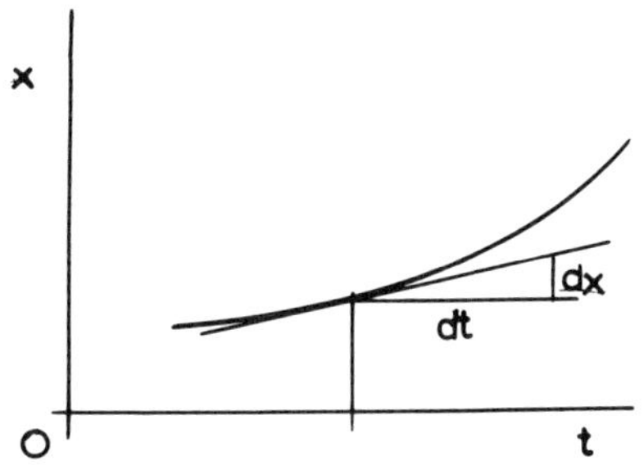

Fig. 1.4.

A velocity–time curve is shown in Fig. 1.5. The slope of the tangent is $\mathrm{d}V/\mathrm{d}t = a$; the acceleration at any time may thus be found by graphical differentiation. The area under the curve between time values t_1 and t_2 is

$$\int_{t_1}^{t_2} V\,\mathrm{d}t = \int_{t_1}^{t_2} \frac{\mathrm{d}x}{\mathrm{d}t}\,\mathrm{d}t = x_2 - x_1$$

The area then gives the distance moved in the time interval $t_2 - t_1$; if the function $V = h(t)$ can be established, the integration may sometimes be performed analytically, otherwise the area may be found by using a planimeter or by other means of graphical integration.

Fig. 1.6 shows the acceleration–time diagram. The slope is of no particular interest, but the area under the curve from time value t_1 to t_2 is

$$\int_{t_1}^{t_2} a\,\mathrm{d}t = \int_{t_1}^{t_2} \frac{\mathrm{d}V}{\mathrm{d}t}\,\mathrm{d}t = V_2 - V_1$$

or the difference in speed between the two times or positions.

Another curve that is sometimes useful is the acceleration–displacement curve shown in Fig. 1.7. The area under the curve from x_1 to x_2 is

$$\int_{x_1}^{x_2} a\,\mathrm{d}x = \int_{x_1}^{x_2} \mathrm{d}V\frac{\mathrm{d}x}{\mathrm{d}t} = \int_{V_1}^{V_2} V\,\mathrm{d}V = \tfrac{1}{2}(V_2^2 - V_1^2)$$

If the velocity is known at one position, it may be determined at another position by graphical integration of the $a - x$ curve, or by using a planimeter.

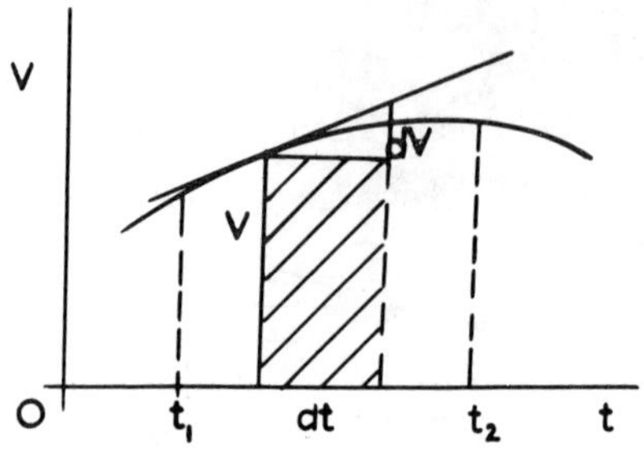

Fig. 1.5.

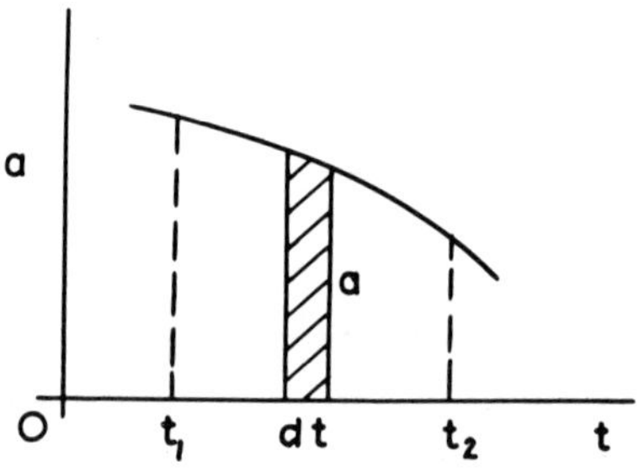

Fig. 1.6.

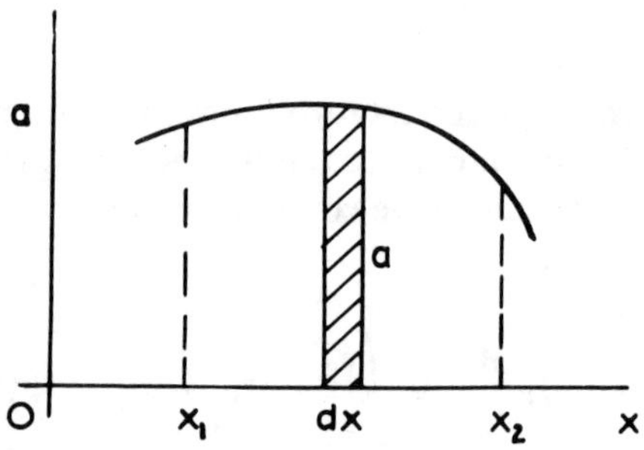

Fig. 1.7.

1.4 CURVILINEAR MOTION

1.4.1 Displacement, velocity and acceleration in curvilinear motion

The motion of a point P in a coordinate system (x, y) (Fig. 1.8) is completely defined

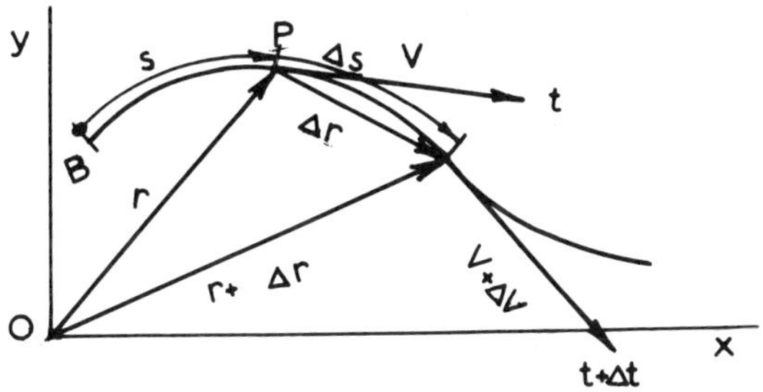

Fig. 1.8.

if $x = f_1(t)$ and $y = f_2(t)$ are given functions of time t. The functions define a curve, the *path* of P.

The position of P may be given by the radius vector $\mathbf{r} = \mathbf{r}(t)$, or position vector **OP**. At time t the position vector is $\mathbf{r}$ and at the time $t + \Delta t$ the point P has moved to a new position given by the position vector $\mathbf{r} + \Delta\mathbf{r}$. We define now the *instantaneous velocity* of P at time t by the vector

$$\mathbf{V} = \lim_{\Delta t \to 0} \frac{\Delta \mathbf{r}}{\Delta t} = \frac{\mathrm{d}\mathbf{r}}{\mathrm{d}t} = \dot{\mathbf{r}}$$

The velocity is, therefore, directed *along the tangent to* the path at the position of P; it is *not* perpendicular to the position vector. The scalar magnitude, the *speed* of P, may be expressed by considering the function $s = h(t)$, where the position of P on the curve is determined by the distance s measured along the curve from a fixed point B on the curve; the displacement along the curve in the time Δt is then Δs as shown, and the speed is

$$V = \lim_{\Delta t \to 0} \frac{\Delta s}{\Delta t} = \frac{\mathrm{d}s}{\mathrm{d}t} = \dot{s}$$

If the velocity vectors of Fig. 1.8 are drawn from the same point, the vector diagram looks as shown in Fig. 1.9. The *instantaneous acceleration* of P is now

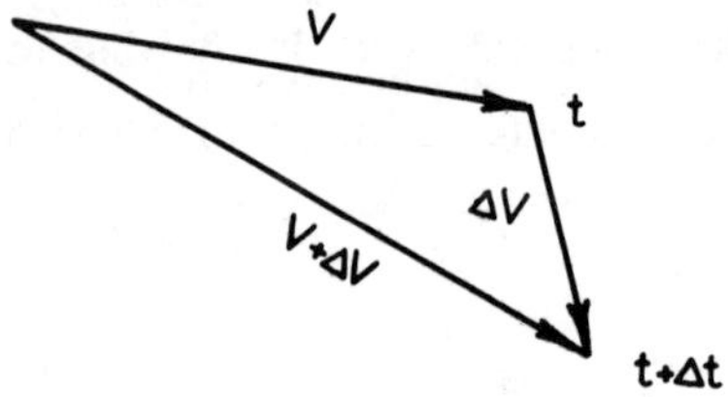

Fig. 1.9.

defined by the vector

$$\mathbf{A} = \lim_{\Delta t \to 0} \frac{\Delta \mathbf{V}}{\Delta t} = \frac{d\mathbf{V}}{dt} = \dot{\mathbf{V}} = \frac{d^2\mathbf{r}}{dt^2} = \ddot{\mathbf{r}}$$

This vector is *not* tangential to the path. To have $\mathbf{A} = \mathbf{0}$ in a time interval, it is necessary that *both* the *magnitude* and the *direction* of **V** are unchanged in that time interval.

These definitions of velocity and acceleration may be seen to be in agreement with the definitions previously given for the simple case of rectilinear motion.

It is usually most convenient to work with *components* of the acceleration vector **A**. The following three sets of components have been found to be most useful in practical problems: rectangular or Cartesian, normal and tangential, and radial and transverse components.

1.4.2 Rectangular components of velocity and acceleration

Fig. 1.10 shows a rectangular coordinate system (x,y) with unit vectors **i** and **j** along

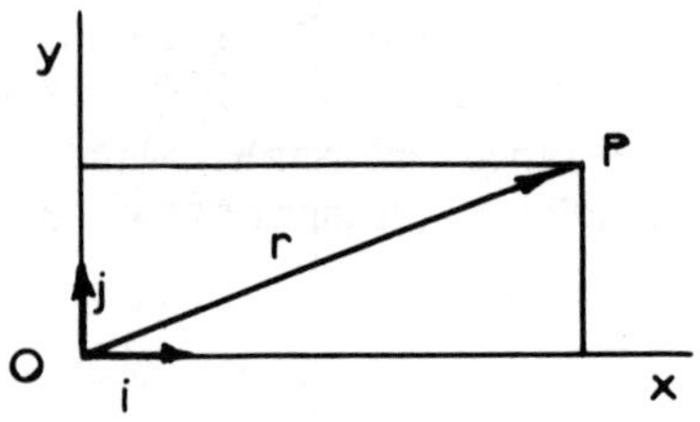

Fig. 1.10.

the axes. A point P is shown with position vector $\mathbf{r} = x\mathbf{i} + y\mathbf{j}$, where x and y are functions of time.

The *velocity* of P, from the previous definition, is

$$\mathbf{V} = \dot{\mathbf{r}} = \dot{x}\mathbf{i} + \dot{y}\mathbf{j} \tag{1.1}$$

The speed of the projections of P on the axes are

$$V_x = \dot{x} \quad V_y = \dot{y}$$

These are the magnitudes of the components of the velocity **V**.

The acceleration of P is

$$\mathbf{A} = \dot{\mathbf{V}} = \ddot{x}\mathbf{i} + \ddot{y}\mathbf{j} \tag{1.2}$$

The magnitudes of the acceleration components are

$$A_x = \ddot{x} \quad A_y = \ddot{y}$$

Example 1.4

The displacement of a point P moving in the xy-plane is given by the functions $x = r \cos\theta$ and $y = r \sin\theta$, where r(m) is a *constant* length and θ(rad) is a function of time $\theta = f(t)$. Determine the path of the point, its velocity and acceleration components in rectangular coordinates, and the magnitude of the velocity and acceleration.

Solution

The angle θ may be directly eliminated to give the path $x^2 + y^2 = r^2$, or a *circle* with radius r. The point P may be visualised as the end point of a radial line OP of length r which rotates about O in the xy-plane, as shown in Fig. 1.11. The angle between the

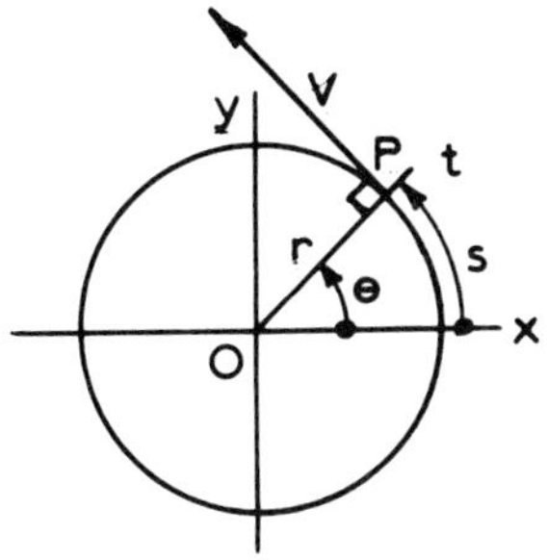

Fig. 1.11.

x-axis and OP is seen to be θ from the expressions for x and y. The angular velocity and acceleration of OP are then $\dot{\theta} = \omega$ and $\ddot{\theta} = \dot{\omega}$.

The velocity components are

$$\dot{x} = -r\omega \sin \theta \qquad \dot{y} = r\omega \cos \theta$$

The magnitude of **V** is the speed $V = \sqrt{(\dot{x}^2 + \dot{y}^2)} = r\omega$ (m/s). The velocity **V** is along the tangent to the path, as shown in Fig. 1.11; its components, found by projection on the x- and y-axes, are seen to be in agreement with the expressions for $\dot{x}$ and $\dot{y}$. The speed may also be determined from $s = r\theta$, so that $V = \dot{s} = r\dot{\theta} = r\omega$.

The acceleration components are

$$\ddot{x} = -r\omega^2 \cos \theta - r\dot{\omega} \sin \theta$$

$$\ddot{y} = -r\omega^2 \sin \theta + r\dot{\omega} \cos \theta$$

The magnitude of the acceleration is

$$A = \sqrt{(\ddot{x}^2 + \ddot{y}^2)} = r\sqrt{(\omega^4 + \dot{\omega}^2)} \text{ (m/s}^2\text{)}$$

Example 1.5

Fig. 1.12 shows a circular wheel of radius r which rolls, *without slipping*, along the

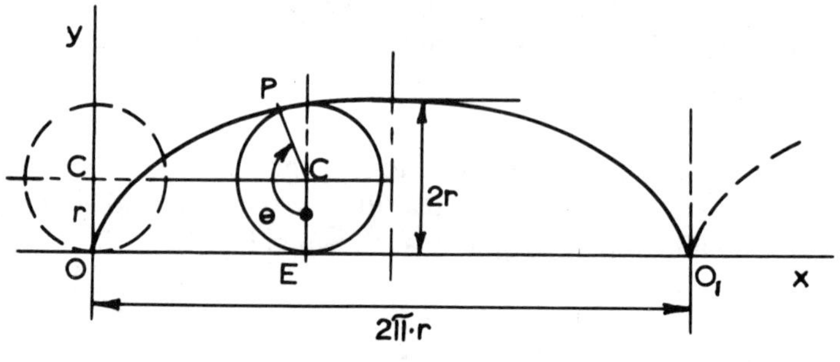

Fig. 1.12.

x-axis in the vertical xy-plane. Determine the displacement, velocity and acceleration of a point P on the rim of the wheel.

Solution

Introducing the xy-coordinate system as shown, so that the point P starts at O, the position of the radial line CP may be given by the angle $\theta = f(t)$ between CP and the vertical direction. The distance $\text{OE} = x_C$ is equal to the arc EP of the wheel, so that $x_C = r\theta$, while $Y_C = r$; the velocity components of C are then $\dot{x}_C = r\dot{\theta}$ and $\dot{y}_C = 0$; the centre C of the wheel moves on a horizontal line with speed $V_C = r\dot{\theta}$.

From the geometry of the figure, we find the coordinates of P:

$$x = r(\theta - \sin \theta)$$

$$y = r(1 - \cos\theta)$$

Differentiation with respect to time gives the velocity components:

$$\dot{x} = r\dot{\theta}(1 - \cos\theta)$$

$$\dot{y} = r\dot{\theta}\sin\theta$$

The acceleration components are

$$\ddot{x} = r\dot{\theta}^2\sin\theta + r\ddot{\theta}(1 - \cos\theta)$$

$$\ddot{y} = r\dot{\theta}^2\cos\theta + r\ddot{\theta}\sin\theta$$

The shape of the path is given by the x- and y-expressions, and has the form indicated in Fig. 1.12. It is called a *cycloid*, and clearly repeats itself indefinitely, so that we need consider only the part for $0 \leqslant \theta \leqslant 2\pi$.

When $\theta = 0$, 2π etc., that is when P is the *point of contact* between the wheel and the x-axis, we find $\dot{x} = 0$ and $\dot{y} = 0$, which means that the *instantaneous velocity of the point of contact is zero*; it may be shown that the cycloid has vertical tangents at these points.

When $\theta = \pi$, P is at the top point of the wheel, $\dot{x} = 2r\dot{\theta} = 2V_C$ and $\dot{y} = 0$: the velocity of the top point is horizontal and twice the magnitude of the velocity of the centre of the wheel.

For the contact points, $\theta = 0$, 2π etc., the acceleration components are $\ddot{x} = 0$, $\ddot{y} = r\dot{\theta}^2$: the acceleration of the contact point is of magnitude $r\dot{\theta}^2$, and always vertical, that is always *directed towards the geometrical centre* of the wheel.

1.4.3 Normal and tangential components of acceleration

1.4.3.1 Radius of curvature of a plane curve

Fig. 1.13 shows a plane curve $y = F(x)$. The tangents at points P and P_1 on the curve

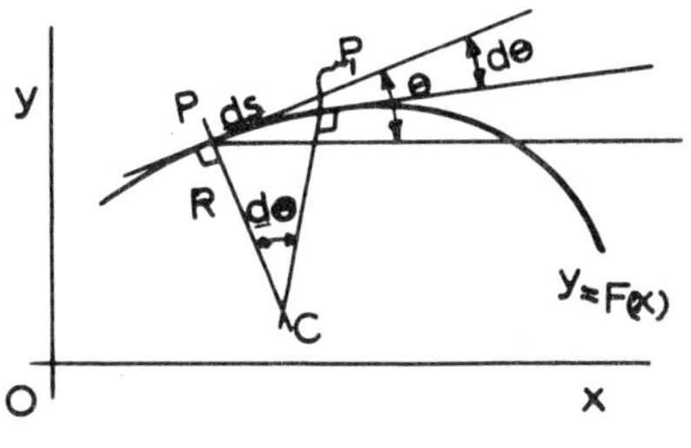

Fig. 1.13.

have been drawn, the distance between P and P_1 being ds; the tangent at P is inclined at an angle θ to the horizontal and the angle between the tangents is dθ.

The radius of curvature of the curve at P is defined as the ratio ds/dθ. A circle with centre C on the normal at P and with radius R has been introduced, and this circle has the same tangent and the same radius of curvature as the curve at P. We have $ds = R\, d\theta$, where R is the radius of curvature at P, C is the centre of curvature, and $R = ds/d\theta$. The slope of the tangent is $\tan\theta = dy/dx = y'$, and differentiating with respect to x gives $(1/\cos^2\theta)d\theta/dx = y''$, or $d\theta/dx = y''\cos^2\theta$. Now $\sin^2\theta/\cos^2\theta = \tan^2\theta = (y')^2$, and substituting in $\sin^2\theta + \cos^2\theta = 1$ gives $\cos^2\theta = 1/[1+(y')^2]$, so that

$$\frac{d\theta}{dx} = \frac{y''}{1+(y')^2}$$

Now

$$\frac{d\theta}{dx} = \frac{d\theta}{ds}\frac{ds}{dx} = \frac{1}{R}\frac{ds}{dx}$$

Using $ds^2 = dx^2 + dy^2$, or $ds/dx = (1+(y')^2)^{1/2}$, gives

$$\frac{1}{R}(1+(y')^2)^{1/2} = \frac{y''}{1+(y')^2} \quad \text{or} \quad R = \frac{[1+(y')^2]^{3/2}}{y''} \tag{1.3}$$

If the curve is given by $x = f_1(t)$ and $y = f_2(t)$, we find

$$y' = \frac{dy}{dx} = \frac{dy}{dt}\frac{dt}{dx} = \frac{\dot{y}}{\dot{x}}$$

$$y'' = \frac{d}{dx}\left(\frac{dy}{dx}\right) = \frac{d}{dt}\left(\frac{dy}{dx}\right)\frac{dt}{dx} = \frac{d}{dt}\left(\frac{\dot{y}}{\dot{x}}\right)\frac{1}{\dot{x}} = \frac{\dot{x}\ddot{y} - \dot{y}\ddot{x}}{\dot{x}^3}$$

Substituting in (1.3) gives

$$R = \frac{(\dot{x}^2 + \dot{y}^2)^{3/2}}{\dot{x}\ddot{y} - \dot{y}\ddot{x}} \tag{1.4}$$

For a circle, the radius of curvature is constant and equal to the radius of the circle.

1.4.3.2 Normal and tangential components of acceleration

Figure 1.14 shows a point P of a plane curve. Unit vectors $\mathbf{e}_t$ and $\mathbf{e}_n$ have been introduced on the tangent and the normal to the curve at P. The velocity vectors along the tangents at point P and a second point P_1 on the curve are shown as $\mathbf{V}$ and $\mathbf{V}_1$.

The point C is the centre of curvature of the curve at P, and the radius of curvature is $R = ds/d\theta$. The instantaneous speed of point P is $V = ds/dt = \dot{s} = R$

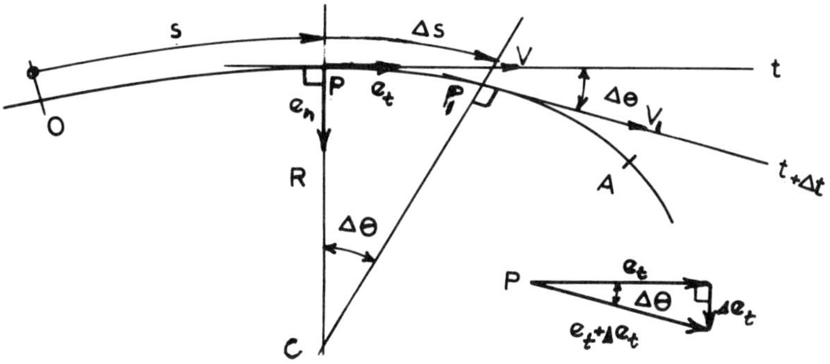

Fig. 1.14.

$\mathrm{d}\theta/\mathrm{d}t = R\dot{\theta}$, from which $\dot{V} = \ddot{s} = R\ddot{\theta} + \dot{R}\dot{\theta}$.

The velocity of P is $\mathbf{V} = V\,\mathbf{e}_t$, from which the acceleration of P is $\mathbf{A} = \dot{\mathbf{V}} = \mathrm{d}(V\mathbf{e}_t)/\mathrm{d}t = \dot{V}\,\mathbf{e}_t + V\,\dot{\mathbf{e}}_t$.

From the figure we find

$$\dot{\mathbf{e}}_t = \frac{\mathrm{d}\,\mathbf{e}_t}{\mathrm{d}t} = \lim_{\Delta t \to 0} \frac{\Delta\,\mathbf{e}_t}{\Delta t}$$

The length of $\Delta\,\mathbf{e}_t$ is $1 \times \Delta\theta$ and for $\Delta t \to 0$ its direction is perpendicular to $\mathbf{e}_t$, that is along $\mathbf{e}_n$. The result is

$$\dot{\mathbf{e}}_t = \lim_{\Delta t \to 0} \frac{1 \times \Delta\theta\,\mathbf{e}_n}{\Delta t} = \dot{\theta}\,\mathbf{e}_n = \frac{V}{R}\,\mathbf{e}_n$$

The acceleration is now

$$\mathbf{A} = \dot{V}\,\mathbf{e}_t + \frac{V^2}{R}\,\mathbf{e}_n = (R\ddot{\theta} + \dot{R}\dot{\theta})\mathbf{e}_t + R\dot{\theta}^2\mathbf{e}_n = \mathbf{A}_t + \mathbf{A}_n \tag{1.5}$$

The tangential component of **A** has the magnitude $A_t = \dot{V} = \ddot{s} = \dot{R}\dot{\theta} + R\ddot{\theta}$, and gives the change in magnitude of the velocity only. The normal component of A has the magnitude $A_n = V^2/R = \dot{s}^2/R = R\dot{\theta}^2$, and is directed towards the centre of curvature C; it gives the change in direction of **V** only.

If the path is a straight line, $R \to \infty$ so that $\mathbf{A}_n = \mathbf{0}$ and $\mathbf{A} = \mathbf{A}_t = \mathrm{d}V/\mathrm{d}t$ along the line.

If $\mathbf{V}$ is of *constant magnitude* along a curved path, we have $V =$ constant and $\mathrm{d}V/\mathrm{d}t = 0$, so that $\mathbf{A}_t = \mathbf{0}$ and $\mathbf{A} = \mathbf{A}_n = V^2\mathbf{e}_n/R$ towards the centre of curvature.

Example 1.6

A point P is moving in a circle as shown in Fig. 1.15. Determine the velocity and the normal and tangential components of its acceleration.

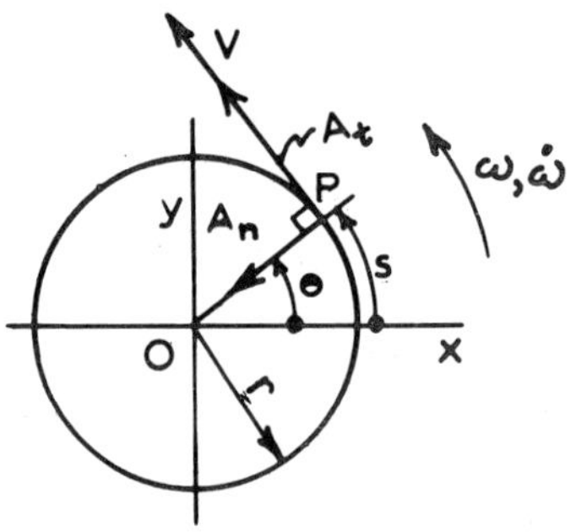

Fig. 1.15.

Solution

We have directly $s = r\theta$ so that the speed of P is $V = \dot{s} = r\dot{\theta} = r\omega$, where ω is the angular velocity of the radial line OP. The velocity is $\mathbf{V}$ as shown; this is directed along the tangent at P and is therefore, in this special case, perpendicular to the radius vector OP.

The tangential component of the acceleration is $\mathbf{A}_t = \dot{V}\,\mathbf{e}_t$ as shown, the magnitude is $\dot{V} = \ddot{s} = r\ddot{\theta} = r\dot{\omega}$, where $\dot{\omega}$ is the angular acceleration of the radial line OP.

The normal component of the acceleration is $\mathbf{A}_n = (V^2/R)\mathbf{e}_n$ as shown; the magnitude of $\mathbf{A}_n$ is $A_n = V^2/R = V^2/r = r\omega^2$.

These two components are of great importance for all work in rigid-body kinematics. Projection on the x- and y-axes gives the rectangular components already found in Example 1.4. The magnitude of the acceleration is $A = \sqrt{(A_t^2 + A_n^2)} = r\sqrt{(\dot{\omega}^2 + \omega^4)}$ as found before.

Example 1.7

A point P is moving with constant speed V on the ellipse $x^2/a^2 + y^2/b^2 = 1$, as shown in Fig. 1.16. The direction of motion is anti-clockwise. Determine the acceleration components when P passes through the point B.

Solution

The $\mathbf{e}_t$ and $\mathbf{e}_n$ vectors are shown in Fig. 1.16 at B. The velocity at B is $\mathbf{V} = V\mathbf{e}_t$, and the acceleration is $\mathbf{A} = \dot{V}\,\mathbf{e}_t + (V^2/R)\mathbf{e}_n$. Since V is constant, we have $\dot{V} = 0$ so that $\mathbf{A}_t =$

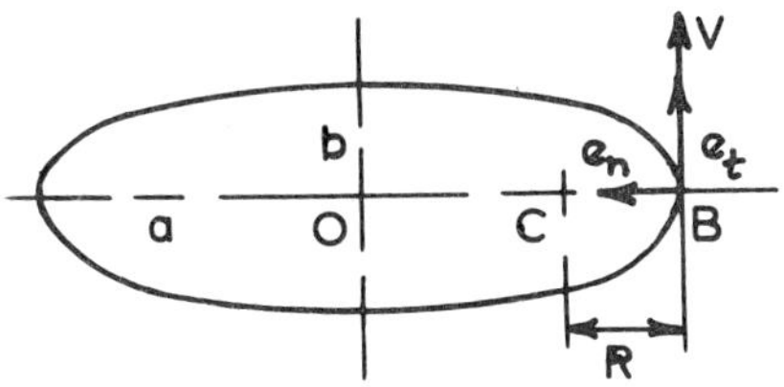

Fig. 1.16.

$\dot{V}\mathbf{e}_t = \mathbf{0}$. The radius of curvature is given by (1.3): $R = [1 + (y')^2]^{3/2}/y''$. Since $y' \to \infty$ at point B, the formula cannot be used directly for point B; it becomes necessary to develop a general formula for R for this particular curve, or to use the formula (1.4) for R.

In the present case we have

$$y = \frac{b}{a}\sqrt{(a^2 - x^2)}$$

$$y' = -\frac{b}{a}\frac{x}{\sqrt{(a^2 - x^2)}}$$

$$y'' = -\frac{ab}{(a^2 - x^2)^{3/2}}$$

so that

$$R = \frac{\left[a^2 + x^2\left(\frac{b^2}{a^2} - 1\right)\right]^{3/2}}{-ab}$$

At position B, $x = a$ and $R = |-b^2/a|$, so that $\mathbf{A}_n = (V^2/R)\mathbf{e}_n = V^2(a/b^2)\mathbf{e}_n$.

The ellipse may also be given by the equations: $x = a\cos\omega t$ and $y = b\sin\omega t$, with ω constant. Differentiating gives $\dot{x} = -a\omega\sin\omega t$, $\dot{y} = b\omega\cos\omega t$, $\ddot{x} = -a\omega^2\cos\omega t$ and $\ddot{y} = -b\omega^2\sin\omega t$. At point B, we have $x = a$, $y = 0$ so that $\omega t = 0$, with the result that $\dot{x} = 0$, $\dot{y} = b\omega$; $\ddot{x} = -a\omega^2$ and $\ddot{y} = 0$. The formula (1.4) gives

$$R = \left|\frac{b^3\omega^3}{-ab\omega^3}\right| = \frac{b^2}{a} \text{ as above.}$$

1.4.4 Polar coordinates. Radial and transverse components

The position of a point P moving in a plane may be given by the *polar coordinates* (r, ϕ), as shown in Fig. 1.17. The coordinates r and ϕ are functions of time t; if the

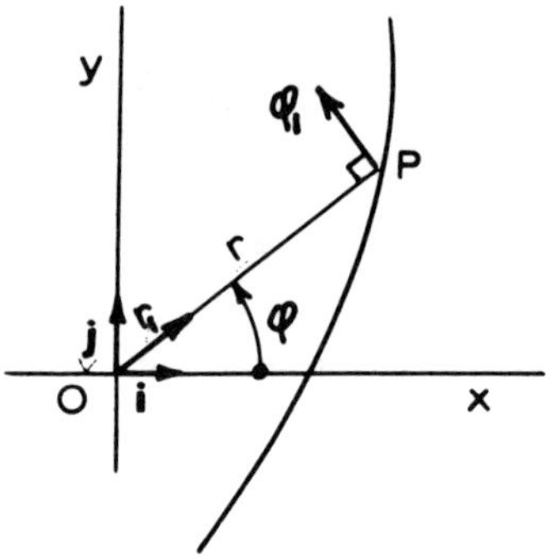

Fig. 1.17.

time is eliminated we find the path of P in polar coordinates $r = f(\phi)$.

Introducing the unit vectors $\mathbf{r}_1$ and $\boldsymbol{\phi}_1$ as shown, we may take the components of velocity and acceleration of the point in the directions of these unit vectors.

The position vector of P is $\mathbf{OP} = \mathbf{r} = r\mathbf{r}_1$. From the figure we find that

$$\mathbf{r}_1 = \cos\phi\mathbf{i} + \sin\phi\mathbf{j} \quad \text{and} \quad \boldsymbol{\phi}_1 = -\sin\phi\mathbf{i} + \cos\phi\mathbf{j}$$

Differentiating gives

$$\dot{\mathbf{r}}_1 = -\sin\phi\dot{\phi}\mathbf{i} + \cos\phi\dot{\phi}\mathbf{j} = \dot{\phi}\boldsymbol{\phi}_1$$

and

$$\dot{\boldsymbol{\phi}}_1 = -\cos\phi\dot{\phi}\mathbf{i} - \sin\phi\dot{\phi}\mathbf{j} = -\dot{\phi}\mathbf{r}_1$$

The velocity $\mathbf{V} = \mathrm{d}(r\mathbf{r}_1)/\mathrm{d}t = \dot{r}\mathbf{r}_1 + r\dot{\mathbf{r}}_1$. Substituting $\dot{\mathbf{r}}_1 = \dot{\phi}\boldsymbol{\phi}_1$ in the expression for $\mathbf{V}$ gives

$$\mathbf{V} = \dot{r}\mathbf{r}_1 + r\dot{\phi}\boldsymbol{\phi}_1 \tag{1.6}$$

We may write this as $\mathbf{V} = \mathbf{V}_r + \mathbf{V}_\phi = V_r\mathbf{r}_1 + V_\phi\boldsymbol{\phi}_1$, where $V_r = \dot{r}$ and $V_\phi = r\dot{\phi}$.

The acceleration of P is $\mathbf{A} = \dot{\mathbf{V}} = \dot{r}\dot{\mathbf{r}}_1 + \ddot{r}\mathbf{r}_1 + r\dot{\phi}\dot{\boldsymbol{\phi}}_1 + (\ddot{r} + \dot{r}\dot{\phi})\boldsymbol{\phi}_1$; by substituting $\dot{\mathbf{r}}_1 = \dot{\phi}\boldsymbol{\phi}_1$ and $\dot{\boldsymbol{\phi}}_1 = -\dot{\phi}\mathbf{r}_1$, we finally obtain the acceleration:

$$\mathbf{A} = (\ddot{r} - r\dot{\phi}^2)\mathbf{r}_1 + (r\ddot{\phi} + 2\dot{r}\dot{\phi})\boldsymbol{\phi}_1 \tag{1.7}$$

We may write this as $\mathbf{A} = \mathbf{A}_r + \mathbf{A}_\phi = A_r\mathbf{r}_1 + A_\phi\boldsymbol{\phi}_1$, where $\mathbf{A}_r$ is the radial component and $\mathbf{A}_\phi$ is the transverse component of acceleration, with

$$A_r = \ddot{r} - r\dot{\phi}^2 \quad A_\phi = r\ddot{\phi} + 2\dot{r}\dot{\phi} = \frac{1}{r}\frac{\mathrm{d}}{\mathrm{d}t}(r^2\dot{\phi})$$

If r is of *constant length*, the motion is circular, and $\dot{r} = \ddot{r} = 0$, $\dot{\phi} = \omega$, $\ddot{\phi} = \dot{\omega}$, so that

$$\mathbf{V} = r\omega\boldsymbol{\phi}_1 \quad \text{and} \quad \mathbf{A} = -r\omega^2\mathbf{r}_1 + r\dot{\omega}\boldsymbol{\phi}_1$$

The direction and magnitude of the velocity and the components of the acceleration are in agreement with the expressions developed in Examples 1.4 and 1.6.

Example 1.8

A point P moves in a plane in such a way that its polar coordinates r (m) and ϕ (rad) are given by the time functions $r = t^2 + 2t$, and $\phi = t$. Determine the path in polar coordinates and the position, velocity and acceleration of P at the instant when $t = 2$ s.

Solution

We find by eliminating t that the path is determined by $r = \phi^2 + 2\phi$. When $t = 2$, we find $r = 8$ m and $\phi = 2$ rad, which determines the position of P at this instant.

The velocity $\mathbf{V} = \dot{r}\mathbf{r}_1 + r\dot{\phi}\boldsymbol{\phi}_1$ from (1.6). We have $\dot{r} = 2t + 2$ and $\dot{\phi} = 1$ (constant), so when $t = 2$ s, $\dot{r} = 6$ m/s and $\dot{\phi} = 1$ rad/s, the instantaneous velocity is then $\mathbf{V} = 6\mathbf{r}_1 + 8\boldsymbol{\phi}_1$ m/s.

The instantaneous speed is $V = \sqrt{(36 + 64)} = 10$ m/s.

The acceleration is determined from (1.7):

$$\mathbf{A} = (\ddot{r} - r\dot{\phi}^2)\mathbf{r}_1 + (r\ddot{\phi} + 2\dot{r}\dot{\phi})\boldsymbol{\phi}_1$$

We have $\ddot{r} = 2$ (constant), and $\ddot{\phi} = 0$, so that

$$\mathbf{A} = -6\mathbf{r}_1 + 12\boldsymbol{\phi}_1 \text{ m/s}^2$$

The magnitude of the acceleration is

$$A = \sqrt{(36+144)} = 13{\cdot}4\,\text{m/s}^2$$

PROBLEMS

1.1 Water drips from a faucet at the uniform rate of n drops per second. Find the distance x between any two adjacent drops as a function of the time t that the trailing drop has been in motion. Neglect air resistance.

1.2 A car starting from rest increases its speed from 0 to V with a constant acceleration a_1, runs at this speed for a time, and finally comes to rest with constant deceleration a_2. Given that the total distance travelled is s, find the total time t required. If the greatest possible acceleration and deceleration that the automobile may have is a and maximum speed is V, what is the minimum time required to cover a distance s from rest to rest?

1.3 A particle moves in the xy-plane according to the law $\dot{x} = 2t - 6$ and $\dot{y} = 3t^2 - 18t + 27$. If the particle is at $(9, -27)$ m when $t = 0$, determine in *rectangular coordinates* (a) the acceleration of the particle when $t = 4$ s, and (b) the equation of the path.

1.4 The position of a particle moving on a circular path with a radius of 32 m varies according to $s = 3t^2 + 4t$, where s is the distance in metres from a fixed point to the particle measured along the path. The radial line from the centre to the particle is turning counterclockwise when $t = 1$ s, and the particle is at the top of the path when $t = 2$ s. Determine the acceleration when $t = 2$ s in *normal and tangential components*.

1.5 A rod (Fig. 1.18) rotates in the xy-plane about O with a *constant angular*

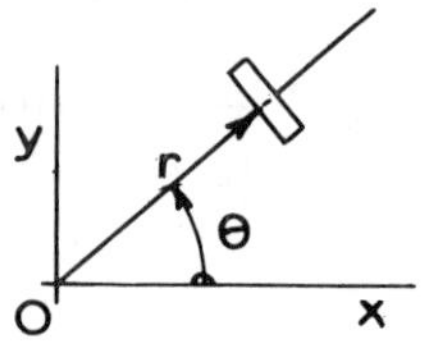

Fig. 1.18.

acceleration of 1 rad/s^2. A washer slides out along the rod from O. The distance r from O to the washer *increases uniformly* at the rate of 2 m/s. Determine the velocity and acceleration of the washer when the angular velocity of the rod is

3 rad/s, $r = 2$ m and $\theta = 90°$ (radial and transverse components).

1.6 AB ≡ BC ≡ 6 m in Fig. 1.19. Block C has a *constant* velocity of 12 m/s to the right.

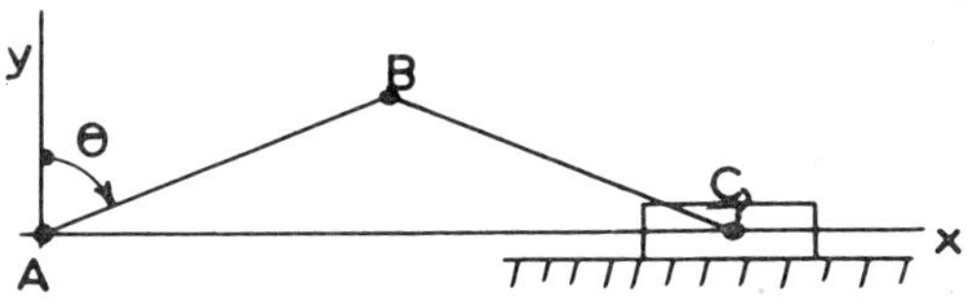

Fig. 1.19.

Determine the angular velocity and acceleration of the rod AB for $\theta = 40°$.

1.7 A point moves in *rectilinear motion* with velocity–time function $V = 3 \sin(\pi t/6)$ m/s. Determine the maximum velocity. Determine the displacement and acceleration at $t = 6$ s, given that $x = 0$ when $t = 0$.

2

Dynamics of a particle

2.1 NEWTON'S LAWS

2.1.1 Newton's laws of motion

The basic SI units of length (m) and time (s) were used extensively in kinematics in Chapter 1. In dynamics it soon becomes apparent that further concepts and units must be introduced to describe adequately the motion of bodies. To move a body, a certain action must be taken; the body must be pushed or pulled. This action is called to apply a *force* to the body.

A force may, in statics, be defined by the change in configuration it produces in a standard body. In dynamics a force also produces changes in a body, but of far greater importance is the fact that a force changes the motion of a body; this change of motion is used to define a force, and the change is accounted for in Newton's three laws of motion; these laws are the foundations of classical dynamics and were first stated in Newton's famous book *Principia* in 1687; the laws are axioms in dynamics. The first law may be stated as follows:

Newton's first law. A body continues in a state of rest or uniform motion, unless it is acted upon by a resultant force.

This law is really a special case of the second law, but it was included by Newton in his statements and has therefore been retained in works on dynamics. It defines a force as an action which changes the motion of a body.

It is common knowledge that to prevent a body from falling to the ground, a force must be exerted on the body, and if the body is moved about, it resists any change in its motion; experiments also show that two bodies attract each other with a force, and that the force necessary to accelerate a body is proportional to the acceleration. This property of the matter of which a body consists, to resist any change in motion and attract other bodies, is called the *inertia* or the *mass* of the body.

The mass of a body is defined in Newton's second law as the proportionality factor between the applied resultant force and the acceleration it produces.

The masses of two bodies may be compared directly by attaching them to identical springs on the same location of the earth's surface; if the extension of the springs is the same, the bodies have the same mass. If the experiment is repeated high above the surface of the earth, the extension of the springs is smaller, but it is still the same for both bodies if this is so at the surface; we assume from this that the mass of a body is independent of its location and is constant throughout our solar system.

To give a body a certain acceleration, the same resultant force must be applied anywhere in our solar system.

The unit of mass in the SI system is the mass of a prototype body kept in Paris; this mass is one *kilogram* (kg). If a body of mass m is moving with a velocity $\mathbf{v}$ it is said to have a *linear momentum* $m\mathbf{v}$.

Newton's second law. This is the most important in dynamics. Newton stated that the motion is proportional to the natural force impressed and in the same direction.

This statement needs some clarification. The law was stated for a particle only, and was later extended to include a collection of particles and a rigid body.

In Newtonian mechanics the mass of a particle is assumed constant. For a system of particles the total mass of the system may change, as in the case of a rocket burning fuel or a snowball rolling down a hill.

Newton's statement has been interpreted in two ways: in the first case the statement has been taken to mean that the resultant force is proportional to the time rate of change of the momentum and in the same direction. In the second interpretation the resultant force is taken to be proportional to the acceleration and in the same direction, the scalar mass being the proportionality factor.

In the first formulation the law may be expressed as follows:

$$\mathbf{F} = \frac{\mathrm{d}}{\mathrm{d}t}(m\mathbf{V}) = m\frac{\mathrm{d}\mathbf{V}}{\mathrm{d}t} + \mathbf{V}\frac{\mathrm{d}m}{\mathrm{d}t}$$

For a constant mass m, the result is

$$\mathbf{F} = m\frac{\mathrm{d}\mathbf{V}}{\mathrm{d}t} = m\ddot{\mathbf{r}} \tag{2.1}$$

and (2.1) expresses the second interpretation of the law. This is a *vector equation*, for which the *three corresponding scalar equations* are

$$F_x = m\ddot{x} \qquad F_y = m\ddot{y} \qquad F_z = m\ddot{z} \tag{2.2}$$

The fact that the acceleration is involved in Newton's law explains why it is necessary to discuss kinematics at considerable length before dealing with dynamics.

The acceleration produced by a force is independent of the previous motion of the body and other forces acting; the resultant acceleration is proportional to the resultant force and in the same direction.

Newton's third law. This is the law of action and reaction. It states that the forces acting between two bodies are always equal in magnitude, opposite in direction and directed along the same line. The law holds at any instant for all forces, whether constant or variable and stationary or moving; it means that forces always occur in equal and opposite pairs, one of the forces being called the *action*, the other the *reaction*.

Newton's laws are claimed to be valid only in a *primary inertial system*, that is a coordinate system with axis fixed in space without rotation, or moving in a parallel translation with constant speed. It may be shown that the acceleration measured in a coordinate system attached to the earth's surface is very nearly the same as the absolute acceleration measured in a primary inertial system.

A coordinate system fixed on the earth may therefore be accepted as a *secondary inertial system* in which Newton's law may be applied without correction in the great majority of cases. The main exceptions to this are orbital motion, space travel, long-range rocket flight and certain problems in fluid and air flow.

The unit for force in the SI system is now derived from the basic units — mass, length and time — by using Newton's second law; the unit is the *newton* (*N*), which is defined as the force which, applied to a mass of 1 kg, gives it an acceleration of 1 m/s^2. A newton is thus equivalent to 1 kg m/s^2.

In *classical or Newtonian dynamics* the mass of a particle is assumed to be constant, and the particle must be large compared to subatomic particles; it is also assumed that velocities are small compared to the velocity of light (about 300 000 km/s).

In the case of subatomic particles, a new branch of dynamics — *quantum mechanics* — has been developed in this century, and for particles moving with velocities approaching the speed of light, the laws of *relativistic mechanics* must be applied. In this book we consider only those situations in which Newtonian dynamics may be assumed valid.

2.1.2 Newton's universal law of gravitation

The law was also given in 1687, and states that the gravitational force of attraction between two particles of mass m_1 and m_2 (Fig. 2.1) and at a distance r is along the

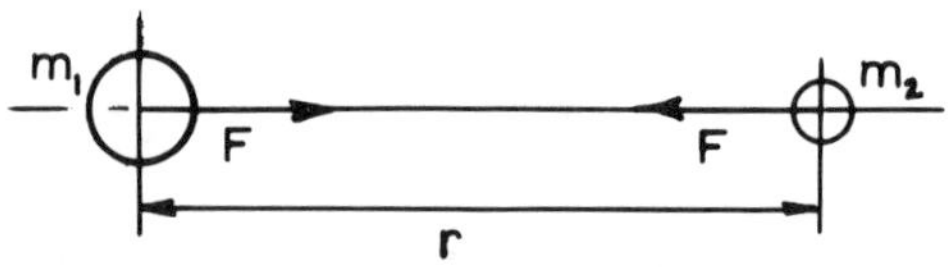

Fig. 2.1.

line connecting them and equal to

$$F = \gamma \frac{m_1 m_2}{r^2} \tag{2.3}$$

This is the famous *inverse square law of gravitation*. The *universal gravitational constant* γ has been found by experiment to be equal to $6{\cdot}673 \times 10^{-11}$ $m^3/kg\ s^2$.

The gravitational attraction on a particle from a homogeneous sphere may be found in the following manner: a thin spherical shell of radius a and thickness da is shown in Fig. 2.2. The shell attracts a particle of mass m at P at a distance r from the

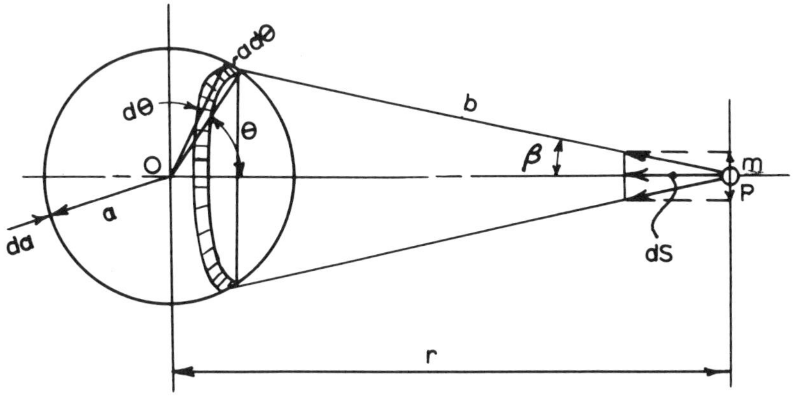

Fig. 2.2.

centre O of the shell ($r > a$).

The gravitational forces from the ring-shaped element of the shell are directed along the generators of the cone shown; taking components along PO and in the plane perpendicular to PO through P, the components in this plane form a star of concurrent equal forces at P; these components cancel in the summation, and only the components along PO give contributions to the resultant force dS; this force is then directed towards the centre of the shell, and since this is the case for all such ring-shaped elements, the resultant force from the spherical shell is towards its centre O.

If the density of the shell is ρ, the mass of the ring-shaped element is $(2\pi a \sin\theta)(a\,d\theta)(da)\rho = 2\pi\rho a^2 \sin\theta\, d\theta da$. If the universal gravitational constant is γ, Newton's gravitational law gives

$$dS = \gamma \frac{(2\pi\rho a^2 \sin\theta\, d\theta da) m \cos\beta}{b^2}$$

From the figure

$$\cos \beta = \frac{r - a \cos \theta}{b} \quad \text{and} \quad b^2 = a^2 + r^2 - 2ar \cos \theta$$

Therefore the total force from the shell is

$$S = 2\pi\gamma\rho m a^2 \, \mathrm{d}a \int_{\theta=0}^{\pi} \frac{\sin \theta (r - a \cos \theta)}{(a^2 + r^2 - 2ar \cos \theta)^{3/2}} \, \mathrm{d}\theta$$

towards O. Using $b^2 = a^2 + r^2 - 2ar \cos \theta$, we have

$$\sin \theta \, \mathrm{d}\theta = \frac{b}{ar} \, \mathrm{d}b \quad \text{and} \quad r - a \cos \theta = \frac{r^2 - a^2 + b^2}{2r}$$

Substitution now gives

$$S = \frac{\pi\gamma\rho m a \, \mathrm{d}a}{r^2} \int_{b=r-a}^{r+a} \left(1 + \frac{r^2 - a^2}{b^2}\right) \mathrm{d}b$$

$$= \frac{\pi\gamma\rho m a \, \mathrm{d}a}{r^2} \left[b - \frac{(r^2 - a^2)}{b} \right]_{r-a}^{r+a} = \frac{4\pi\gamma\rho m a^2}{r^2} \, \mathrm{d}a$$

For a solid homogeneous sphere of centre O and radius R, the gravitational attraction may be found by dividing it into thin concentric shells; the attraction from each shell is towards O, so that the total gravitational force on m is towards the centre of the sphere.

The magnitude of the force may be found from the above expression by summing up for all the shells:

$$F = \frac{4\pi\gamma\rho m}{r^2} \int_{a=0}^{R} a^2 \mathrm{d}a = \left(\frac{4}{3}\pi R^3\right) \frac{\gamma\rho m}{r^2}$$

The mass of the sphere is

$$M = \left(\frac{4}{3}\pi R^3\right) \rho$$

Hence $F = \gamma M m / r^2$. This shows that the total gravitational force from a homogeneous sphere acts towards the centre of the sphere and as if the total mass of the sphere were concentrated there. It may be seen that the same result is found if the density ρ is a function of the radius only. This is the actual situation in the case of the earth.

If the earth is considered to be a sphere of radius R_e and mass M_e, the gravitational force on a mass m at a distance $r(r > R_e)$ is towards the centre of the earth and of magnitude

$$F = \frac{\gamma M_e m}{r^2} \equiv \left(\frac{\gamma M_e}{R_e^2}\right) \frac{m}{r^2} R_e^2$$

Introducing a new factor $g = \gamma M_e/R_e^2$, we have $F = gm(R_e/r)^2$; at the earth's surface $r = R_e$, so

$$F = mg \tag{2.4}$$

We call g the *acceleration due to gravity*; it is not a constant, since the earth is not a sphere and its density varies; g depends on altitude, location and local geography.

The mass of the earth is about $M_e = 5.98 \times 10^{24}$ kg, and the radius is about 6378 km at the equator and about 6357 km at the poles, giving a mean value of about 6368 km; this gives a value of g of about 9.8 m/s^2.

The total variation in g over the earth's surface up to an altitude of 5000 m is about $\frac{3}{4}$%; this variation is too small to be of importance in most practical engineering problems, and a standard value of g from 45° latitude is used. This value is $g = 9.80665$ m/s^2; for practical work this is rounded off to $g = 9.81$ m/s^2.

The gravitational force from the earth is called the *weight force* or the weight of a body. If the mass of a body is m kg, the weight is $W = mg$ N, where $g = 9.81$ m/s^2.

For a sphere of steel with a radius of 1 m, the weight force is about 325 000 N; the gravitational attraction from a similar sphere on the earth's surface at a centre distance of 10 m is about 0.00073 N; for comparison, the attraction from the moon is about 1.1 N and from the sun about 196 N; consequently the gravitational force from the earth is the only gravitational force that need be considered in all practical problems.

Since the gravitational forces acting on the mass particles of a body are all directed towards the earth's centre about 6 368 000 m distant, we can consider these forces as parallel forces, so that the gravitational force field may be considered to be a *uniform parallel force field* in most practical problems.

2.2 DIFFERENTIAL EQUATIONS OF MOTION

Newton's second law, $\mathbf{F} = m\ddot{\mathbf{r}}$, for the motion of a particle, is a vector equation and we may, therefore, take components of the vectors in the equation along any line, to find the scalar equation of the projected motion along that line. Generally coordinate axes are introduced to give the simplest possible expressions for the scalar equations of motion; these equations are of the form $F_x = m\ddot{x}$, and may be used to solve two types of problem.

In the first type the position x is given as a function of time or may be established as a function of time. If this function can be differentiated twice, we can directly determine the velocity $\dot{x}$ and acceleration $\ddot{x}$ along the line, the direct substitution in $F_x = m\ddot{x}$ gives the force as a function of time. In the second type of problem, the acting forces are given and the problem is to determine the resultant motion. This involves integration and cannot always be done directly.

2.2.1 Displacements given as functions of time

When the displacements are given, or may be determined as functions of time, the forces necessary to produce the motion may be determined directly from Newton's second law: the procedure is best illustrated by some examples.

Example 2.1

Fig. 2.3 shows a so-called Scotch Yoke Mechanism. The crank arm AB is rotated

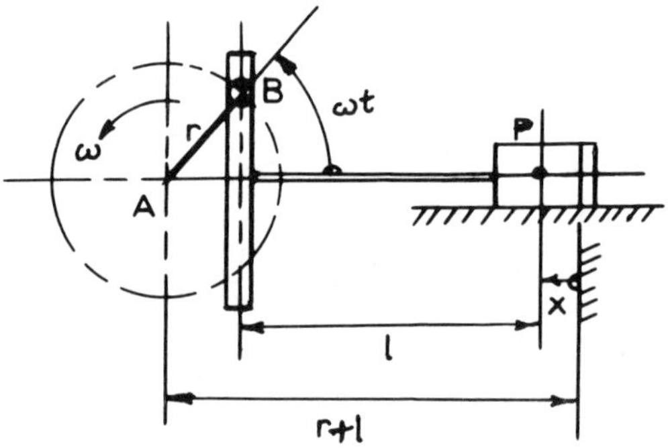

Fig. 2.3.

about A by an external torque with constant angular velocity ω. The point B slides in the yoke, and the yoke and piston P move in rectilinear motion. This mechanism was used in the early steam pumps and is now used in vibration test equipment and as a sine or cosine function generator. Determine the displacement, velocity and acceleration of the piston in terms of ω and the given lengths r and l. Determine also the maximum force in the piston rod, if the mass of the piston is $M = 10$ kg, $r = 10$ cm, and the angular velocity of the crank arm is 120 rpm. Neglect friction and wind resistance.

Solution

Taking the displacement x of the piston as shown from the top dead centre, we obtain $x = r + l - r\cos\omega t - l = r(1\text{-}\cos\omega t)$; since this expression contains the function $\cos\omega t$, and no higher order terms, it is called simple harmonic motion. Differentiating gives the velocity $\dot{x} = r\omega\sin\omega t$ and acceleration $\ddot{x} = r\omega^2\cos\omega t$. Neglecting friction and wind resistance, the force on the piston from the piston rod is $F = M\ddot{x} = Mr\omega^2\cos\omega t$; the maximum acceleration is when $\cos\omega t = 1$; therefore $\omega t = 0$ and therefore $x = 0$. The maximum force on the piston is then $F_m = Mr\omega^2$; this is in the positive direction of x and therefore the force in the piston rod is tension. We have $\omega = 2\pi \times \frac{120}{60} = 4\pi$ rad/s, and $F_m = 10 \times \frac{10}{100} \times 16\,\pi^2 = 157.91$ N.

Example 2.2
Fig. 2.4 shows a horizontal turntable which is rotating about a vertical axis through O

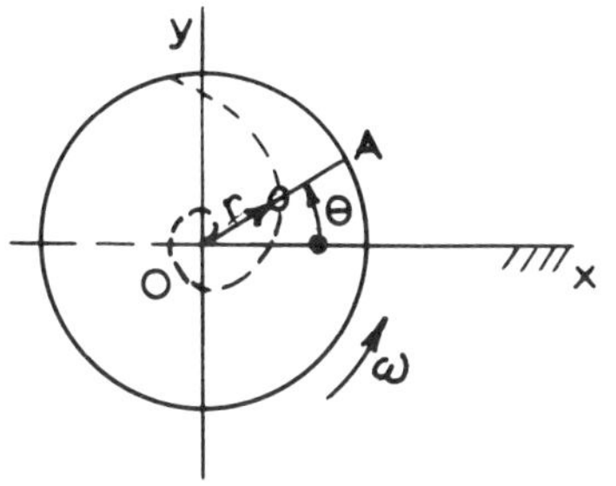

Fig. 2.4.

at a *constant* angular velocity $\omega = 0.5$ rad/s. A man of mass 85 kg walks from O towards A along a radial line drawn on the disc; the velocity of the man relative to the disc is $V = 2$ m/s *constant*.

Assuming that the man starts from O when $t = 0$, determine the *horizontal* force on his feet at the instant when $t = 2$ s. Use radial and transverse components.

Solution
Fixing a coordinate system (x, y) as shown, and determining the position of the man by the polar coordinates (r, θ), we have $r = Vt = 2t$ m and $\theta = \omega t = 0.5t$ rad. The path of the man in the horizontal plane is then the Archimedes spiral $r = 4\theta$, as indicated on the figure.

The components of acceleration in polar coordinates are of magnitudes $A_r = \ddot{r} - r\dot{\theta}^2$ and $A_\theta = r\ddot{\theta} + 2\dot{r}\dot{\theta}$; we have now $\dot{r} = 2$, $\ddot{r} = 0$, $\dot{\theta} = 0.5$ and $\ddot{\theta} = 0$, so that $A_r = -r/4 = -t/2$ *m/s², and* $A_\theta = 2 \times 2 \times 0.5 = 2$ m/s².

When $t = 2$ s, the position of the man is found by $r = 2t = 4$ m, and $\theta = 0.5t = 1$ rad; the acceleration components are $A_r = -1$ m/s², and $A_\theta = 2$ m/s²; the resultant acceleration is $A = \sqrt{(A_r^2 + A_\theta^2)} = \sqrt{5}$ m/s², and the force on the man is $F = ma = 85\sqrt{5} = 190.1$ N.

2.2.2 Dry friction. Motion with constant force

Fig. 2.5 shows a body of mass m resting on a rough, horizontal plane. The vertical forces on the body are the weight force mg and reaction from the plane N. Since there is no motion, these two forces are equal and opposite as shown. If the body is now pushed with a small force P as shown, the body will not move owing to the contact between the rough surfaces which creates *a resistant force F called the friction force*. As long as the body does not move we have $F = P$ in magnitude. Increasing P

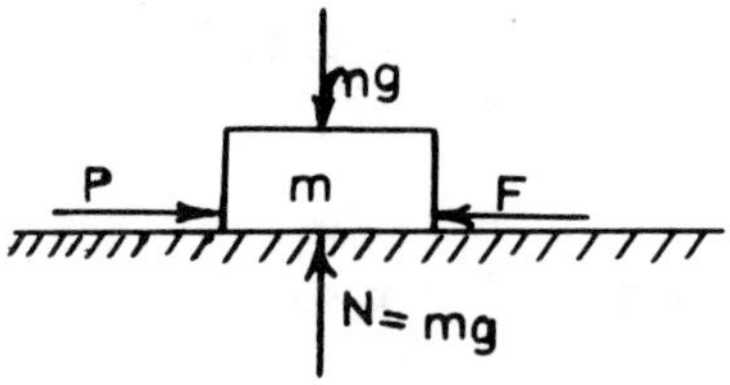

Fig. 2.5.

increases F until the contact surfaces break down and the body moves. The largest friction force F_{max} is called the limiting friction and occurs when motion impends.

Experiments by Coulomb in 1781 showed that $F_{max} = \mu_s N$, where μ_s is a dimensionless constant called the coefficient of static friction. It is noticeable that the surface area is not involved in this expression; only the surface quality determined by μ_s, and the normal pressure between the surfaces count. The law is an empirical law, since it is based on experiments.

If the body moves with constant velocity, so that no acceleration is involved, it is found that $F'_{max} = \mu_k N$, where μ_k is called the coefficient of kinetic friction; this coefficient varies somewhat with the velocity; in most cases there is a sudden drop in the coefficient when motion starts, so that $\mu_k < \mu_s$, and for low velocities μ_k is approximately constant.

The difference between μ_s and μ_k is generally small and we do not distinguish between them, but take $\mu_s = \mu_k = \mu$. The law for dry friction is called Coulomb's law and gives the dry friction force F as

$$F = \mu N \tag{2.5}$$

Some useful values of μ are the following:

Steel	on steel	0.4 –0.8
Wood	— wood	0.2 –0.5
Metal	— leather	0.3 –0.6
Cast iron	— cast iron	0.3 –0.4
Rubber	— concrete	0.6 –0.8
Rubber	— ice	0.05–0.2

The simplest case of motion occurs when the *resultant force is constant in magnitude and in a fixed direction*; if the initial velocity of the particle is zero, or in the direction of the force, the motion is rectilinear, along the line of action of the force; this situation is shown in Fig. 2.6, where the x-axis has been taken along the line of action of the resultant force F.

The equation of motion is $m\ddot{x} = F$ (constant), so that $\ddot{x} = F/m = a =$ constant acceleration; two successive integrations give as functions of time the velocity $\dot{x} = at + C_1$ and the displacement

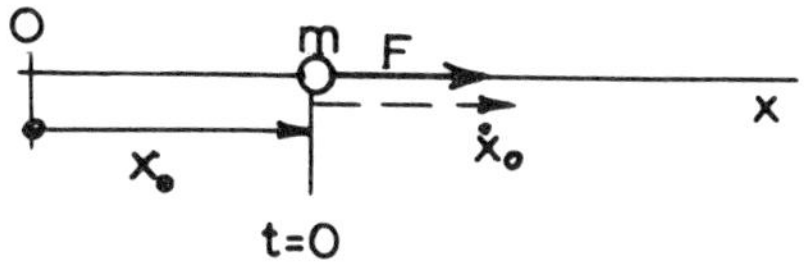

Fig. 2.6.

$$x = \tfrac{1}{2}at^2 + C_1 t + C_2$$

The integration constants C_1 and C_2 must be determined from the initial conditions; if the displacement and velocity at the time $t = 0$ are $x = x_0$ and $\dot{x} = \dot{x}_0$, we find $C_1 = \dot{x}_0$ and $C_2 = x_0$; the complete solution of the problem is then

$$x = \tfrac{1}{2}at^2 + \dot{x}_0 t + x_0$$

$$\dot{x} = at + \dot{x}_0$$

$$\ddot{x} = a = F/m \text{ (constant)}$$

Example 2.3

A body is dropped in a free fall starting from rest; determine the displacement as a function of time.

Solution

Assuming that the height above the earth's surface is relatively small, we can take g as a constant; if the body has a mass that is large compared to the same volume of air and is of a compact shape, we can neglect the air resistance for small velocities; the only force acting is then the constant weight force $F = W = mg$.

Taking the positive x-direction to be vertically downwards, we have $m\ddot{x} = mg$, or $\ddot{x} = g =$ constant. With $x = 0$ and $\dot{x} = 0$ at $t = 0$, we find $\dot{x} = gt$ and $x = \tfrac{1}{2}gt^2$ for a free fall starting from rest; these expressions were first established by Galileo during his investigations of falling bodies about the year 1600.

Example 2.4

A body of mass m (Fig. 2.7) is released from the rest on a plane inclined at α to the horizontal. The coefficient of friction between the mass and the plane is μ; determine the equation of motion of the body, and the distance moved as a function of time.

Solution

Introducing a coordinate system (x, y) as shown, the body is in a rectilinear motion along the x-axis. The acting forces are the gravity force mg and a reaction from the plane, which is resolved into a normal pressure N and a force F along the plane; F is the friction force and is of magnitude $F = \mu N$.

Since there is no motion in the y-direction, $\ddot{y} = 0$, so that the forces in this direction are balanced. We have then $N - mg \cos \alpha = 0$, or $N = mg \cos \alpha$, so that $F = \mu mg \cos \alpha$.

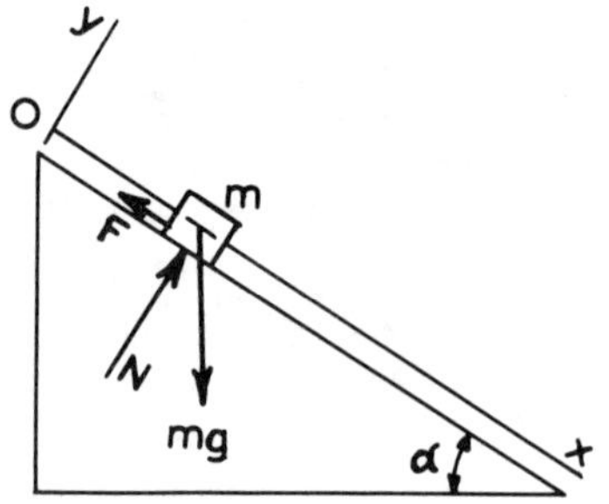

Fig. 2.7.

The equation of motion in the x-direction is $m\ddot{x} = mg \sin \alpha - F = mg \sin \alpha - \mu mg \cos \alpha$, so that $\ddot{x} = g \sin \alpha - \mu \cos \alpha = \text{constant}$. To enable the mass to slide down, we must have $\ddot{x} > 0$:

$$g \sin \alpha - \mu \cos \alpha > 0 \quad \text{or} \quad \mu < g \tan \alpha$$

If $\mu > g \tan \alpha$, the mass will not start to move; if $\mu < g \tan \alpha$, we find $\dot{x} = g(\sin \alpha - \mu \cos \alpha)t$, and $x = \frac{1}{2}g(\sin \alpha - \mu \cos \alpha)t^2$, where the integration constants have been taken as zero since the body starts from rest.

Example 2.5
Fig. 2.8 shows a small stone of mass m which is attached to a string of length l. The

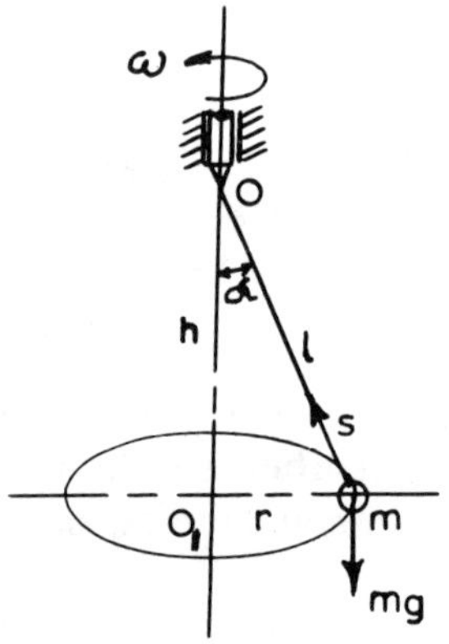

Fig. 2.8.

string is attached to a fixed point O which is the end point of a vertical shaft. The shaft is rotating with constant angular velocity ω. The stone moves in a horizontal circle and the string generates the surface of a cone. The system is called a conical pendulum. Determine the string tension and the velocity of the stone in terms of the system constants l, m and ω. Neglect wind resistance.

Solution
Introducing the radius r and height h of the cone and the angle α as shown in the figure, the only forces acting on the stone are the string tension s and gravity mg as shown. There is no motion in the vertical direction so the vertical forces must be balanced; this gives the equation $s \cos \alpha = mg$, with $\cos \alpha = h/l$; we find $s = mgl/h$.

The acceleration in the horizontal direction towards O is $A_n = r\omega^2$. Newton's second law in the horizontal direction gives $s \sin \alpha = mA_n = \mathrm{m}r\omega^2$. Substituting $\sin \alpha = r/l$ gives the result that $s = ml\omega^2$, which gives s in terms of the system constants. Equating the two expressions for s leads to $ml\omega^2 = mgl/h$, or $h = g/\omega^2$. From the right triangle in the figure we have $r^2 + h^2 = l^2$, or $r = \sqrt{l^2 - h^2} = \sqrt{l^2\omega^4 - g^2}/\omega^2$. We have now $V = r\omega\sqrt{l^2\omega^4 - g^2}/\omega$.

It is not possible to make the string rotate in a horizontal plane; to do this would require $h \to 0$, which would mean that ω and s would be infinite.

As an example of motion under a constant force, where the initial velocity is not in the direction of the force, consider the case of *projectile motion*.

Example 2.6
Fig. 2.9 shows a projectile of mass m which starts with an initial velocity V_0 inclined at

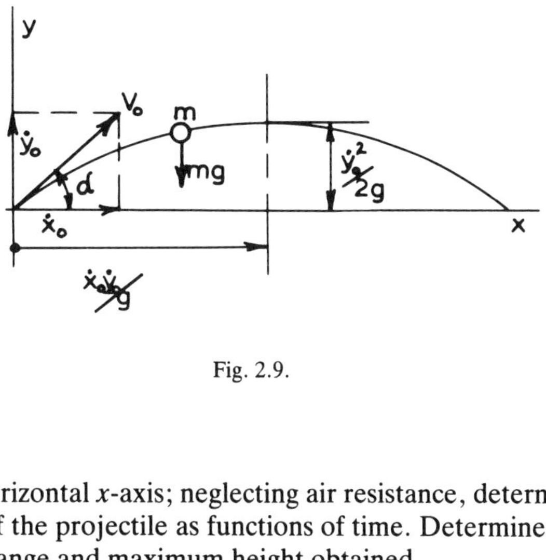

Fig. 2.9.

an angle α to the horizontal x-axis; neglecting air resistance, determine the velocity and displacement of the projectile as functions of time. Determine also the path of the projectile, the range and maximum height obtained.

Solution
The equations of motion are

$$m\ddot{x} = 0 \qquad m\ddot{y} = -mg$$

Integrating these equations and taking $\dot{x} = \dot{x}_0$, $\dot{y} = \dot{y}_0$ when $t = 0$, we find the velocity components:

$$\dot{x} = \dot{x}_0 \qquad \dot{y} = -gt + \dot{y}_0$$

The component of velocity in the x-direction is constant; seen from above, the projectile moves along the x-axis with constant velocity $\dot{x}_0$ until it suddenly stops.

Integrating the velocity equations and taking $x = y = 0$ when $t = 0$ gives the displacements:

$$x = \dot{x}_0 t \qquad y = -\tfrac{1}{2}gt^2 + \dot{y}_0 t$$

Eliminating t gives the equation of the path

$$y = \frac{x}{\dot{x}_0}\left(\dot{y}_0 - \frac{gx}{2\dot{x}_0}\right)$$

This is a parabola; substituting $\dot{x}_0 = V_0 \cos\alpha$, $\dot{y}_0 = V_0 \sin\alpha$ gives a second form for the path:

$$y = x\left(\tan\alpha - \frac{gx}{2V_0^2\cos^2\alpha}\right)$$

The vertex is found when

$$\frac{\mathrm{d}y}{\mathrm{d}x} = \frac{\dot{y}_0}{\dot{x}_0} - \frac{gx}{\dot{x}_0^2} = 0 \qquad \text{or} \qquad x = \frac{\dot{x}_0\dot{y}_0}{g}$$

This value gives $y = \dot{y}_0^2/2g$ as the maximum height; this result may also be found from $\dot{y} = 0$.

The range, from symmetry, is $r = 2\dot{x}_0\dot{y}_0/g$; this may also be found by setting $y = 0$ in the equation for the parabola. The time to impact is $t = x/\dot{x}_0$, where $x = 2\dot{x}_0\dot{y}_0/g$, so that $t = (2\dot{y}_0/g)$ s. The range may also be given as

$$r = \frac{2\dot{x}_0\dot{y}_0}{g} = \frac{2}{g}V_0\cos\alpha\, V_0\sin\alpha = \frac{V_0^2}{g}\sin 2\alpha$$

For a given velocity V_0, the maximum range is found for $\sin 2\alpha = 1$, or $\alpha = 45°$.

The solution given is quite unrealistic, since the air resistance is of considerable importance at the velocities of normal projectiles. Introducing air resistance results in equations of motion that can only be solved numerically by computer calculations.

2.2.3 Force as function of time

If the resultant force is given as a function of time that can be integrated twice, the motion produced by the force may be found directly by integration.

Example 2.7
Fig. 2.10 shows a mass m in rectilinear motion under the action of a force $F = F_0 \cos$

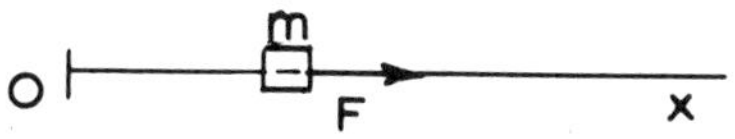

Fig. 2.10.

ωt, where F_0 and ω are given constants. Determine the equation of motion and the displacement–time function, if the displacement and velocity at $t = 0$ are $x = x_0$ and $\dot{x} = \dot{x}_0$.

Solution
The equation of motion, from Newton's second law, is $m\ddot{x} = F_0 \cos \omega t$, so that $d\dot{x} = (F_0/m) \cos \omega t \, dt$, or

$$\dot{x} = \frac{F_0}{m} \int \cos \omega t \, dt + C_1 = \frac{F_0}{m\omega} \sin \omega t + C_1$$

When $t = 0$, $\dot{x} = \dot{x}_0$, so $C_1 = \dot{x}_0$; we have now

$$dx = \frac{F_0}{m\omega} \sin \omega t \, dt + \dot{x}_0 \, dt$$

and integration gives

$$x = -\frac{F_0}{m\omega^2} \cos \omega t + \dot{x}_0 t + C_2$$

When $t = 0$, $x = x_0$, or $x_0 = F_0/m\omega^2 + C_2$, so that $C_2 = x_0 + F_0/m\omega^2$. This finally gives

$$x = \frac{F_0}{m\omega^2} (1 - \cos \omega t) + \dot{x}_0 t + x_0$$

2.2.4 Resistant force proportional to the displacement

Consider the system shown in Fig. 2.11; it consists of a mass m which slides without friction on a horizontal plane; the mass is connected by a spring to a vertical wall. The mass is shown in the static-equilibrium position where there is no force in the spring, and the displacements x are measured from this position, positive to the right.

The force in the spring is determined by the force law $F = Kx$, where K is a

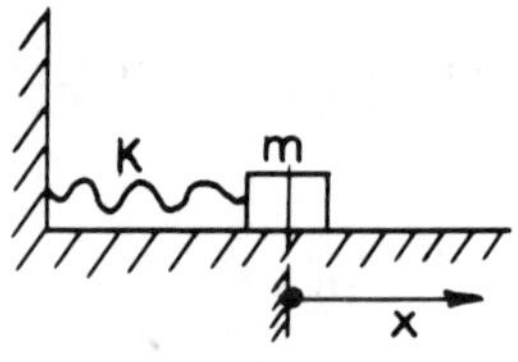

Fig. 2.11.

constant, called the spring constant, which is the force necessary to extend or compress the spring one unit of length. The units of K are N/m or N/cm. If the mass is displaced a distance x to the right, the spring pulls back on it with a force of magnitude Kx to the left; the force on the mass must then be stated as $F = -Kx$. Note that if x is negative, the force is to the right, so that this expression holds for x positive or negative.

The equation of motion of the mass is now

$$m\ddot{x} = -Kx \qquad \text{or} \qquad \ddot{x} + (K/m)x = 0$$

Introducing a new constant $\omega_0 = \sqrt{K/m}$ rad/s, we have

$$\ddot{x} + \omega_0^2 x = 0 \tag{2.6}$$

The general solution to this equation is $x = A \cos \omega_0 t + B \sin \omega_0 t$, where A and B are arbitrary constants; this may be seen by direct substitution.

From the general mathematical solution to equation (2.6), we must now find the one solution that fits the physical situation; the constants A and B must be found from the starting conditions of the motion. Let us assume that the mass is started in motion by giving it a displacement x_0 followed by a blow to give it an initial velocity $\dot{x}_0$; counting time t from this instant, we have the starting conditions at $t = 0$: $x = x_0$ and $\dot{x} = \dot{x}_0$; *substituting* $x = x_0$ and $t = 0$ in the general solution gives $A = x_0$; differentiating gives $\dot{x} = -A\omega_0 \sin \omega_0 t + B\omega_0 \cos \omega_0 t$, so that $\dot{x}_0 = B\omega_0$; for these starting conditions the solution is then

$$x = x_0 \cos \omega_0 t + \frac{\dot{x}_0}{\omega_0} \sin \omega_0 t$$

Writing this as

$$x = A_0 \cos (\omega_0 t - \phi) = A_0 \cos \omega_0 t \cos \phi + A_0 \sin \omega_0 t \sin \phi$$

we have $A_0 \cos \phi = x_0$, and $A_0 \sin \phi = \dot{x}_0/\omega_0$, so that

$$A_0 = \sqrt{\left(x_0^2 + \frac{\dot{x}_0^2}{\omega_0^2}\right)} \quad \text{and} \quad \tan\phi = \frac{\dot{x}_0}{x_0\omega_0}$$

the solution may then be given as

$$x = A_0 \cos(\omega_0 t - \phi)$$

The maximum displacement is then A_0, and this is called the *amplitude* of the motion. The angle ϕ is called the *phase angle*.

The acceleration is $\ddot{x} = -A_0\omega_0^2 \cos(\omega_0 t - \phi) = -\omega_0^2 x$, so the acceleration is proportional to the displacement, but in the opposite direction.

The simplest solution is found for $\dot{x}_0 = 0$, which gives $A_0 = x_0$ and $\phi = 0$, so that $x = x_0 \cos \omega_0 t$; this motion is shown in Fig. 2.12; it is clearly a vibratory motion along

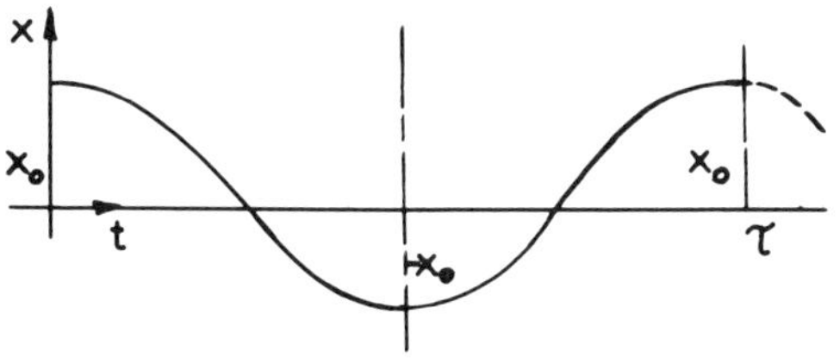

Fig. 2.12.

the x-axis, which completely repeats itself after a certain time τ; the motion is called a *periodic motion* with *period* $\tau = 2\pi/\omega_0$ s. This is the time for one complete vibration; the constant $\omega_0 = \sqrt{(K/m)}$ rad/s is called the *natural circular frequency* of the motion. The *frequency* f is the number of complete vibrations in one second:

$$f = \frac{1}{\tau} = \frac{1}{2\pi}\omega_0 = \frac{1}{2\pi}\sqrt{\left(\frac{K}{m}\right)} \text{ cycles/s}$$

The motion just described is called *simple harmonic motion*, and is of great importance in physics and engineering. The so-called *restoring force* $-Kx$ may be found in many different forms in various systems. Because of its importance, a general discussion of vibratory motion will be deferred to a special chapter, Chapter 9, where further examples may be found.

2.2.5 Resistant force proportional to the velocity

For motion, with small velocities, in a resisting fluid medium, it is found that the resisting force may be taken proportional to the velocity. The force may then be expressed by $-cv$, where the constant c depends on the size and shape of the body, and must be found experimentally.

Taking a simple case of this motion of a body moving down through a fluid, the driving force is the gravity force mg. Taking the x-axis vertical, the equation of motion is

$$m\frac{\mathrm{d}v}{\mathrm{d}t} = mg - cv \quad \text{or} \quad \frac{\mathrm{d}v}{\mathrm{d}t} = g - \frac{c}{m}v = g - \beta v$$

where $\beta = c/m$. The resisting force increases with the velocity and eventually reaches the same magnitude as the constant driving force. When this happens, the forces on the mass are balanced, so that the body continues with a constant velocity and zero acceleration. We have then $\mathrm{d}v/\mathrm{d}t = 0$, or $g - \beta v = 0$, so that $v = g/\beta = v_0$, and this is the *limiting* or *terminal* velocity.

Introducing v_0 in the equation of motion gives

$$\frac{\mathrm{d}v}{\mathrm{d}t} = \beta v_0 - \beta v = \beta(v_0 - v) \quad \text{or} \quad \mathrm{d}t = \frac{\mathrm{d}v}{\beta(v_0 - v)}$$

$$\mathrm{t} = \frac{1}{\beta}\int \frac{\mathrm{d}v}{v_0 - v} + C_1 = -\frac{1}{\beta}\int \frac{\mathrm{d}(1 - v/v_0)}{1 - v/v_0} + C_1$$

$$= -\frac{1}{\beta}\log\left(1 - \frac{v}{v_0}\right) + C_1$$

*I*f the body starts at rest, we have $v = 0$ at $t = 0$, so that $C_1 = 0$, and

$$\log\left(1 - \frac{v}{v_0}\right) = -\beta t, \qquad 1 - \frac{v}{v_0} = \mathrm{e}^{-\beta t}$$

or $v = v_0(1 - e^{-\beta t})$. We find from this that

$$\mathrm{d}x = v_0(1 - e^{-\beta t})\mathrm{d}t$$

$$x = v_0\int (1 - \mathrm{e}^{-\beta t})\mathrm{d}t + C_2$$

$$= v_0 t - \frac{v_0}{\beta}\int \mathrm{e}^{-\beta t}\,\mathrm{d}\beta t + C_2 = v_0\left[\frac{1}{\beta}\mathrm{e}^{-\beta t} + t\right] + C_2$$

If $x = 0$ when $t = 0$, we obtain $C_2 = -v_0/\beta$, so that $x = (v_0/\beta)\,[\mathrm{e}^{-\beta t} + \beta t - 1]$. An example of this motion is small particles settling in a fluid, for instance silt settling in water.

A series of examples has been given in which the solution to the equation of motion can be found by differentiation or integration. In more complicated problems, the solution to the equation of motion cannot be found in this way and we must then find solutions by graphical or numerical differentiation or integration, or by computer calculations. This type of problem is, however, outside the scope of this book. In some cases the equation of motion may be solved if *displacements are small*; a simple example of this is the following.

Example 2.8
Fig. 2.13 shows a *simple pendulum* consisting of a concentrated mass m suspended on

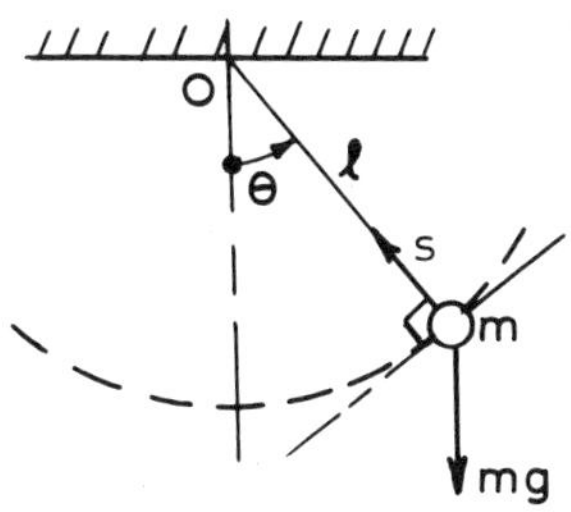

Fig. 2.13.

a string of length l. The pendulum is swinging in a vertical plane under the action of gravity. Air resistance may be neglected. Determine the equation of motion; simplify the equation for small angular displacements and solve the equation.

Solution
The mass m is in a circular motion with centre O and radius l; the normal and tangential acceleration components are $l\omega^2$ and $l\dot{\omega}$, so that the corresponding forces are $F_n = ml\omega^2 = ml\dot{\theta}^2$ and $F_t = ml\ddot{\theta}$.

The forces acting are the string tension s and the gravity force mg; projecting these forces on the radial line and the tangent at m, we find $F_n = s - mg \cos\theta$ and $F_t = -mg \sin\theta$; the minus sign for F_t is due to the fact that the force is in the opposite direction to θ. The equations of motion are now

$$s - mg \cos\theta = ml\dot{\theta}^2$$

$$-mg \sin\theta = ml\ddot{\theta}$$

The first equation gives $s = mg \cos\theta + ml\dot{\theta}^2$, which determines s if θ can be found as a function of time. The second equation gives the equation of motion:

$$\theta + \frac{g}{l} \sin\theta = 0$$

This is a *non-linear* differential equation; the time t of the swing depends on the angle θ and can only be stated as a function of θ in terms of elliptic integrals, which require numerical methods for solution.

The equation may be given a much simpler form if we restrict the motion to *small angles* θ; if $\theta = 20° = 0.34907$ rad, $\sin\theta = 0.34202$, so that θ is about 2% bigger than $\sin\theta$; for $\theta = 10° = 0.17453$ rad, $\sin\theta = 0.17365$, so that θ is only about $\frac{1}{2}$% bigger than $\sin\theta$; up to about 20° of swing, we can, therefore, expect reasonable accuracy if we take $\sin\theta = \theta$; the equation then takes the form $\theta + (g/l)\theta = 0$; this is a *linear* differential equation. Comparing this with the equation $\ddot{x} + \omega_0^2 x = 0$ shows that the equation gives simple harmonic motion of the pendulum with a frequency

$$f = \frac{\omega_0}{2\pi} = \frac{1}{2\pi}\sqrt{\frac{g}{l}}$$

2.3 WORK AND POWER

It is common knowledge that to move or lift a body, certain forces must be applied and act through the distances moved, that is a certain amount of *work* must be done on the body; to use this important concept in dynamics, we need a concise definition that can be used in the simplest possible way, so we define the work done by a force F (Fig. 2.14). with a point of application that moves through a distance ds, as the

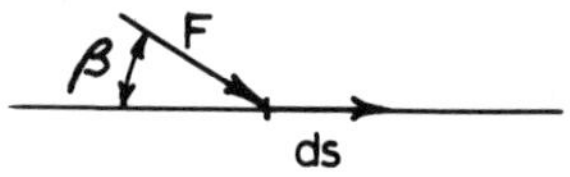

Fig. 2.14.

product $|\mathbf{F}|\,|\mathbf{ds}|\cos\beta$ = work done (W.D.).

The work done is thus defined as the product of the magnitude of the force component in the direction of the displacement, and the magnitude of the displacement; work is therefore a *scalar quantity*. In vector language, the work is W.D. = $\mathbf{F}\cdot\mathbf{ds}$, or the dot product of the force and the displacement.

It follows from the definition that the component of the force perpendicular to the displacement *does not do any work*.

If the force moves from a position 1 to a position 2, the work done is

$$\text{W.D.} = \int_1^2 \mathbf{F}\cdot\mathbf{ds} = \int_1^2 (F_x \mathrm{d}x + F_y \mathrm{d}y + F_z \mathrm{d}z)$$

The unit for work is the Newton metre (Nm) = kg m^2/s^2, and this unit is called the *joule* (J). It is the work done when the point of application of a constant force of 1 N is displaced through a distance of 1 m in the direction of the force.

An important concept associated with work is *power*, which is the *rate* of work done or the *instantaneous work* d*W*/d*t*. This is

$$\mathbf{F}\cdot\frac{\mathrm{d}\mathbf{s}}{\mathrm{d}t} = \mathbf{F}\cdot\mathbf{V}$$

The unit for power is the *watt* (W) = 1 J/s = 1Nm/s.

The *mechanical efficiency* of a system or machine is defined by the ratio

$$\frac{\text{work output}}{\text{work input}} \quad \text{or} \quad \frac{\text{useful work}}{\text{expended work}}$$

2.4 KINETIC ENERGY

If a particle of mass m is acted upon by a resultant force $\mathbf{F}$ of constant magnitude and in a fixed x-direction, Newton's second law gives the equation of motion: $\ddot{x} = F/m$ (constant). If we count time from the instant of application of the force and assume that the particle is initially at rest at position $x = 0$, the velocity V at time t is

$$V = \dot{x} = \int_0^t (F/m)\mathrm{d}t = (F/m)t$$

The distance moved is

$$x = \int_0^t (F/m)t \; \mathrm{d}t = \frac{1}{2}(F/m)t^2$$

The work done by the force is $Fx = \frac{1}{2}(F^2/m)t^2 = \frac{1}{2}mV^2 = T$; the expression $\frac{1}{2}mV^2$ is called the *kinetic energy* of the particle at the time t; this is the work required to increase its velocity from rest to a velocity V.

The unit for kinetic energy is the same as for work, and is the kg m^2/s^2 = Nm = J. The concept of kinetic energy is of fundamental importance in dynamics.

For a rigid body moving in such a way that all particles have the same velocity V_c at any instant, we find the total kinetic energy T by summing up for all the particles: $T = \frac{1}{2}V_c^2\,\Sigma m = \frac{1}{2}MV_c^2$, where M is the total mass of the body.

2.5 WORK–ENERGY EQUATION

If the resultant instantaneous force on a particle of mass m is $\mathbf{F}$ and the instantaneous velocity is $\mathbf{V}$, the displacement of the particle in the time dt is $\mathbf{V}$ dt; the work done on the particle is then d(W.D.) = $\mathbf{F}\cdot\mathbf{V}$ dt, and the total work in moving from a position 1 to a position 2 is W.D. = $\int_1^2 \mathbf{F}\cdot\mathbf{V}\,\mathrm{d}t$. Introducing Newton's second law

$$\mathbf{F} = m\,\frac{\mathrm{d}\mathbf{V}}{\mathrm{d}t}$$

we obtain

$$\begin{aligned}\text{W.D.} &= \int_1^2 m\,\frac{\mathrm{d}\mathbf{V}}{\mathrm{d}t}\cdot\mathbf{V}\,\mathrm{d}t = m\int_1^2 \mathbf{V}\cdot\mathrm{d}\mathbf{V}\\ &= \frac{1}{2}m\int_1^2 \mathrm{d}(\mathbf{V}\cdot\mathbf{V}) = \frac{1}{2}\,m\Big[V^2\Big]_1^2\\ &= \frac{1}{2}m\,(V_2^2 - V_1^2) = T_2 - T_1 \qquad (2.7)\end{aligned}$$

This shows that the total work done, in moving from a position 1 to a position 2, is equal to the total *change* in kinetic energy between the two positions; this equation is called the *work–energy equation*, and is an alternative form of Newton's second law. Although no new information is introduced by this equation, it is in a form that is convenient in many problems. It is particularly useful in problems where time is not involved in the solution.

In terms of its components, for instance in the x-direction, the equation takes the form

$$\int_2^x x_1\,F_x\,\mathrm{d}x = \frac{1}{2}\,m\,(\dot{x}_2^2 - \dot{x}_1^2) \qquad (2.8)$$

Example 2.9
A body (Fig. 2.15) of mass m is released from rest and falls through a distance h to the

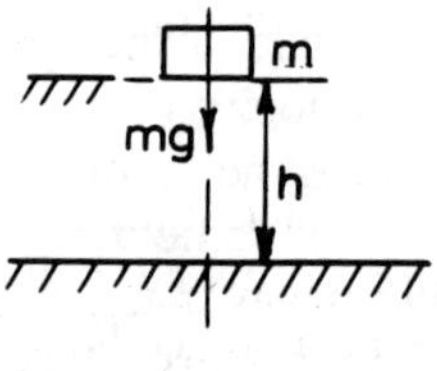

Fig. 2.15.

ground. Determine its impact velocity.

Solution
The work done on the body is mgh and the change in its kinetic energy is $\frac{1}{2}mV^2$; the work–energy equation now states that $mgh = \frac{1}{2}mV^2$, so that $V = \sqrt{(2gh)}$.

Example 2.10
Fig. 2.16 shows a body of mass m, which is sliding down a plane inclined at an angle α

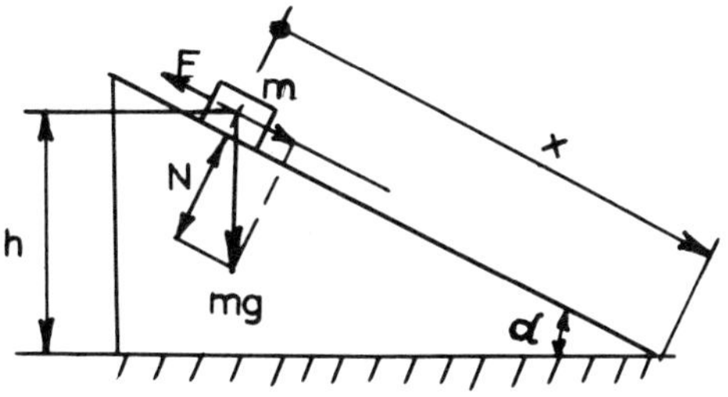

Fig. 2.16.

to the horizontal direction; the body starts from rest, and the coefficient of friction between the body and the plane is μ.

Determine the velocity of the body as a function of the distance x that it has moved along the plane.

Solution
The resultant force in the x-direction is $F_x = mg \sin \alpha - \mu mg \cos \alpha$; this is a constant force, and the work done in moving through a distance x is W.D. $= F_x x$; the component $mg \cos \alpha$ of the weight force is perpendicular to the motion and does no work.

The change in kinetic energy is $T = \frac{1}{2}mV^2$; so that $\frac{1}{2}mV^2 = F_x x$, and

$$V = \sqrt{\frac{2}{m}\left(F_x x\right)} = \sqrt{\left[2g(\sin \alpha - \mu \cos \alpha)x\right]}$$

For motion down the plane, we must have $\sin \alpha - \mu \cos \alpha > 0$; therefore $\tan \alpha > \mu$.

If $\mu = 0$, $V = \sqrt{[2g(\sin \alpha)x]} = \sqrt{(2gh)}$; this is the same velocity that the body would reach in a free fall through the distance h.

Example 2.11
Fig. 2.17 shows a block of mass M which starts from rest at A and slides in a vertical plane along AB on a smooth circular cylinder; the block leaves the cylinder at B. Determine the velocity at B and the angle ϕ. Determine also the distance a defining the point of impact, if the radius of the cylinder is $r = 2$ m.

Solution
The equation of motion in the normal direction for motion between A and B is

$$Mg \cos \theta - N = Mr\omega^2 = M\frac{V^2}{r}$$

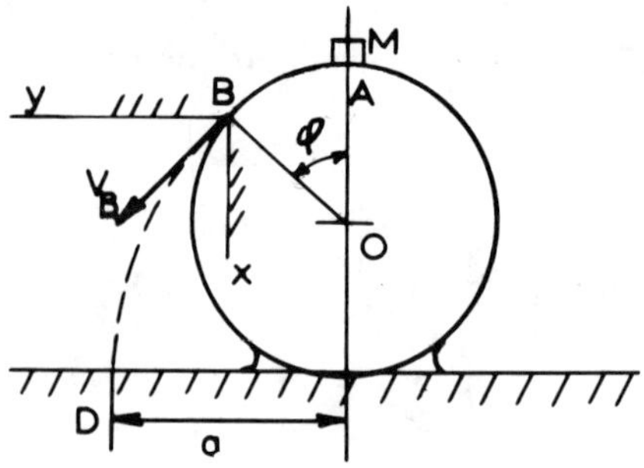

Fig. 2.17.

where N is the normal force on the block and $0 < \theta < \phi$. The block leaves the cylinder at B, where $\theta = \phi$ and $N = 0$, so that $Mg \cos \phi = MV_B^2/r$; therefore $\cos \phi = V_B^2/rg$.

During the motion from A to B, the normal force does no work, and

$$\text{W.D.} = Mgr(1 - \cos \phi)$$

The change in kinetic energy is $\frac{1}{2}MV_B^2$, so that $Mgr(1 - \cos \phi) = \frac{1}{2}MV_B^2$. Substituting $\cos \phi = V_B^2/rg$ gives the result $V_B^2 = 2rg/3$, so that $V_B = 3.62$ m/s, and $\cos \phi = \frac{2}{3}$.

Introducing a coordinate system (x, y) as shown, we have the equations of motion $M\ddot{x} = Mg$, and $M\ddot{y} = 0$, so that

$$\dot{x} = gt + V_0 \sin \phi \qquad x = \tfrac{1}{2}gt^2 + V_B(\sin \phi)t$$

$$\dot{y} = V_0 \cos \phi \qquad y = (V_B \cos \phi)t$$

Substituting the values for V_B and ϕ gives $x = 4.905t^2 + 2.70\text{t}$ and $y = 2.41t$.

The impact occurs when $x = r + r \cos \phi = 3.334$ m, so that $4.905t^2 + 2.70t - 3.334 = 0$, or $t = 0.594$ s, which gives $y = 1.43$ m, and

$$a = r \sin \phi + y = 2.92 \text{ m}$$

2.6 CONSERVATIVE FORCES. POTENTIAL ENERGY

Fig. 2.18 shows a force **F** which is moving in space along a curve from a point A to a point B.

By definition, the W.D. in this motion is

$$\text{W.D.} = \int_A^B \mathbf{F} \cdot \mathbf{ds} = \int_A^B (F_x \mathrm{d}x + F_y \mathrm{d}y + F_z \mathrm{d}z)$$

The W.D. generally depends on the curve followed by the force from A to B;

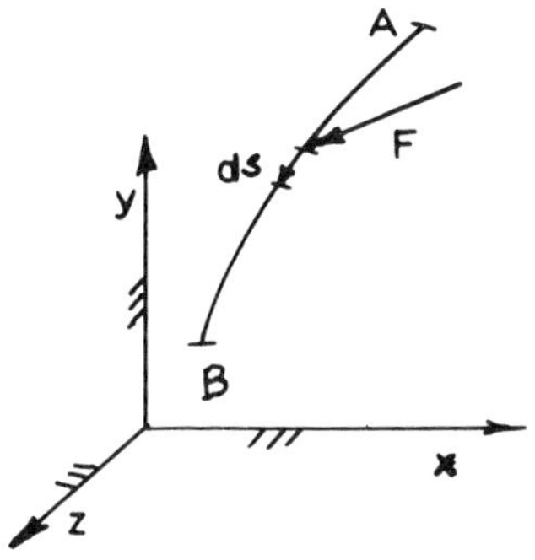

Fig. 2.18.

however, for some forces the work is independent of this curve, and depends only on the positions of A and B. An important example of such a force is the gravity force W shown in Fig. 2.19. By dividing the curve into an infinite number of horizontal and

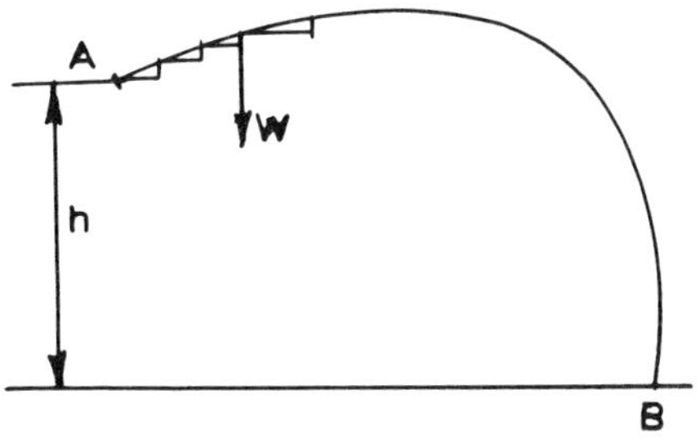

Fig. 2.19.

vertical steps, it may be seen that the horizontal movement of the gravity force produces no work, while the vertical motion of the force produces work that adds up algebraically to Wh, so that the work done depends only on the vertical difference between A and B. As a second example consider the system shown in Fig. 2.20. A

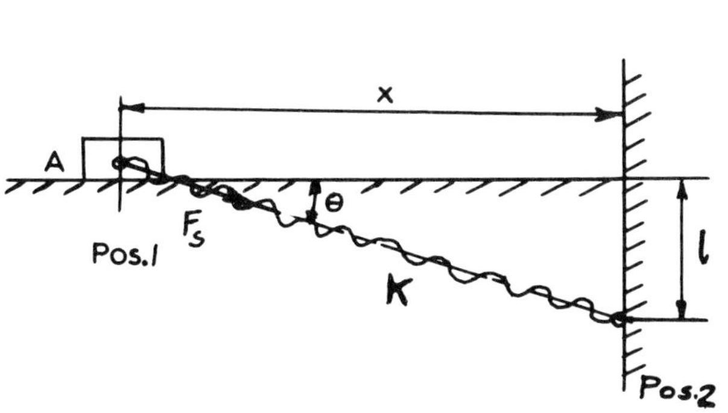

Fig. 2.20.

body A is acted upon by a massless spring of spring constant K and unstressed length l. The body slides without friction along a horizontal plane; during this motion only the spring force F_s does work on the body — in fact only the horizontal component does any work; the work done by F_s from position 1 to position 2 is W.D. $|_1^2 = \int_0^x F_s \cos\theta \, dx$.

If the extension is e_x, the force is $F_s = Ke_x = K(\sqrt{x^2+l^2} - l)$, while $\cos\theta = x/\sqrt{(x^2+l^2)}$. The integration for the work done is straightforward and leads to W.D. $|_1^2 = \frac{1}{2}Kx^2 - Kl[\sqrt{x^2+l^2} - l] = \frac{1}{2}K[x^2 - 2l\sqrt{x^2+l^2} + 2l^2] = \frac{1}{2}K[\sqrt{x^2+l^2} - l]^2$. The work done is then independent of the angle θ, that is independent of the position of the spring in the plane, since the result is $\frac{1}{2}Ke_x^2$ which is the work done by a spring released from an extension e_x

Forces like the gravity force and the elastic force are called *potential* or *conservative forces*; other examples are the central gravity field and forces with constant magnitude parallel to themselves. The electrostatic field is also conservative. Forces for which the work done depends on the path of the force or its velocity are called *non-conservative forces*. The most important examples are dry-friction, air and fluid resistance forces, and the magnetic field forces.

Considering now the force in Fig. 2.18, we *define* the *potential energy difference* between the two positions of the force by

$$V_B - V_A = -\int_A^B \mathbf{F}\cdot d\mathbf{s}$$

The potential energy difference is evidently the *negative* of the work done by the force in the motion from A to B. The definition is given this form to make it most useful. Since we do not want to specify the path of the force, we define a potential energy only for conservative forces. Since we always use the *difference* in potential energy between two positions, an arbitrary constant may be taken as the value of the potential energy in one particular position called the *datum* position. We usually take $V=0$ in this position, which is chosen as the *most convenient* position for the particular situation at hand.

If we take point B as datum position, $V_B = 0$ and we obtain $V_A = \int_A^B \mathbf{F}\cdot d\mathbf{s}$, so that V_A = W.D. by the conservative forces when the system is moved *from position A to the datum position*. In practical problems it is simplest to use the concept of potential energy as follows: the P.E. in any position is the W.D. by potential forces when the system is moved *from that position back to the datum position*. (How the system is moved there and the forces necessary to do this do not enter the definition.)

Example 2.12

Determine the potential-energy function for the systems shown in Fig. 2.21 (a) and (b).

Solution

(a) The two forces acting on the mass m are the string tension s and the gravity force mg. The string tension is always perpendicular to the motion and so produces no

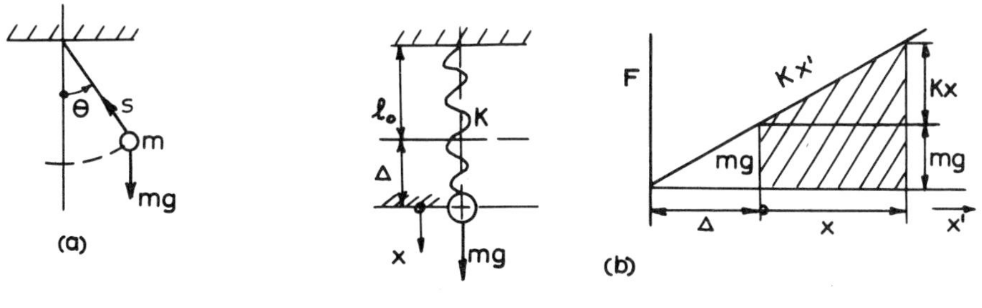

Fig. 2.21.

work, and the gravity force moves through a vertical distance $l - l\cos\theta$; taking the vertical position of the pendulum as datum position with $V = 0$, we find $V = mgl(1 - \cos\theta)$.

(b) In system (b), it is convenient to take the static equilibrium position as datum position with $V = 0$. The unstressed length of the spring is l_0; in the static position the spring is extended a length Δ, and the spring tension is mg to balance the gravity force. From the spring force diagram in Fig. 2.21(b) it follows that the potential energy at position x is $mgx + \frac{1}{2}Kx^2$ for the spring, and $-mgx$ for the gravity force, so that $V = \frac{1}{2}Kx^2$.

Using the static equilibrium position as datum position, the gravity force does not appear in the final expression for V.

2.7 PRINCIPLE OF CONSERVATION OF MECHANICAL ENERGY

If a particle is acted upon by a system of conservative forces with resultant **F**, the difference in potential energy between two positions 1 and 2 is by definition.

$$V_2 - V_1 = -\int_1^2 \mathbf{F}\cdot\mathbf{ds}$$

From the work-energy equation we find that the work done is

$$\text{W.D.} = \int_1^2 \mathbf{F}\cdot\mathbf{ds} = T_2 - T_1$$

Introducing the potential energy gives $-(V_2 - V_1) = T_2 - T_1$, or $T_1 + V_1 = T_2 + V_2$, so that

$$V + T = \text{constant} \tag{2.9}$$

This means that the sum of the potential and kinetic energy (the mechanical energy) is constant for a particle moving under the action of conservative forces; this is the *principle of conservation of mechanical energy* for a particle.

The principle is sometimes useful in establishing an equation of motion by differentiation with respect to time, since $dV/dt + dT/dt = 0$.

If friction forces or other non-conservative forces are acting, the more powerful work–energy equation may be employed instead of the principle of conservation of mechanical energy.

Example 2.13

Fig. 2.22 shows two bodies of mass M_1 and M_2, which are connected by a string over a

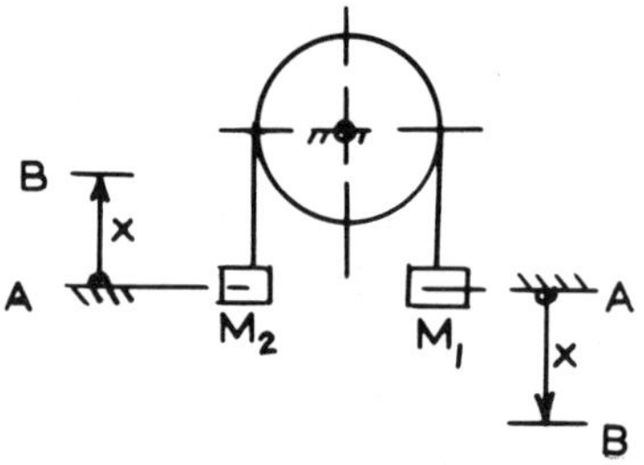

Fig. 2.22.

pulley which can rotate about a fixed axis. The inertia of the pulley and friction may be neglected; given that $M_1 > M_2$ and that the system starts from rest in position A, determine the velocity of M_1 in position B, the acceleration of M_1 and the string tension.

Solution

In position A we have the kinetic energy of the system $T_A = 0$; taking position A as our datum position, we also have $V_A = 0$. In position B the kinetic energy is $T_B = \frac{1}{2}(M_1 + M_2)\dot{x}^2$, and $V_B = (M_2 - M_1)gx$. The principle of conversation of mechanical energy now gives the equation

$$T_B + V_B = T_A + V_A$$

therefore

$$\tfrac{1}{2}(M_1 + M_2)\dot{x}^2 + (M_2 - M_1)gx = 0$$

so that

$$\dot{x} = \sqrt{\left[\frac{2(M_1 - M_2)}{M_1 + M_2}\, gx\right]}$$

Differentiating the equation leads to

$$(M_1 + M_2)\dot{x}\ddot{x} + (M_2 - M_1)g\dot{x} = 0 \quad \text{or} \quad \ddot{x} = \frac{M_1 - M_2}{M_1 + M_2}g$$

This result may be seen to be correct dimensionally. An investigation of the result for the so-called *logical extremes* shows that as $M_1 \to \infty$, $\ddot{x} \to g$, so that M_1 is in a free fall; if $M_2 = 0$, we find again $\ddot{x} = g$, or M_1 in a free fall as it should be.

Since there is no friction on the pulley, there are no tangential friction forces around the pulley; since these forces are necessary to change the string tension, it may be seen that *for a frictionless pulley the string forces are the same on both sides of the pulley*.

Calling the string force s, we may write Newton's law for the body of mass M_1, which gives the result $M_1g - s = M_1\ddot{x}$, or $s = M_1(g - \ddot{x})$; this shows that a mass accelerating downwards does not pull in the string with its total weight; part of the weight force goes to accelerate the body. $s = M_1g$ only if $\ddot{x} = 0$, or when $\dot{x}$ is constant, including $\dot{x} = 0$. Substituting $\ddot{x}$ from the previous result leads to

$$s = \frac{2M_1M_2}{M_1 + M_2}g$$

The same result is found if we write Newton's law for the body of mass M_2.

Example 2.14

Fig. 2.23 shows a vertical glass U-tube with a uniform bore of cross-sectional area A.

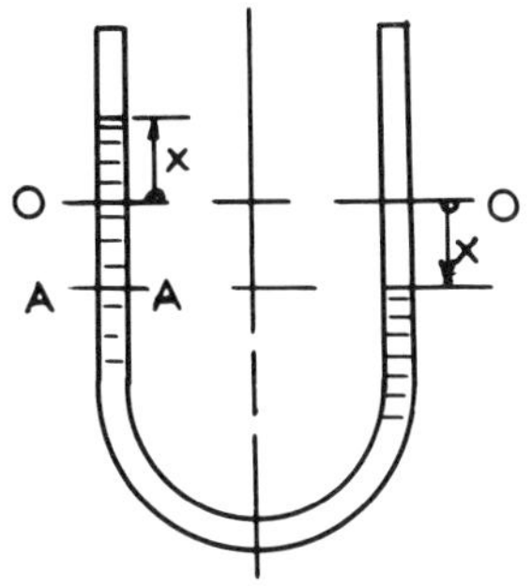

Fig. 2.23.

The tube is open at both ends, and contains a column of liquid of total length l and density ρ. Neglecting friction and the motion of air, determine the equation of motion of the liquid after a displacement x from the equilibrium position.

Solution

All the particles of the liquid move with the same displacement x and velocity $\dot{x}$. The total mass is $M = Al\rho$, and the kinetic energy is $T = \frac{1}{2}M\dot{x}^2 = \frac{1}{2}(Al\rho)\dot{x}^2$.

For a displacement x, the unbalanced force is $(2A\rho g)x$, which is proportional to the displacement. The action is, therefore, the same as for a spring constant $K = 2A\rho g$, so that the potential energy is

$$V = \tfrac{1}{2}Kx^2 = A\rho g x^2$$

We have now

$$\begin{aligned} \tfrac{1}{2}Al\rho\dot{x}^2 + \rho A g x^2 &= \textit{constant} \\ Al\rho\dot{x}\ddot{x} + 2\rho A g x\dot{x} &= 0 \end{aligned}$$

or $\ddot{x} + (2g/l)x = 0$ as equation of motion; this is S.H.M. with frequency $f = \dfrac{1}{2\pi}\sqrt{2g/l}$.

2.8 IMPULSE–MOMENTUM EQUATION

Newton's second law for the motion of a particle: $\mathbf{F} = m\mathrm{d}\mathbf{v}/\mathrm{d}t$ may be integrated directly to give

$$\int_1^2 \mathbf{F}\,\mathrm{d}t = \int_1^2 m\,\mathrm{d}\mathbf{v} = m(\mathbf{v}_2 - \mathbf{v}_1) \tag{2.10}$$

The term $\mathbf{F}\,\mathrm{d}t$ or $\int_1^2 \mathbf{F}\,\mathrm{d}t$ is called the *impulse* of the force $\mathbf{F}$ in the time interval, and the vector $m\mathbf{v}$ is called the *momentum* of the particle. Equation (2.10) then states that the impulse is equal to the total change in momentum in the same time; the equation is called the *impulse–momentum equation* and is a vector equation; scalar component equations may be taken along any line. In the x-axis direction, the equation is

$$\int_1^2 F_x\mathrm{d}t = m(\dot{x}_2 - \dot{x}_1)$$

If the *force is constant* (2.10) takes the form

$$\mathbf{F}(t_2 - t_1) = m(\mathbf{v}_2 - \mathbf{v}_1) \tag{2.11}$$

The unit of impulse is the newton second = kg m/s; that for momentum is, of course, the same.

If there is no resultant force acting on a particle in a certain time interval, the impulse is zero, and (2.10) states that there is no change in the momentum in that time interval, so that the velocity is constant in magnitude and direction. In that case the momentum is conserved, and this is a special case of the so-called *principle of conservation of momentum*.

In the case of two interacting particles, the forces between the particles are equal and opposite and act for the same time; their impulses are then equal and opposite anmd make no contribution to the total impulse on the system. If there are no external forces acting, the total momentum of the system is conserved.

The impulse–momentum equation is based directly on Newton's second law, and contains exactly the same information. It is, however, convenient in certain problems where the impulse of a force is known and the variation of the force with time is not.

Example 2.15

Fig. 2.24 shows the approximate variation of the gas pressure in a rifle barrel after

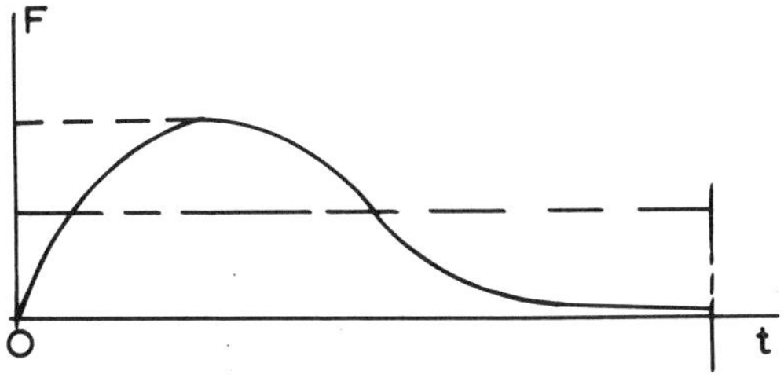

Fig. 2.24.

firing. The mass of the bullet is m and that of the rifle M; the muzzle velocity of the bullet is v; neglecting friction, determine the recoil velocity of the rifle at the moment when the bullet leaves the barrel.

Solution

The total impulse on the bullet is $\int_0^t F\,\mathrm{d}t = mv$, which is the area under the curve. If a curve is available, the impulse may be found by measuring the area with a planimeter; in the present case the curve is not known in detail and is very difficult to find experimentally owing to the high pressure and extremely short time. Incidentally, the same impulse is obtained from a constant force indicated by the horizontal line, if the area under the line is made the same as that under the curve.

The gas pressure on the rifle is equal and opposite to the pressure on the bullet and acts for the same time. The impulse on the rifle is then $\int_0^t F\,\mathrm{d}t = Mv_\mathrm{r}$, where v_r is the recoil velocity of the rifle. We have now $Mv_\mathrm{r} = mv$, or $v_\mathrm{r} = (m/M)v$, so that v_r has been determined without knowledge of the forces acting.

Example 2.16

In order to reduce the effect of the recoil forces, a horizontal gun barrel of mass 363 kg is arranged to slide axially on guides against the action of a compression spring with spring constant $K = 35 \times 10^4$ N/m. The spring is initially compressed to hold the barrel against its forward stop, with a force of 26 700 N. The gun fires a shell of mass 8.16 kg with a muzzle velocity of 549 m/s.

Determine the distance through which the barrel recoils, if friction and the time taken for the shell to traverse the barrel are neglected.

Solution

The momentum of the system is conserved, so that $MV - mv = 0$, where $M = 363$ kg, $m = 8.16$ kg and $v = 549$ m/s. This gives $V = mv/M = 12.35$ m/s.

The barrel is now in rectilinear motion, with initial velocity 12.35 m/s against a resisting force $F_x = 26\,700 + Kx$; taking $x = 0$ and $\dot{x} = 12.35$ when $t = 0$ at the initial position, we have the equation of motion.

$$M\ddot{x} = -F_x$$

so that

$$\ddot{x} + 965x = -73.6$$

As particular solution to this equation is $x = C =$ constant; by substitution, the value of C is found to be $C = -0.0761$. The solution to the homogeneous part of the equation $\ddot{x} + 965x = 0$ is $x = A \sin \omega t + B \cos \omega t$, where $\omega = \sqrt{965} = 31.1$ rad/s. The total solution is then

$$x = A \sin \omega t + B \cos \omega t - 0.0761$$

$$\dot{x} = A\omega \cos \omega t - B\omega \sin \omega t$$

When $t = 0$, $x = 0$, so $B = 0.0761$. Since $\dot{x}_0 = 12.35$, $A\omega = 12.35$, or $A = 12.35/31.1 = 0.397$. When the barrel stops, $\dot{x} = 0$, so that $A\omega \cos \omega t - B\omega \sin \omega t = 0$, or tan $\omega t = A/B = 5.21$, $\omega t = 79.14°$. Substituting this gives $x = 0.397 \cdot 0.982 + 0.0761 \cdot 0.1884 - 0.0761 = 0.328$ m recoil

The impulse–momentum equation is particularly useful in problems with large forces acting for very short times, as shown in the examples. The same situation occurs in impact problems to be considered later, and a similar type of problem occurs when a continuous stream of particles changes direction abruptly; the forces between the particles are then equal and opposite, and only external forces need be considered.

Example 2.17

Fig. 2.25 shows a horizontal water jet, which is stopped by a vertical wall. The jet is of cross-section A, the density is ρ and the constant velocity of the water is V. Determine the force on the wall.

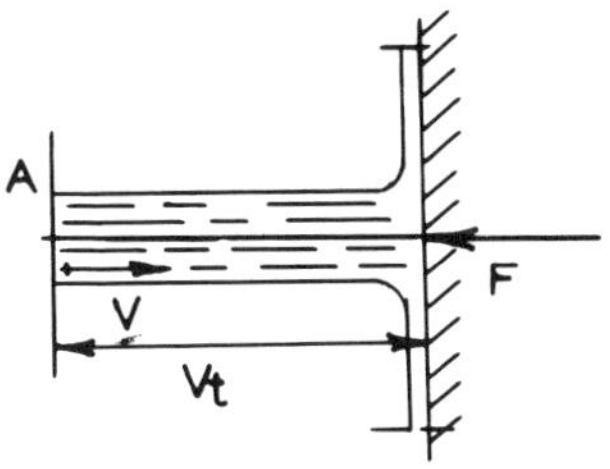

Fig. 2.25.

Solution

The volume of water passing in t s is VtA. The mass is $M = VtA\rho$ and the momentum is $MV = V^2tA\rho$. The force on the wall is F and on the water $-F$; the impulse–momentum equation gives the result

$$-Ft = M(V_2 - V_1) = M(0 - V) = -MV$$

so that

$$F = \frac{MV}{t} = V^2A\rho$$

2.9 IMPACT

An impact may be defined as a sudden contact between two bodies, involving large contact forces acting for a short time. This definition implies that at least one of the bodies must be moving before the impact; the large forces occur because of the big change in velocities in a very short time.

If a light steel hammer is used to strike a blow on a large piece of steel, an impression is made, indicating that large forces act at the impact; pressing the hammer against the steel piece produces no impression on the steel.

To investigate this further, experiments may be performed as shown in Fig. 2.26.

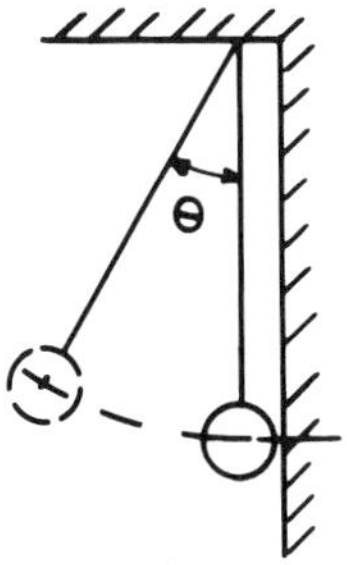

Fig. 2.26.

A pendulum with a 3 cm diameter brass sphere as a bob is released at an angle θ from the vertical position. It will be noticed that the pendulum, after impact against the wall, swings out to an angle slightly smaller than θ, because its starting velocity after impact is slightly less than its impact velocity. The time of the impact may be measured electrically, and is found to be of the order 1.5×10^{-4} s. In this extremely short time, the bob has stopped, and regained an almost equal velocity in the opposite direction. The maximum value of the impact force may be found by measuring the diameter of the impact area on the bob, after covering the wall with soot or dye, and a static compression test, giving the same compression area, enables us to measure the maximum force; some idea of the magnitude of the force may be found by calculating the *constant* force necessary to give the same impulse: the mass of the sphere is $M = 0.122$ kg, the impact velocity $V = 0.3$ m/s, and the impulse–momentum equation states that $\int_0^t F\,dt = M(V_2 - V_1)$. Taking force and velocities positive to the right, this gives

$$-Ft = M(-V - V) = -2MV$$

so that

$$F = \frac{2MV}{t} = \frac{2 \times 0.122 \times 0.3}{1.5 \times 10^{-4}} = 488 \text{ N}$$

The weight W of the sphere is $W = 1.20$ N, so that $F \sim 400\ W$; the actual maximum force may be 2–3 times bigger. These observations enable us to deal with a series of practical problems in an *approximate* way by the following two rules for impact of this nature.

1. The impact forces are so great that other forces such as gravity, friction, etc., *may be neglected* during impact.
2. The time of impact is so short that no appreciable motion can take place; the configuration just after impact may therefore be assumed to be the same as the configuration just before impact.

Consider now two spheres (Fig. 2.27) of mass m_1 and m_2 and in rectilinear motion

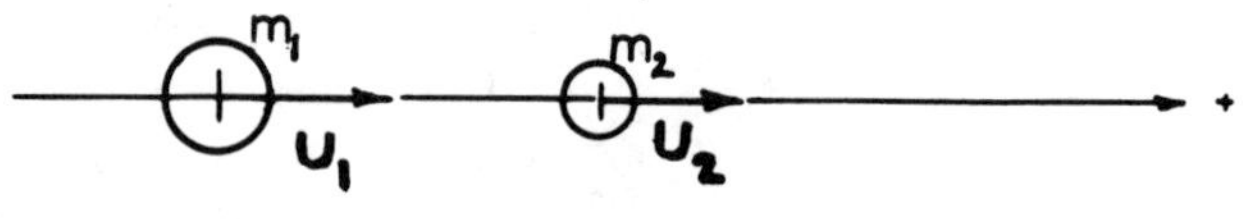

Fig. 2.27.

with velocities U_1 and U_2; with $U_1 > U_2$ an impact will occur; this is the simplest form of impact and is called a *direct central impact*. Neglecting all external forces during impact, taking velocities as positive to the right, and calling the velocities after impact V_1 and V_2, we can apply the *law of conservation of momentum*, since the impact forces are equal and opposite and act through the same time. This gives the equation.

$$m_1U_1 + m_2U_2 = m_1V_1 + m_2V_2$$

which holds for all impacts, independent of the elasticity of the bodies. This equation may also be established by writing the impulse–momentum equation for each body and adding the equations.

In any impact there is a loss of kinetic energy; this is due to friction on the impact surfaces, possible vibration of the bodies and, most important, inelastic behaviour of the bodies, so that the bodies do not return to the same shape as they had before impact. Calling the loss Δ we may now relate the kinetic energy before and after impact by the equation

$$\tfrac{1}{2}m_1U_1^2 + \tfrac{1}{2}m_2U_2^2 = \tfrac{1}{2}m_1V_1^2 + \tfrac{1}{2}m_2V_2^2 + \Delta$$

Considering now a *perfectly elastic impact* and ignoring other losses of energy, we have $\Delta = 0$ and the energy equation for this case is

$$\tfrac{1}{2}m_1U_1^2 + \tfrac{1}{2}m_2U_2^2 = \tfrac{1}{2}m_1V_1^2 + \tfrac{1}{2}m_2V_2^2$$

The equations of conservation of momentum and the kinetic energy equation may be stated as follows:

$$\begin{aligned} m_1(U_1 - V_1) &= m_2(V_2 - U_2) \\ m_1(U_1^2 - V_1^2) &= m_2(V_2^2 - U_2^2) \end{aligned}$$

The ratio of these equations gives $U_1 + V_1 = V_2 + U_2$, or $V_1 - V_2 = -(U_1 - U_2)$, so that the relative velocity after impact is the negative of the relative velocity before impact.

In any practical case there is a loss in kinetic energy, since the material is not perfectly elastic. This may be taken into account by writing the equation in the form $V_1 - V_2 = -e(U_1 - U_2)$, where the factor e is called the *coefficient of restitution* of the material, and $0 \leqslant e \leqslant 1$. In this form the equation is called *Newton's empirical law of impact*, since it is based on experiments.

We now have the following two equations for *direct central impact*:

$$\begin{aligned} m_1U_1 + m_2U_2 &= m_1V_1 + m_2V_2 \\ V_1 - V_2 &= -e(U_1 - U_2) \end{aligned} \qquad (2.12)$$

Assuming that U_1, U_2 and e are known, the velocities V_1 and V_2 may be determined from (2.12).

If $e = 0$, we find $V_2 = V_2$ and the first equation gives the solution; the two bodies continue as one body and in a distorted shape, and there is no tendency to rebound; this is called a *plastic impact*; a material like putty has a value of e close to zero. If $e = 1$, we have the same relative velocity before and after impact. This is called a *perfectly elastic impact*, and the total energy is constant during impact. For $0 < e < 1$, the impact is called *semi-elastic*. For glass or polished hardened steel the value of e is about 0.9; for ivory, $e \sim 0.8$; and for lead, $e \sim 0.1$.

Example 2.18
Fig. 2.28 shows a pile of mass M_2 which is driven into the ground by blows from a

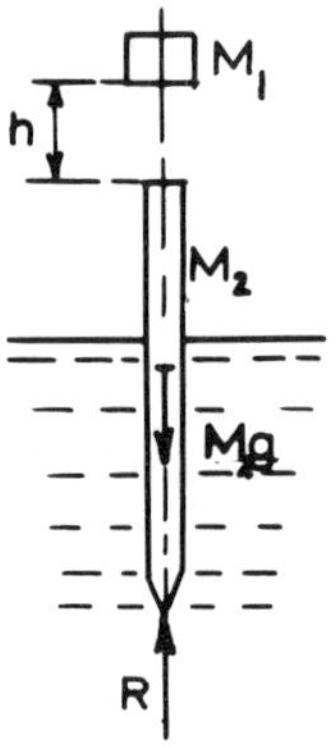

Fig. 2.28.

hammer of mass M_1 which falls freely through a height of h. The impact may be assumed plastic, and one blow moves the pile a distance b into the ground. Determine the resistance R from the ground if it is assumed constant.

Solution
The impact velocity of the hammer is $U_1 = \surd(2gh)$, while the initial velocity of the pile is $U_2 = 0$; *taking* $e = 0$ in equation (2.12) gives $V_2 = V_1 = V$, so that M_1 and M_2 move as one body after impact; the first of the equations (2.12) now gives $M_1\surd(2gh) = (M_1 + M_2)V$, or $V = M_1\surd(2gh)/(M_1 + M_2)$. The kinetic energy right after impact is $T_1 = \frac{1}{2}(M_1 + M_2)V^2 = ghM_1^2/(M_1 + M_2)$; the pile and hammer move through a distance b and stop; the work done during this motion is $(M_1 + M_2)gb - Rb$; the work-energy equation states that W.D. $= T_2 - T_1$, which gives

$$(M_1 + M_2)gb - Rb = 0 - \frac{ghM_1^2}{M_1 + M_2}$$

so that

$$R = \frac{ghM_1^2}{b(M_1 + M_2)} + (M_1 + M_2)g$$

Example 2.19
Fig. 2.29 shows a ballistic pendulum consisting of a bag of sand of mass M suspended

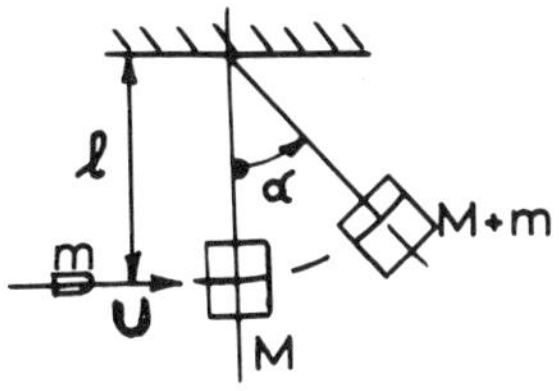

Fig. 2.29.

from a string of length l. A bullet of mass m is fired into the sand in a horizontal direction at a speed U, while the pendulum is at rest; the pendulum swings out an angle α. Assuming that the bullet is retained by the sand, develop a formula for the velocity U of the bullet in terms of the ratio M/m, l and α. Determine U if $M/m = 402.5$, $l = 0.954$ m and $\alpha = 28.4°$.

Solution
The velocities before impact are $U_1 = U$ and $U_2 = 0$; after impact the velocity is V for the sand and the bullet. Conservation of momentum, equation (2.12), gives

$$mU = (M+m)V \quad \text{or} \quad U = \frac{M+m}{m}V$$

The kinetic energy just after impact is $T_1 = \frac{1}{2}(M+m)V^2$, while $T_2 = 0$; the work done is W.D. $= -(M+m)gl(1-\cos\alpha) = T_2 - T_1 = -\frac{1}{2}(M+m)V^2$, so that $V = \sqrt{[2gl(1-\cos\alpha)]}$; the final result is

$$U = \left(\frac{M}{m}+1\right)\sqrt{[2gl(1-\cos\alpha)]}$$

With the numerical values given

$$U = (402.5+1)\sqrt{[2\times 9.81\times 0.954(1-\cos 28°.4)]} = 604 \text{ m/s}$$

2.10 D'ALEMBERT'S PRINCIPLE

Newton's second law for a particle, $\mathbf{F} = m\ddot{\mathbf{r}}$, may be stated in the form

$$F + (-m\ddot{\mathbf{r}}) = \mathbf{0} \tag{2.13}$$

The term $-m\ddot{\mathbf{r}}$ has the dimension of a force, the minus sign indicating a direction always opposite to the acceleration $\ddot{\mathbf{r}}$. The expression $-m\ddot{\mathbf{r}}$ may be considered as a fictitious force, convenient in certain applications, although it is not a force in the usual meaning of the word, and cannot be applied by external means. It has been given the name *inertia force*.

Writing equation (2.13) obviously corresponds to writing an equation in dynamics in the same form as a static equilibrium equation: this form of the equation of motion was first given by the French mathematician D'Alembert in his book *Traité de Dynamique* in 1743. Using D'Alembert's principle, we apply a force $-m\ddot{\mathbf{r}}$ to the particle in question, so that the forces may be considered to be in equilibrium. This situation is called *dynamic equilibrium*.

Once the inertia forces have been applied, the equilibrium equations of *statics* may be applied, and this may sometimes result in simplifications in the solution.

Example 2.20

Fig. 2.30 shows a light frictionless pulley of radius r, which carries two masses M_1 and

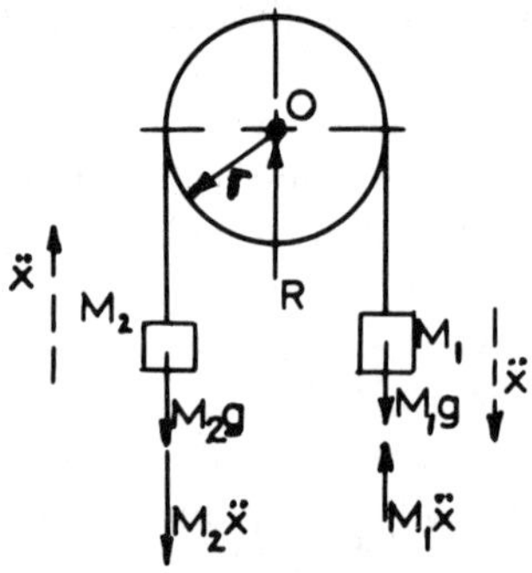

Fig. 2.30.

M_2 connected by a string; assuming that $M_1 > M_2$, determine the acceleration of the masses, by applying D'Alembert's principle.

Solution

The external forces R, M_1g and M_2g are first applied to the system. The acceleration of M_1 is $\ddot{x}$ downwards; the inertia force is $-M_1\ddot{x}$ and this force is applied, as shown, to M_1; M_2 has an acceleration $\ddot{x}$ upwards, so the inertia force $-M_2\ddot{x}$ is then applied downwards on M_2. The system is now in dynamic equilibrium, and equations of statics may be used. Taking moments about point O, to eliminate the unknown reaction R, gives the result

$$M_2gr + M_2r\ddot{x} + M_1r\ddot{x} - M_1gr = 0$$

from which

$$\ddot{x} = \frac{M_1 - M_2}{M_1 + M_2} g$$

The same result was found in Example 2.13. The advantage of the present method is that the fixed reaction and the string tension need not be considered.

2.11 PRINCIPLE OF VIRTUAL WORK

A virtual displacement $\delta\mathbf{r}$ is an assumed infinitesimal displacement of a particle, in which the forces and constraints are unchanged; the notation $\delta\mathbf{r}$ is used to distinguish a virtual displacement from a real displacement $\mathrm{d}\mathbf{r}$, which takes place in a time $\mathrm{d}t$ in which forces and constraints may change.

We may write $\delta\mathbf{r}$ in terms of its components as $\delta x\mathbf{i} + \delta y\mathbf{j} + \delta z\mathbf{k}$. Since $\delta\mathbf{r}$ is infinitesimal, it follows the same rules as $\mathrm{d}\mathbf{r} = \mathrm{d}x\mathbf{i} + \mathrm{d}y\mathbf{j} + \mathrm{d}z\mathbf{k}$: for instance, if a displacement y at a point of a mechanism is related to the displacement x of another point by $y = \sin x$, we obtain $\delta y = \cos x\, \delta x$; the determination of virtual displacements then becomes a problem in geometry.

Any static force system acting on a particle may be reduced to a single resultant force $\mathbf{F}$; if the particle is given a virtual displacement $\delta\mathbf{r}$, the work done by $\mathbf{F}$ is the *virtual work* $\delta(\text{W.D.}) = \mathbf{F}\cdot\delta\mathbf{r}$. If the force system is in equilibrium, it will also be in equilibrium during this virtual displacement, following the definition of virtual displacements, and we have $\delta(\text{W.D.}) = 0$, for any virtual displacement. This is the *principle of virtual work* in statics.

In dynamics, the acting forces are generally not in equilibrium, but if D'Alembert's principle is applied, we have dynamic equilibrium, and the principle of virtual work may then be applied as in a static case.

The combination of D'Alembert's principle and the principle of virtual work is a powerful method in many dynamics problems. The principle of virtual work in dynamics may now be stated in the form

$$(\mathbf{F} - m\ddot{\mathbf{r}})\cdot\delta\mathbf{r} = 0 \tag{2.14}$$

Equation (2.14) is called *D'Alembert's equation*, and is of fundamental importance in further developments in dynamics; it was used by Lagrange, about 1790, to develop his famous equations in dynamics, which will be discussed in Chapter 3. For more complicated problems, Lagrange's equations largely supersede D'Alembert's equation.

Example 2.21

Determine the acceleration of the masses in Example 2.20, by using the principle of virtual work.

Solution
Applying all external forces and inertia forces as shown in Fig. 2.30, we have created dynamic equilibrium. Giving the mass M_1 a virtual displacement δx, the principle of virtual work states that $\delta(\text{W.D.}) = 0$.

Now $\delta(\text{W.D.}) = M_1 g\ \delta x - M_1\ddot{x}\ \delta x - M_2 g\ \delta x - M_2\ddot{x}\ \delta x = 0$, and cancelling δx, we find

$$\ddot{x} = \frac{M_1 - M_2}{M_1 + M_2} g$$

as in Example 2.20

The great advantage of the principle is that internal forces, occurring in equal and opposite pairs, produce no work in the work summation; fixed reactions and normal forces also produce no work. All of these forces may therefore be omitted in the calculation of the work done.

PROBLEMS

2.1 In Fig. 2.31, M_1 represents a mass which slides on a plane which is inclined at an

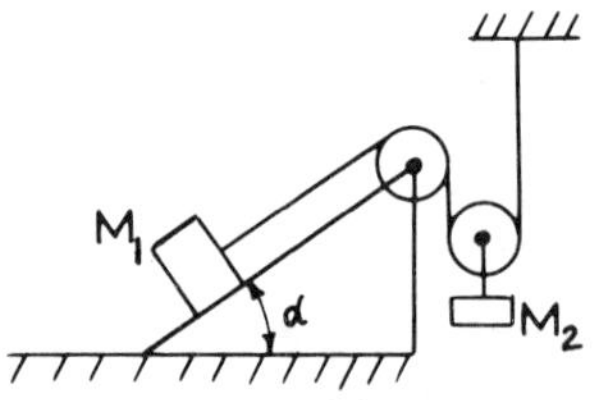

Fig. 2.31.

angle α to the horizontal. A string attached to M_1 passes over a fixed pulley at the top of the slope, and then around a free pulley before being anchored to 'ground'. The free pulley carries a mass m_2.

Neglecting the inertia and friction of the pulleys, derive a formula for the tension in the string, for the case when M_1 *is moving up the plane*. Coefficient of friction is μ.

Calculate the magnitude of the tension when $\alpha = 45°$, $M_2 = 200$ kg, $M_1 =$ 100 kg and the coefficient of friction between M_1 and the plane is 0.2. Determine the acceleration of M_2.

2.2 A system of weights and pulleys is arranged in a vertical plane as shown in Fig. 2.32. Neglecting friction, inertia and weight of the pulleys, find the acceleration

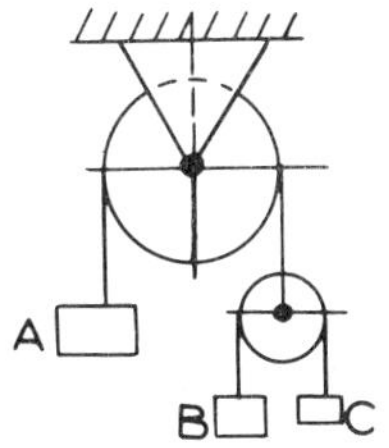

Fig. 2.32.

of each weight if their mass magnitudes are in the ratios $A{:}B{:}C = 4{:}2{:}1$. The string tension stays constant across a pulley without inertia and friction.

2.3 The mass m (Fig. 2.33) is supported by wires subjected to tensile force T.

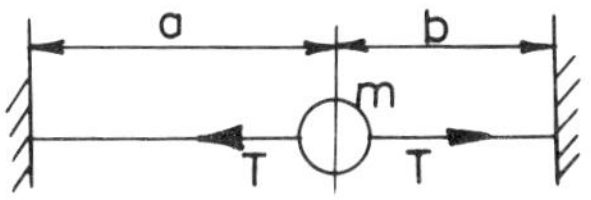

Fig. 2.33.

Assuming small amplitudes, determine the natural frequency of vibration in a plane perpendicular to the wire. Show that the period of vibration is greatest when $a = b$. Assume that T is large and unchanged for small displacements. (Neglect gravity.)

2.4 In Fig. 2.34 a small ball of mass 0.5097 kg starts from rest at O and slides down

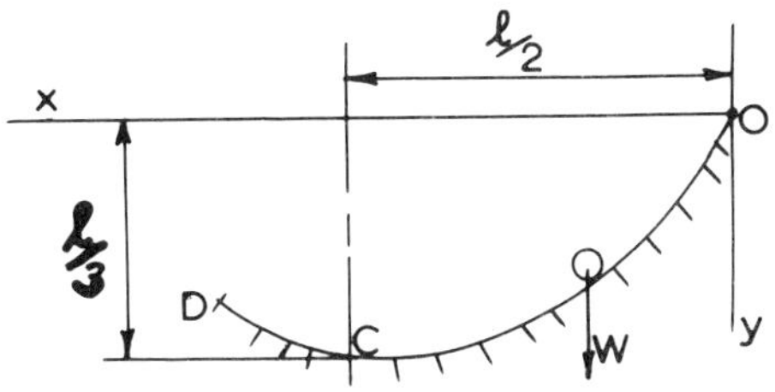

Fig. 2.34.

the smooth track OCD under the influence of gravity. Find the *reaction* on the ball at C if the curve OCD is defined by the equation $y = \frac{l}{3}\sin(\pi x/l)$. Repeat for position $x = \frac{l}{4}$.

2.5 An aircraft is to be launched by catapult at a speed of 241 km/h, at a constant acceleration of $3g$. Determine the minimum distance for take-off.

When landing, the aircraft is arrested by a heavy rubber band acting as a spring with spring constant K N/m. The aircraft weighs 80 000 N and has a landing speed of 193 km/h; during landing, a constant braking force from the wheels and propellers of 22 000 N is applied.

Determine the value of K, if the plane must be stopped in a distance of 45.7 m.

2.6 Find the velocity of escape for a projectile fired from the surface of the moon in a vertical direction. Repeat for the earth, neglecting air resistance. $g_m = 1.62$ m/s^2; $r_m = 1738$ km; $r_e = 6368$ km.

2.7 A weight W_1 falls through a height h onto a block W_2 supported by a spring of constant K (Fig. 2.35). Assuming *plastic* impact, find the maximum compression

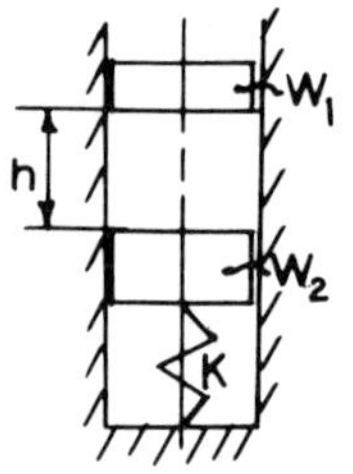

Fig. 2.35.

Δ of the spring from the equilibrium position shown. $W_1 = W_2 = 10$ N, $K =$ 10 N/cm, $h =$ 3 cm.

2.8 For the pile and piledriver shown in Fig. 2.36, $W_1 = 2000$ N, $W_2 =$ 1000 N, $h =$

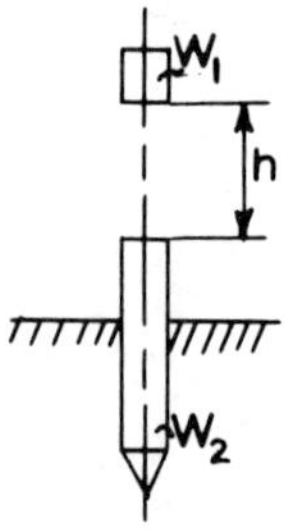

Fig. 2.36.

10 m and $e = \frac{1}{4}$. If the resistance to penetration is constant $= 60\,000$ N, find the minimum number of blows necessary to drive the pile to a depth of more than 1 m. (Neglect the possibility of further impacts from rebound.)

3

Lagrange's equations

Lagrange's equations were developed by the great French mathematician J. L. Lagrange and published in his celebrated book *Analytical Mechanics* (Paris, 1788). Lagrange's equations are of great importance and are widely used in engineering, physics and mathematics.

For complicated dynamics systems these equations give the simplest means of establishing the equations of motion of the system. Before Lagrange's equations can be developed it is necessary to discuss certain concepts and definitions involved in the equations.

3.1 DEGREES OF FREEDOM AND EQUATIONS OF CONSTRAINT

The number of degrees of freedom n of a particle is defined as the number of independent coordinates that are necessary to determine the position of the particle.

A fixed particle may be said to have $n = 0$; a particle moving in rectilinear motion has $n = 1$, since one coordinate is sufficient to define the position of the particle; if the particle is in plane curvilinear motion, we also find $n = 1$; the two coordinates x and y, to specify the position of the particle in the plane, are connected by an equation $y = f(x)$, which gives the path of the particle. In this case x and y are therefore not independent; the equation $y = f(x)$ is called an equation of constraint, and each equation of constraint lowers the number of degrees of freedom by one.

If the particle is free to move in a plane, $n = 2$. A particle moving on a space curve has $n = 1$; for a surface in space, $n = 2$. A particle free to move in space has $n = 3$; this is then the maximum possible number of degrees of freedom of a particle.

The concepts of degrees of freedom and equations of constraint are of fundamental importance in more advanced dynamics.

3.2 GENERALISED COORDINATES

In most problems the position of a body may be given in rectangular coordinates. In general these are not independent but connected by equations of constraints, and to avoid the complications of these equations, it is a great advantage if some independent quantities can be found to describe the system. These quantities are called generalised coordinates, and there are just enough of them to specify the position of the system. The number of generalised coordinates is therefore equal to the number of degrees of freedom of the system.

In most cases several different sets of generalised coordinates may be found; often the usual rectangular coordinates and angles of rotation may be used. Quantities other than lengths or angles may also be used: for instance the pressure in a cylinder may be taken as a generalised coordinate for the piston, if it uniquely determines the position of the piston.

Because different sets of generalised coordinates may be found, it is customary to use the notation $(q_1, q_2, \ldots, q_n)$ for the generalised coordinates of a system with n degrees of freedom.

The usual rectangular coordinates (x, y, z) may be expressed in terms of the generalised coordinates by certain functions:

$$\begin{aligned} x_i &= G_i\,(q_1, \ldots, q_n) \\ y_i &= H_i\,(q_1, \ldots, q_n) \qquad (3.1) \\ z_i &= K_i\,(q_1, \ldots, q_n) \end{aligned}$$

As an example, consider the system shown in Fig. 3.1. This is a double pendulum

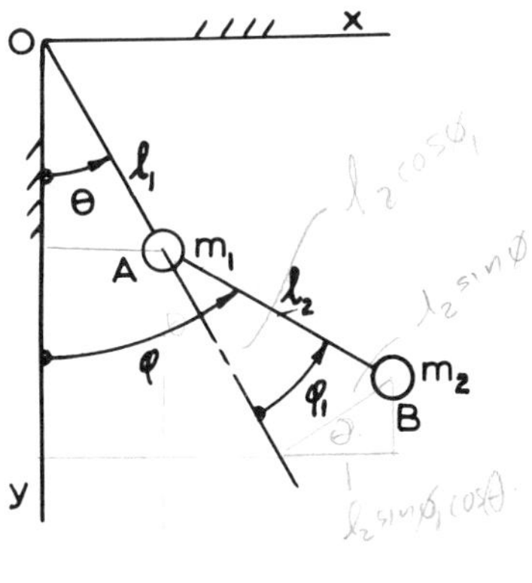

Fig. 3.1.

swinging in the vertical plane. The system has two degrees of freedom. The position of the two bobs A and B may be given by rectangular coordinates x_1, y_1, x_2 and y_2; these are not independent, but are connected by the equations of constraint $x_1^2 + y_1^2 = l_1^2$ and $(x_2 - x_1)^2 + (y_2 - y_i)^2 = l_2^2$. Only two of the coordinates may be used as independent generalised coordinates. A simpler set of generalised coordinates are the angles θ and ϕ as shown, which are absolute coordinates. We may also use the angles θ and ϕ_1 measured as shown on the figure, and in this case ϕ_1 is a relative coordinate. A mixture of absolute and relative coordinates is often most useful for use as generalised coordinates.

Using θ and ϕ as coordinates, we find for the functions (3.1) in this case:

$$x_1 = G_1(\theta,\phi) = l_1 \sin \phi \qquad x_2 = G_2(\theta,\phi) = l_1 \sin \theta + l_2 \sin \phi$$

$$y_1 = H_1(\theta,\phi) = l_1 \cos \theta \qquad y_2 = H_2(\theta,\phi) = l_1 \cos \theta + l_2 \cos \phi$$

3.3 GENERALISED FORCES

Consider a general case of a system with n degrees of freedom. The position of the system may be given by a set of generalised coordinates $(q_1, \ldots, q_n)$, and these must all be independent, which means that we may give a small increment δq_i to the coordinate q_i, without changing any of the other coordinates.

To this change δq_i corresponds a certain displacement of the system, and the forces acting will move through certain distances, dependent on δq_i, and perform an amount of work δ(W.D.).

We now express the work done as $\delta(\text{W.D.}) = \delta q_i[Q_i]$, where Q_i is an expression involving the forces acting. The expression Q_i is called the generalised force corresponding to the generalised coordinate q_i. The generalised force is thus that quantity by which we must multiply δq_i, to obtain the work done by all the forces acting on a displacement δq_i of the system, with all the other coordinates kept constant. All the forces may be assumed to be of constant magnitude in working out δ(W.D.).

As an example, consider the double pendulum in Fig. 3.1. Using the angles θ and ϕ as generalised coordinates, we find the generalised force Q_ϕ corresponding to the coordinate ϕ by keeping θ constant and give a small increment $\delta\phi$ to ϕ as shown in Fig. 3.2.

The point B moves up a distance $\delta h_1 = l_2\, \delta\phi \sin \phi$, and the total work done is $\delta(\text{W.D.}) = -m_2 g l_2\, \delta\phi \sin \phi = \delta\phi[Q_\phi]$. We thus find $Q_\phi = -m_2 g l_2 \sin \phi$; this has the dimension of a moment (N m), which agrees with the fact that ϕ is an angle, and $Q_\phi \delta\phi$ must have the dimension of work.

To find the generalised force Q_θ corresponding to θ, we keep ϕ constant and give a small increment $\delta\theta$ to θ. The point A moves up a distance $\delta h_2 = l_1\, \delta\theta \sin \theta$, and since ϕ is unchanged, the line AB moves parallel to itself, so that point B also moves a distance δh_2. The total work done is thus

$$\delta(\text{W.D.}) = -(m_1 g + m_2 g)\delta h_2 = -(m_1 + m_2) g l_1\, \delta\theta \sin \theta = \delta\theta\, [Q_\theta]$$

so that $Q_\theta = -(m_1 + m_2) g l_1 \sin \theta$, which again has the dimensions of a moment.

In general only the external forces are involved in the expressions for generalised forces. Of these external forces, fixed reactions and normal forces do no work and

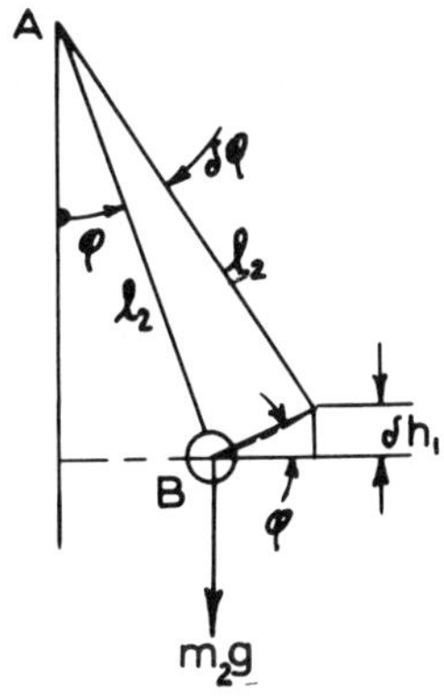

Fig. 3.2.

are not involved; internal forces occur in equal and opposite pairs and again do no work in a summation of work done. There are also cases where no external forces do any work on δq_i, and $Q_i = 0$.

3.4 GENERALISED FORCES AND POTENTIAL ENERGY

The generalised forces may be found in a different, and sometimes simpler, way for forces with a potential. The two most common conservative forces are gravity and elastic.

If the position of a conservative system is determined by the generalised coordinates $(q_1, \ldots, q_n)$, the potential energy V is a function of the coordinates $V = f(q_1, \ldots, q_n)$. If the coordinate q_i is given an increment δq_i, while all the other coordinates are kept constant, the incremental change in V is $\delta V = (\partial V/\partial q_i)\delta q_i$. By the definition of potential energy, this change is also the negative of the work done by all the potential forces, so that $\delta V = (\partial V/\partial q_i)\delta q_i = -\delta(\text{W.D.})$. The generalised force Q_i corresponding to q_i is defined by $Q_i \delta q_i = \delta(\text{W.D.})$; thus $Q_i \delta q_i = -\delta V = -(\partial V/\partial q_i)\delta q_i$, or

$$Q_i = -\partial V/\partial q_i \tag{3.2}$$

For conservative systems, the generalised forces may thus be found by partial differentiation of the potential-energy function expressed in generalised coordinates. We shall find that this way of determining the generalised forces sometimes results in less labour than the definition of generalised forces in section 3.3.

As an example, consider the generalised forces for the system in Fig. 3.1. The potential-energy function is

$$V = m_1 g l_1 (1 - \cos\theta) + m_2 g\,[l_1 (1 - \cos\theta) + l_2 (1 - \cos\phi)]$$

where the datum position with $V=0$ has been taken with both pendulums in the vertical line. Application of (3.2) directly gives the result

$$Q_\phi = -\frac{\partial V}{\partial \theta} = -m_2 g l_2 \sin\phi$$

and

$$Q_\theta = -\frac{\partial V}{\partial \theta} = -(m_1+m_2)\, g l_1 \sin\theta$$

as found before.

3.5 SOME IMPORTANT RELATIONSHIPS FOR PARTIAL AND TIME DERIVATIVES OF FUNCTIONS OF SEVERAL VARIABLES

In the functions $x = G(q_1, q_2)$ and $y = H(q_1, q_2)$, the variables q_1 and q_2 are functions of time.

For the function G we have the usual differential relationship

$$\mathrm{d}x = \frac{\partial x}{\partial q_1}\mathrm{d}q_1 + \frac{\partial x}{\partial q_2}\mathrm{d}q_2$$

from which

$$\frac{\mathrm{d}x}{\mathrm{d}t} = \dot{x} = \frac{\partial x}{\partial q_1}\frac{\mathrm{d}q_1}{\mathrm{d}t} + \frac{\partial x}{\partial q_2}\frac{\mathrm{d}q_2}{\mathrm{d}t} = \frac{\partial x}{\partial q_1}\dot{q}_1 + \frac{\partial x}{\partial q_2}\dot{q}_2$$

Differentiating $\dot{x}$ partially w.r.t. $\dot{q}_1$ gives the result

$$\frac{\partial \dot{x}}{\partial \dot{q}_1} = \frac{\partial x}{\partial q_1} \tag{3.3}$$

The operation is sometimes called 'to remove the dots'. A similar relationship exists for the function $y = H(q_1, q_2)$.

If we differentiate the expression for $\dot{x}$ partially w.r.t q_1, the result is

$$\frac{\partial \dot{x}}{\partial q_1} = \frac{\partial^2 x}{\partial q_1^2}\dot{q}_1 + \frac{\partial^2 x}{\partial q_1 \partial q_2}\dot{q}_2$$

From the function $x = G(q_1, q_2)$ we find in general

$$\frac{\partial x}{\partial q_1} = f(q_1, q_2)$$

and

$$\mathrm{d}\left(\frac{\partial x}{\partial q_1}\right) = \frac{\partial}{\partial q_1}\left(\frac{\partial x}{\partial q_1}\right)\mathrm{d}q_1 + \frac{\partial}{\partial q_2}\left(\frac{\partial x}{\partial q_1}\right)\mathrm{d}q_2 = \frac{\partial^2 x}{\partial q_1^2}\mathrm{d}q_1 + \frac{\partial^2 x}{\partial q_1 \partial q_2}\mathrm{d}q_2$$

from which

$$\frac{\mathrm{d}}{\mathrm{d}t}\left(\frac{\partial x}{\partial q_1}\right) = \frac{\partial^2 x}{\partial q_1^2}\dot{q}_1 + \frac{\partial^2 x}{\partial q_1 \partial q_2}\dot{q}_2$$

Comparing this to the expression above for $\partial \dot{x}/\partial q_1$ shows that

$$\frac{\mathrm{d}}{\mathrm{d}t}\left(\frac{\partial x}{\partial q_1}\right) = \frac{\partial}{\partial q_1}\left(\frac{\mathrm{d}x}{\mathrm{d}t}\right) \tag{3.4}$$

This expression shows that we can determine $\partial x/\partial q_1$ first and then $\mathrm{d}/\mathrm{d}t\,(\partial x/\partial q_1)$, or we can determine $\mathrm{d}x/\mathrm{d}t$ first and then $\partial/\partial q_1(\mathrm{d}x/\mathrm{d}t)$, the final result will be the same. In practical work it is simplest to determine $\partial x/\partial q_1$ first, since $\partial x/\partial q_1$ is generally a simpler expression than $\mathrm{d}x/\mathrm{d}t$.

A similar expression to (3.4) may be obtained for $y = H(q_1, q_2)$.

In the derivation of Lagrange's equations we will find the expression $\ddot{x}(\partial x/\partial q_1)$; this may be expressed in first-order derivatives by using the expression

$$\frac{\mathrm{d}}{\mathrm{d}t}\left(\dot{x}\frac{\partial x}{\partial q_1}\right) = \ddot{x}\frac{\partial x}{\partial q_1} + \dot{x}\frac{\mathrm{d}}{\mathrm{d}t}\left(\frac{\partial x}{\partial q_1}\right)$$

from which

$$\ddot{x}\frac{\partial x}{\partial q_1} = \frac{\mathrm{d}}{\mathrm{d}t}\left(\dot{x}\frac{\partial x}{\partial q_1}\right) - \dot{x}\frac{\mathrm{d}}{\mathrm{d}t}\left(\frac{\partial x}{\partial q_1}\right)$$

Using formulae (3.3) and (3.4) on this, we find

$$\ddot{x}\frac{\partial x}{\partial q_1} = \frac{d}{dt}\left(\dot{x}\frac{\partial \dot{x}}{\partial \dot{q}_1}\right) - \dot{x}\frac{\partial \dot{x}}{\partial q_1} = \frac{d}{dt}\left(\frac{\partial \dot{x}^2/2}{\partial \dot{q}_1}\right) - \frac{\partial(\dot{x}^2/2)}{\partial q_1} \tag{3.5}$$

with a similar expression $\ddot{y}(\partial y/\partial q_1)$.

3.6 DERIVATION OF LAGRANGE'S EQUATIONS

The equations may be derived as follows.

Consider a particle of mass m (Fig. 3.3) in general plane motion.

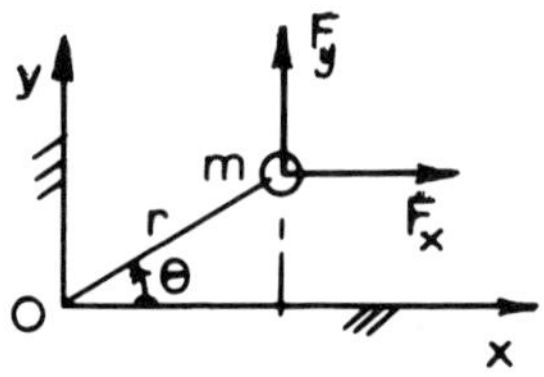

Fig. 3.3.

The particle has two degrees of freedom, and therefore two generalised coordinates q_1 and q_2 are necessary to specify its position; these coordinates may be taken, for instance, as polar coordinates (r, θ), with $x = r\cos\theta$ and $y = r\sin\theta$; taking $r = q_1$ and $\theta = q_2$ we have $x = G(q_1, q_2) = q_1 \cos q_2$ and $y = H(q_1, q_2) = q_1 \sin q_2$. It is possible to establish different sets of generalised coordinates and in the general case we take

$$\begin{cases} x = G(q_1, q_2) \\ y = H(q_1, q_2) \end{cases}$$

Incremental changes δx and δy are determined by the usual differential relationships

$$\delta x = \frac{\partial x}{\partial q_1}\delta q_1 + \frac{\partial x}{\partial q_2}\delta q_2$$

$$\delta y = \frac{\partial y}{\partial q_1}\delta q_1 + \frac{\partial y}{\partial q_2}\delta q_2$$

Since q_1 and q_2 are independent we may now give an increment δq_1 to q_1 and keep q_2 constant, so that $\delta q_2 = 0$; we find then

$$\delta(\text{W.D.}) = F_x\delta x + F_y\delta y = \left[F_x\frac{\partial x}{\partial q_1} + F_y\frac{\partial y}{\partial q_1}\right]\delta q_1 = [Q_1]\delta q_1$$

where Q_1 is the generalised force corresponding to q_1; we have now

$$F_x\frac{\partial x}{\partial q_1} + F_y\frac{\partial y}{\partial q_1} = Q_1$$

Introducing Newton's second law:

$$\begin{cases} F_x = m\ddot{x} \\ F_y = m\ddot{y} \end{cases}$$

gives

$$m\ddot{x}\frac{\partial x}{\partial q_1} + m\ddot{y}\frac{\partial y}{\partial q_1} = Q_1$$

Substituting $\ddot{x}(\partial x/\partial q_1)$ and $\ddot{y}(\partial y/\partial q_1)$ from equation (3.5) in this leads to

$$m\left[\frac{\mathrm{d}}{\mathrm{d}t}\frac{\partial}{\partial \dot{q}_1}\left(\frac{\dot{x}^2}{2}+\frac{\dot{y}^2}{2}\right) - \frac{\partial}{\partial q_1}\left(\frac{\dot{x}^2}{2}+\frac{\dot{y}^2}{2}\right)\right] = Q_1$$

Introducing the kinetic energy $T = \frac{1}{2}mV^2 = m(\dot{x}^2/2 + \dot{y}^2/2)$ gives the result

$$\frac{\mathrm{d}}{\mathrm{d}t}\frac{\partial T}{\partial \dot{q}_1} - \frac{\partial T}{\partial q_1} = Q_1$$

This is Lagrange's equation for coordinate q_1. Similarly by using an increment δq_2 and keeping q_1 constant, so that $\delta q_1 = 0$, we find the same equation in the coordinate q_2. We may expand the result to three-dimensional motion by including the z-terms in the development; the result is *Lagrange's equations for a particle*:

$$\frac{\mathrm{d}}{\mathrm{d}t}\frac{\partial T}{\partial \dot{q}_i} - \frac{\partial T}{\partial q_i} = Q_i \quad i = 1,2,3 \tag{3.6}$$

If the particle has three degrees of freedom, we obtain three Lagrange equations for its motion.

Example 3.1
Fig. 3.4 shows a body of mass m which is in rectilinear motion owing to the action of a force $P(t)$. Neglecting friction, determine the equation of motion by using equation (3.6).

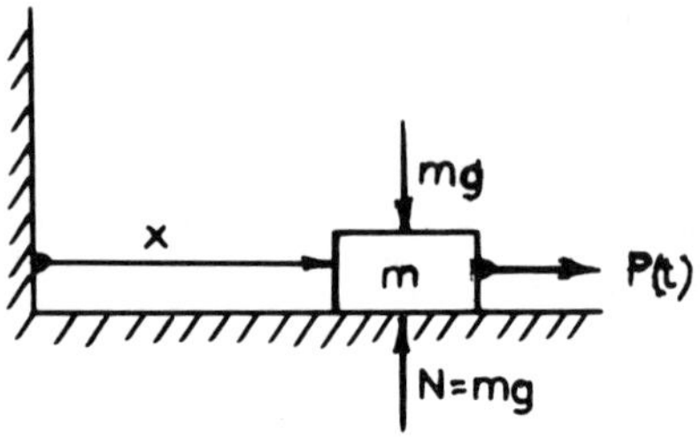

Fig. 3.4.

Solution
The body has one degree of freedom, and taking x as the generalised coordinate the equation (3.6) has the form

$$\frac{\mathrm{d}}{\mathrm{d}t}\frac{\partial T}{\partial \dot{x}} - \frac{\partial T}{\partial x} = Q_x \qquad T = \tfrac{1}{2}m\,V^2 = \tfrac{1}{2}m\dot{x}^2$$

$$\frac{\partial T}{\partial \dot{x}} = m\dot{x} \qquad \frac{\mathrm{d}\partial T}{\mathrm{d}t\partial \dot{x}} = m\ddot{x} \qquad \text{and} \qquad \frac{\partial T}{\partial x} = 0$$

Giving an increment δx to x, we find

$$\delta(\text{W.D.}) = P(t)\,\delta x = Q_x\delta x \qquad \text{or} \qquad Q_x = P(t)$$

and the equation of motion is $m\ddot{x} = P(t)$.

Lagrange's equations are clearly of no advantage in this simple example, since Newton's second law would have given the same result directly. We will find, however, that with growing complexity of the systems, Lagrange's equations are a great advantage.

Example 3.2
Determine the equation of motion of the simple pendulum in Fig. 2.13, by using equation (3.6).

Solution
Taking θ as the generalised coordinate, equation (3.6) has the form

$$\frac{\mathrm{d}}{\mathrm{d}t}\frac{\partial T}{\partial \dot{\theta}} - \frac{\partial T}{\partial \theta} Q_\theta \qquad T = \tfrac{1}{2}m\, l^2\dot{\theta}^2$$

$$\frac{\mathrm{d}\partial T}{\mathrm{d}t\partial\dot{\theta}} = m\, l^2\ddot{\theta} \qquad \frac{\partial T}{\partial\theta} = 0$$

δ(W.D.) is determined as in Fig. 3.2: δ(W.D.) $= -mg\, l \sin\theta\ \delta\theta = Q_\theta\ \delta\theta$, or

$$Q_\theta = -mg\, l\sin\theta$$

The generalised force here is a moment. The result is

$$m\, l^2\ddot{\theta} = -mg\, l\sin\theta \quad \text{or} \quad \ddot{\theta} + (g/l)\sin\theta = 0$$

as determined before.

Example 3.3
Determine the equation of motion of the system in Fig. 3.5 by equation (3.6).

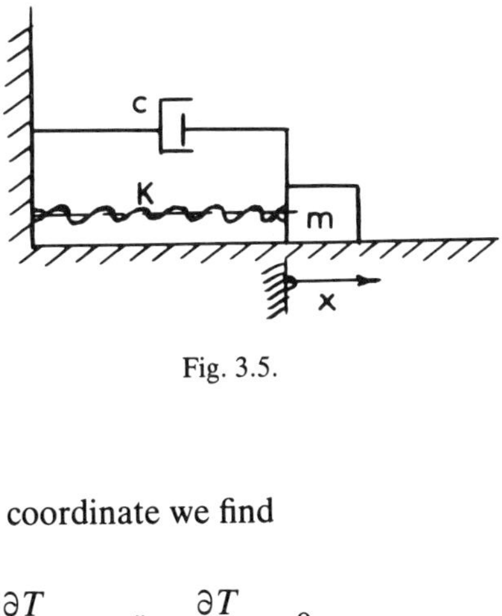

Fig. 3.5.

Solution
With x as the generalised coordinate we find

$$T = \tfrac{1}{2}m\dot{x}^2 \qquad \frac{\mathrm{d}\partial T}{\mathrm{d}t\partial\dot{x}} = m\ddot{x} \qquad \frac{\partial T}{\partial x} = 0$$

$$\delta(\text{W.D.}) = -Kx\ \delta x - c\dot{x}\ \delta x = Q_x\ \delta x\text{: } m\ddot{x} = -Kx - c\dot{x}$$

or the equation of motion:

$$\ddot{x} + \frac{c}{m}\dot{x} + \frac{K}{m}x = 0$$

Example 3.4

Determine the equation of motion of the particle in Fig. 3.6, in polar coordinates.

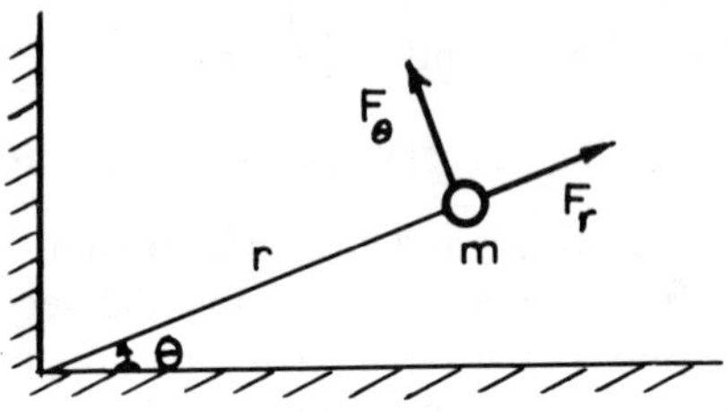

Fig. 3.6.

Solution

The system has two degrees of freedom, and taking (r,θ) as the generalised coordinates, the Lagrange equations are

$$\frac{\mathrm{d}}{\mathrm{d}t}\frac{\partial T}{\partial \dot{r}} - \frac{\partial T}{\partial r} = Q_r \quad \text{and} \quad \frac{d}{dt}\frac{\partial T}{\partial \dot{\theta}} - \frac{\partial \mathrm{T}}{\partial \theta} = Q_\theta$$

From the kinematics of a point in polar coordinates, the velocity has the perpendicular components $\dot{r}$ and $r\dot{\theta}$, so that

$$V^2 = \dot{r}^2 + r^2\dot{\theta}^2 \quad \text{and} \quad T = \tfrac{1}{2}m\,(\dot{r}^2 + r^2\dot{\theta}^2)$$

The differentiations for the first equation are

$$\frac{\mathrm{d}}{\mathrm{d}t}\frac{\partial T}{\partial \dot{r}} = m\ddot{r} \quad \text{and} \quad \frac{\partial T}{\partial r} = mr\dot{\theta}^2$$

To determine Q_r, we keep θ constant and give an increment δr to r; the work done is then $\delta(\mathrm{W.D}) = F_r\delta r = Q_r\,\delta r$, with $Q_r = F_r$.

The gravity force is included in F_r and F_θ. The first equation is now $m(\ddot{r} - r\dot{\theta}^2) = F_r$, which is the same as Newton's equation in the radial direction. The radial acceleration may be seen to agree with previous results in polar coordinates. The differentiations for the second equation are

$$\frac{\partial T}{\partial \dot{\theta}} = mr^2\dot{\theta}$$

It is important to notice that this involves a product, since both r and $\dot{\theta}$ are functions of time.

$$\frac{\mathrm{d}}{\mathrm{d}t}\frac{\partial T}{\partial \dot{\theta}} = mr^2\ddot{\theta} + 2mr\dot{r}\dot{\theta} \quad \text{and} \quad \frac{\partial T}{\partial \theta} = 0$$

To determine Q_θ, we keep r constant and give an increment $\delta\theta$ to θ. The work done is $\delta(\text{W.D.}) = F_\theta(r\delta\theta) = Q_\theta\ \delta\theta$, so that $Q_\theta = F_\theta r$, and the second equation is $m(r\ddot{\theta} + 2\dot{r}\dot{\theta}) = F_\theta$, which is Newton's equation in the direction perpendicular to r. The transverse acceleration is in agreement with previous results.

3.7 LAGRANGE'S EQUATIONS FOR CONSERVATIVE SYSTEMS

For conservative systems, the generalised forces are usually found by using the potential-energy function V in the expression (3.2):

$$Q_i = -\partial V/\partial q_i$$

If we are dealing with a conservative system, this expression may be substituted in the Lagrange equations (3.6), which then takes the form

$$\frac{\mathrm{d}}{\mathrm{d}t}\frac{\partial T}{\partial \dot{q}_i} - \frac{\partial T}{\partial q_i} + \frac{\partial V}{\partial q_i} = 0 \quad i = 1,2,3 \tag{3.7}$$

This gives the form of Lagrange's equations valid for conservative systems only. If we define a certain function $L = T - V$, we have

$$\frac{\partial L}{\partial \dot{q}_i} = \frac{\partial T}{\partial \dot{q}_i} - \frac{\partial V}{\partial \dot{q}_i} = \frac{\partial T}{\partial \dot{q}_i}$$

since the potential energy cannot contain time derivatives, so that $\partial V/\partial \dot{q}_i = 0$. We also find

$$\frac{\partial L}{\partial q_i} = \frac{\partial T}{\partial q_i} - \frac{\partial V}{\partial q_i}$$

Substituting these expressions in equation (3.7) gives an alternative form of Lagrange's equations:

$$\frac{\mathrm{d}}{\mathrm{d}t}\frac{\partial L}{\partial \dot{q}_i} - \frac{\partial L}{\partial q_i} = 0 \quad i = 1,2,3 \tag{3.8}$$

This is the shortest possible form of Lagrange's equation, and it is valid for conservative systems only. The function $L = T - V$ is called the Lagrangian function.

Example 3.5
Determine the equation of motion of the system in Example 3.2, by using equation (3.8)

Solution
Taking the angle θ as the generalised coordinate, the kinetic energy is $T = \frac{1}{2}mV^2 = \frac{1}{2}ml^2\dot{\theta}^2$.

Taking the vertical position of the pendulum as datum position, the potential energy is $V = mgl$ (1-cos θ).

The Lagrangian function $L = T - V = \frac{1}{2}ml^2\dot{\theta}^2 - mgl(1\text{-cos } \theta)$. Hence

$$\frac{\mathrm{d}\partial L}{\mathrm{d}t\partial\dot{\theta}} = ml^2\ddot{\theta} \quad \frac{\partial L}{\partial \theta} = -mgl \sin \theta$$

The equation of motion from equation (3.8) is then $ml^2\ddot{\theta} + mgl \sin\theta = 0$, or $\ddot{\theta} + (g/l) \sin \theta = 0$, as found before for this case.

Example 3.6
Assuming that the damping coefficient $c = 0$, determine the equation of motion of the system in Example 3.3 by using equation (3.8).

Solution
$T = \frac{1}{2}m\dot{x}^2$. Taking the static-equilibrium position as datum position, $V = \frac{1}{2}Kx^2$ and

$$L = \tfrac{1}{2}m\dot{x}^2 - \tfrac{1}{2}Kx^2 \quad \frac{\mathrm{d}\partial L}{\mathrm{d}t\partial\dot{x}} = m\ddot{x} \quad \frac{\partial L}{\partial x} = -Kx$$

and (3.8) gives the result $m\ddot{x} + Kx = 0$.

Although at first glance the equation (3.8) appears more attractive than equation (3.6), the actual difficulties in setting up the potential-energy function may well be greater than the determination of the generalised forces in the equation (3.6).

3.8 LANGRANGE'S EQUATIONS FOR A SYSTEM OF PARTICLES, INCLUDING RIGID BODIES

Suppose we have a system in plane motion. p = *number of particles*, n = degrees of freedom, generalised coordinates $(q_1, \ldots, q_n)$. We have then

$$\begin{cases} x_i = G_i(q_1, \ldots, q_n) \\ y_i = H_i(q_1, \ldots, q_n) \quad i = 1, \ldots, p \end{cases}$$

with

$$\delta x_i = \frac{\partial x_i}{\partial q_1}\delta q_1 + \ldots \frac{\partial x_i}{\partial q_n}\delta q_n$$

$$\delta y_i = \frac{\partial y_i}{\partial q_1}\delta q_1 + \ldots \frac{\partial y_i}{\partial q_n}\delta q_n$$

Giving an increment of δq_1 to coordinate q_1 and keeping $q_2, \ldots, q_n$ constant therefore $\delta q_2 = \ldots = \delta q_n = 0$, we obtain

$$\delta x_i = \frac{\partial x_i}{\partial q_1}\delta q_1 \quad \text{and} \quad \delta y_i = \frac{\partial y_i}{\partial q_1}\delta q_1$$

with

$$\delta(\text{W.D.}) = [Q_1]\delta q_1 = \sum_{i=1}^{p} (F_{x_i}\, \delta x_i + F_{y_i}\, \delta y_i)$$

The rest of the development is exactly the same as for one particle, and we substitute in the end $T = \sum_{i=1}^{p} \frac{1}{2} m_i(\dot{x}_i^2 + \dot{y}_i^2)$; the result is the Lagrange equation

$$\frac{\mathrm{d}}{\mathrm{d}t}\frac{\partial T}{\partial \dot{q}_1} - \frac{\partial T}{\partial q_1} = Q_1$$

determined previously for a particle. As before the same equation may be developed for the rest of the coordinates $q_2, \ldots, q_n$, and the result may be extended to three-dimensional motion by including the z-terms in the development. The final result is

$$\frac{\mathrm{d}\partial T}{\mathrm{d}t\partial \dot{q}_i} - \frac{\partial T}{\partial q_i} = Q_i \qquad i = 1, \ldots, n \tag{3.9}$$

This is a general form for Lagrange's equations, and for a system with n degrees of freedom and, therefore, n generalised coordinates $q_1, \ldots, q_n$, we obtain n Lagrange equations of motion.

Example 3.7

Fig. 3.7 shows a pendulum consisting of a rigid slender bar OA of length l and a bob A

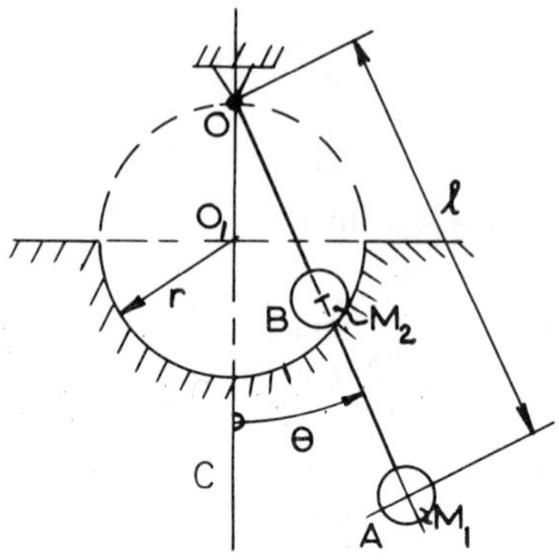

Fig. 3.7.

of mass M_1. The pendulum swings in a vertical plane about a fixed point O as shown. A second bob B of mass M_2 is guided by a circular seat of radius r and centre O_1, and slides freely along the rod OA. The two masses may be assumed to be concentrated, and the mass of the bar OA and all friction may be neglected.

(a) Determine the kinetic- and potential-energy functions of the system in terms of the system constants.
(b) Determine the equation of motion of the system.
(c) Show from the results of (b) that if $M_2 = 0$ the equation of motion takes the usual form for a simple pendulum of length l. Show also that if $M_1 = 0$ *the equation* takes the usual form for a simple pendulum of length r.
(d) Simplify the equation of motion by assuming small displacements and determine the frequency of the system if $l = 20$ cm, $r = 5$ cm, and $M_1 = M_2 = M$.

Solution (*a*)

The system has one degree of freedom, and taking θ as the generalised coordinate the velocity of body A is $l\dot{\theta}$. The angle CO_1B is 2θ, and the velocity of body B is $r(2\dot{\theta})$. The result is

$$T = \tfrac{1}{2}(M_1 l^2 + 4M_2 r^2)\dot{\theta}^2$$

From the triangle O_1OB the distance $OB = 2r\cos\theta$. Taking the position OC as datum position we obtain

$$V = M_1 g l(1 - \cos\theta) + M_2 g 2r\sin^2\theta$$

(b) The system is conservative with

$$L = \tfrac{1}{2}(M_1 l^2 + 4M_2 r^2)\dot{\theta}^2 - M_1 gl(1 - \cos\theta) - M_2 g_2 r \sin^2\theta$$

from which

$$\frac{\mathrm{d}}{\mathrm{d}t}\frac{\partial L}{\partial\dot{\theta}} = (M_1 l^2 + 4M_2 r^2)\ddot{\theta}$$

$$\frac{\partial L}{\partial\theta} = -\mathrm{M}_1 gl \sin\theta - 2M_2 gr \sin 2\theta$$

The equation of motion is

$$(M_1 l^2 + 4M_2 r^2)\ddot{\theta} + M_1 gl\sin\theta + 2M_2\, gr \sin 2\theta = 0$$

(c) Substituting $M_2 = 0$ in the equation of motion leads to

$$\ddot{\theta} + \frac{g}{l}\sin\theta = 0$$

Substituting $M_1 = 0$ leads to

$$(2\ddot{\theta}) + \frac{g}{r}\sin(2\theta) = 0$$

(d) $\sin\theta \simeq \theta$ $\sin 2\theta \simeq 2\theta$ and $M_1 = M_2 = M$.
The equation of motion takes the form

$$(l^2 + 4r^2)\ddot{\theta} + g(l + 4r)\theta = 0$$

This is S.H.M. with $f = \sqrt{[g(l + 4r)/(l^2 + 4r^2)]} = 1.41$ cps.

Examples on the use of Lagrange's equations for systems involving rigid bodies have to be deferred until after the treatment of the kinematics and the moments of inertia of rigid bodies.

3.9 SOME GENERAL REMARKS ON LAGRANGE'S EQUATIONS

The following are some of the advantages of Lagrange's equations:

1. The most suitable coordinates may be used to describe the motion; these may be absolute or a mixture of absolute and relative coordinates.
2. The equations of motion are established by a series of simple differentiations of the kinetic energy expressed in the chosen coordinates. The method is the same for all types of systems.

3. The necessary kinematic analysis involves only the velocities and not the accelerations.
4. Fixed reactions, internal forces and normal forces do not enter the equations.

When a set of generalised coordinates has been established, it should be assumed that the system is moving in the positive direction of all coordinates when the velocities are determined, otherwise mistakes may be made in the direction of the velocity components.

When a set of equations of motion of a system has been determined, we sometimes want the equations of motion of a simplified system where some of the coordinates of the original system are kept constant. Suppose that the equations of motion are established in coordinates x and θ for a system with two degrees of freedom, and we want the equation of motion for a simpler system in which $x = \text{constant}$; the equation of motion for this system is then the original equation of motion in the coordinate θ. This equation may of course be simplified where appropriate by substituting $x = \text{constant}$, $\dot{x} = \ddot{x} = 0$.

The equations for systems with more than two degrees of freedom may be treated in a similar way, if some of the coordinates are fixed, to form a simplified system.

PROBLEMS

3.1 A simple pendulum in plane motion is suspended on an elastic string as shown in Fig. 3.8. Under the action of gravity in the middle position the length is x_0.

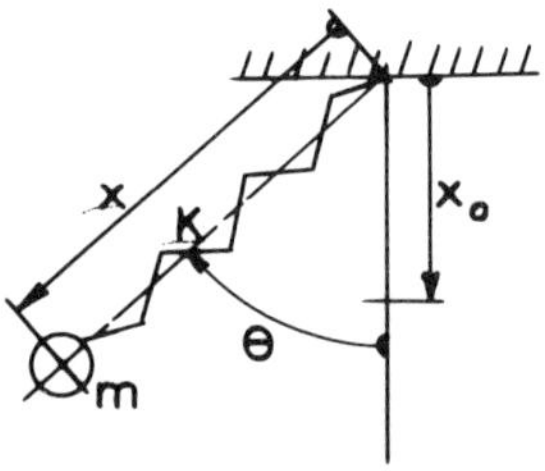

Fig. 3.8.

Taking coordinates as shown, find the equations of motion. Simplify the equations by assuming small vibrations with small $\dot{x}$ and $\dot{\theta}$.

3.2 A spherical pendulum (Fig. 3.9) consists of a particle of mass m supported by a massless string of length l. Using θ and ϕ as generalised coordinates, derive the two differential equations of motion. Show that these equations reduce to previously known results when θ and ϕ are successively held constant.

3.3 Find the equations of motion for the particle in Fig. 3.10. The string has cross-section A and modulus E. The particle is performing vibrations along the horizontal x-axis. Assume that there is no initial tension in the string and that the tension produced by the weight of the particle is negligible.

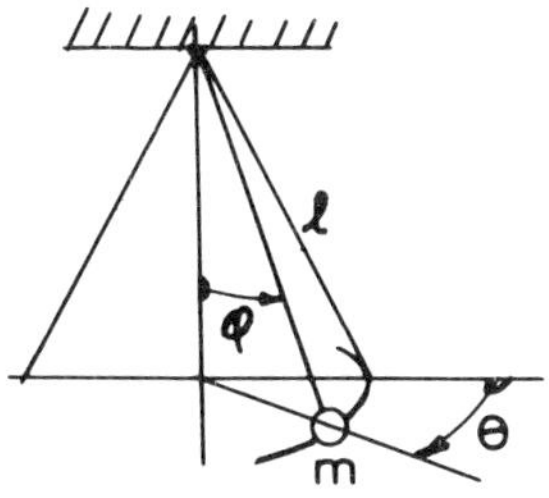

Fig. 3.9.

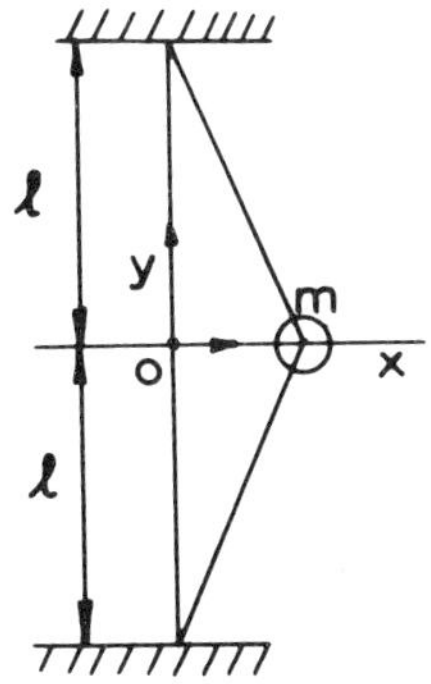

Fig. 3.10.

Use $\sqrt{(l^2+x^2)} \simeq l + x^2/2l$. Find the equation of motion if the initial tension s is large and assumed constant for small vibrations. The spring constant is $K = AE/2l$.

3.4 Fig. 3.11 shows a pendulum which is able to swing in the vertical plane about point O. The pendulum consists of a light rigid rod on which a bob of mass m is sliding without friction. The bob is connected to point O through a spring of spring constant K.

The motion of the bob along the bar is damped by a light dashpot with coefficient of damping c. The length of the spring is l_0 in the vertical statical-equilibrium position. Determine

(a) The kinetic-energy function of the system in terms of the system constants and the coordinates x and θ as shown on the figure.
(b) The equations of motion of the system.
(c) The equation of motion if $\theta = 0$ and constant. The equation of motion if $x = b$ constant.

3.5 Find the equation of motion for the system in Fig. 3.12.

3.6 Fig. 3.13 shows a double pendulum. The string of each pendulum is of length l

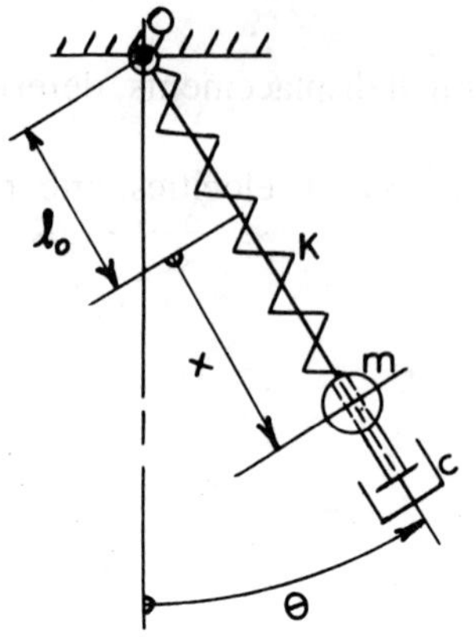

Fig. 3.11.

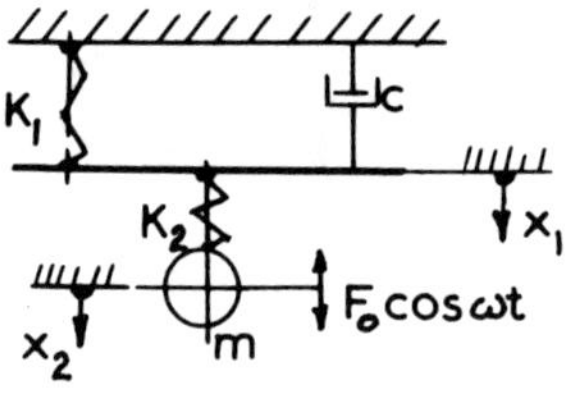

Fig. 3.12.

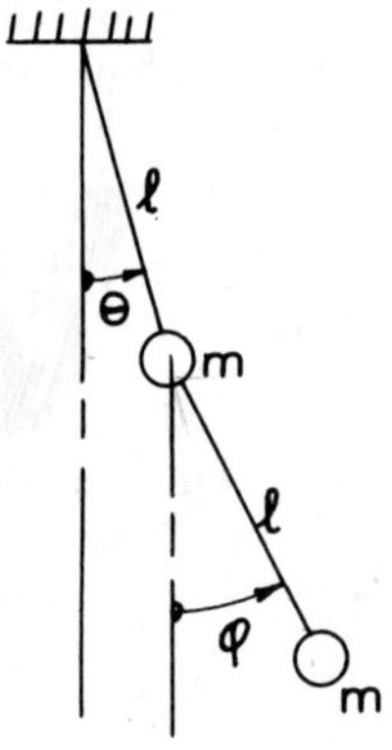

Fig. 3.13.

and the bob of mass m; the pendulum swings in a vertical plane owing to the action of gravity.

(a) Without assuming small displacements, determine the equations of motion of the system.
(b) For small displacements and velocities, linearise the equations under (a).

4

Kinematics of a rigid body

4.1 TYPES OF MOTION. ANGULAR VELOCITY

Some of the various types of motion of a rigid body are illustrated by the mechanism in Fig. 4.1. As the crank arms O_1A and O_2B, of equal length, rotate in the plane of

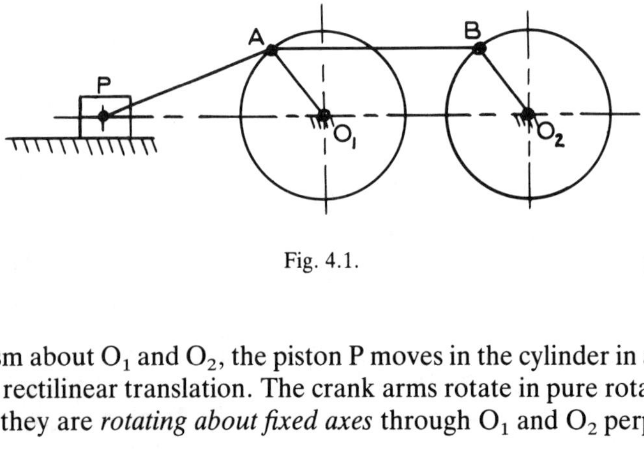

Fig. 4.1.

the mechanism about O_1 and O_2, the piston P moves in the cylinder in a translation, in this case a rectilinear translation. The crank arms rotate in pure rotation in plane motion, and they are *rotating about fixed axes* through O_1 and O_2 perpendicular to the figure.

The side rod AB is moving in the plane, and is always parallel to O_1O_2; it is in plane *curvilinear translation*, while the connecting rod PA is in *general plane motion*.

For the purpose of kinematic and dynamic analysis, the motion of a rigid body may be classified under only six different headings; these are, in increasing order of complexity: (1) rest, (2) translation, (3) rotation about a fixed axis, (4) plane motion, (5) rotation about a fixed point, and (6) general three-dimensional motion.

The first five are all special cases of (6); in fact case (3) is a special case of (5) and (4).

Case (1) is dealt with in statics, while case (5) in general is so-called gyroscopic motion; cases (5) and (6) are much more complicated than the others, and so we shall therefore concentrate on cases (2), (3) and (4); the great majority of engineering problems belong in these categories.

In Fig. 4.1, only the radial arms and the connecting rod change their angular position with respect to O_1O_2. Before the motion of a rigid body can be discussed, the concept of change of angular relationship which we call rotation must be discussed in more detail.

In rotation of a rigid body about a fixed axis, all points on the axis remain stationary, while all other points in the body move in circles in planes perpendicular to the axis, and with centres on the axis of rotation.

Because of the power of the vector concept, it seems reasonable to try to define a rotation as a vector on the axis of rotation, of length equal to the magnitude of the angle of rotation and with sense following the right-hand screw rule. For rotations to be vectors, they must obey the commutative addition law; to investigate whether this is the case, consider rotation in a plane as shown in Fig. 4.2: a point P is rotated about

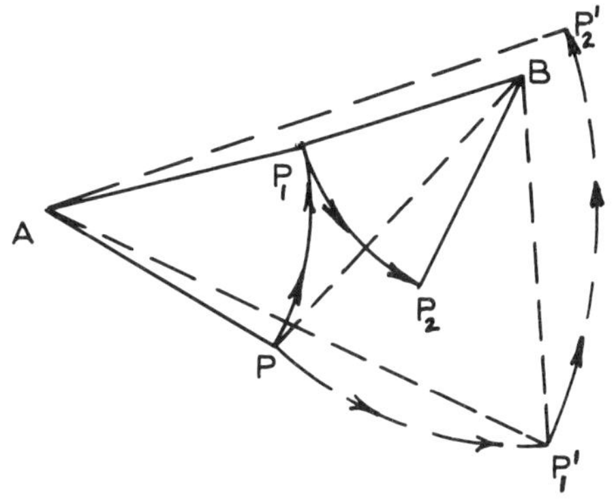

Fig. 4.2.

an axis perpendicular to the paper through A, the angle of rotation is θ_1, and P moves to position P_1. Then P_1 is rotated about an axis through B to position P_2, the angle of rotation being θ_2; if the order of rotation is reversed, P moves first to P_1' and then to P_2'; clearly the final positions P_2 and P_2' are quite different; in fact for angles $\theta_1 = 43.5°$ and $\theta_2 = 47°$, the linear distance P_2P_2' is about 30 mm on the figure.

Finite angular rotations cannot, therefore, be considered vectors, since the order of rotations determines the final position, and this violates the cummutative addition law of vectors.

If we now repeat Fig. 4.2, but this time with angles $\theta_1 = \theta_2 = 5°$, we will find that P_2 and P_2' practically coincide; in fact P_2P_2' is well under 1 mm; clearly if we take infinitesimal angles $\Delta\theta_1$ and $\Delta\theta_2$ and let $\Delta\theta_1 \rightarrow 0$ and $\Delta\theta_2 \rightarrow 0$, the final positions of P_2 and P_2' are identical, so that we have $\boldsymbol{\Delta\theta}_1 + \boldsymbol{\Delta\theta}_2 = \boldsymbol{\Delta\theta}_2 + \boldsymbol{\Delta\theta}_1$. Infinitesimal rotations in a plane are thus vectors.

The instantaneous angular velocity ω, which was introduced in Chapter 1, may now be defined as a vector.

$$\boldsymbol{\omega} = \lim_{\Delta t \to 0} \frac{\Delta \boldsymbol{\theta}}{\Delta t} = \frac{\mathrm{d}\boldsymbol{\theta}}{\mathrm{d}t} = \dot{\boldsymbol{\theta}}$$

This is through O and perpendicular to the rotating radial line, of magnitude equal to the instantaneous angular velocity ω and with sense following the right-hand screw rule. The vector $\boldsymbol{\omega}$ then determines the instantaneous axis of rotation.

The instantaneous angular acceleration $\dot{\omega}$ introduced in Chapter 1 is now the vector.

$$\dot{\boldsymbol{\omega}} = \lim_{\Delta t \to 0} \frac{\Delta \boldsymbol{\omega}}{\Delta t} = \frac{\mathrm{d}\boldsymbol{\omega}}{\mathrm{d}t} = \ddot{\boldsymbol{\theta}}$$

The $\boldsymbol{\omega}$ and $\dot{\boldsymbol{\omega}}$ vectors for a point A in circular motion are shown in Fig. 4.3.

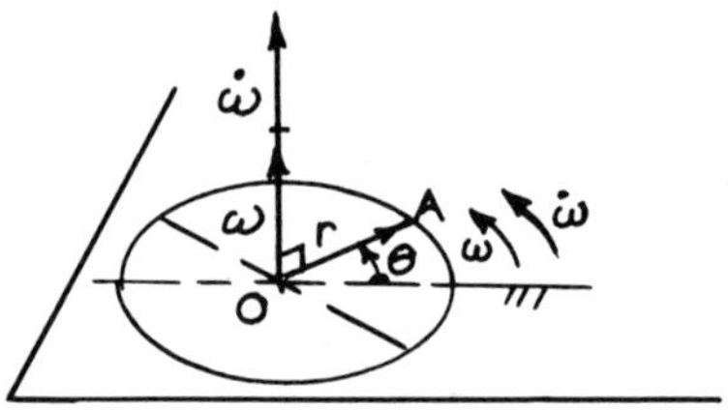

Fig. 4.3.

4.2 DERIVATIVE OF A VECTOR OF CONSTANT LENGTH

Fig. 4.4 shows a vector $\mathbf{r}$ in a position A_1B_1 in a plane. After an interval of Δt, the vector is in a position A_2B_3; the length of the vector is r, and this is a constant length. The vector may be moved to the final position by first translating it to position A_2B_2 and then rotating it an angle $\Delta\theta$ about A_2. The translation of the vector does not change its magnitude or direction and therefore gives no contribution to $\Delta\mathbf{r}$ and therefore no derivative with respect to time.

The total change in $\mathbf{r}$ is the vector $\Delta\mathbf{r}$ as shown, and we have

$$\lim_{\Delta t \to 0} \frac{\Delta \mathbf{r}}{\Delta t} = \lim r \frac{\Delta\theta}{\Delta t}\mathbf{e} = r\frac{\mathrm{d}\theta}{\mathrm{d}t}\mathbf{e} = r\omega\mathbf{e}$$

where $\mathbf{e}$ is a unit vector in the direction $\Delta\mathbf{r}$. The vector $\Delta\mathbf{r}$ is a chord in a circular arc,

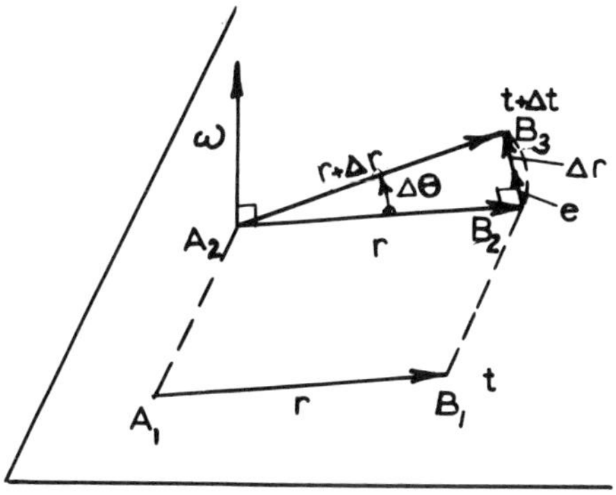

Fig. 4.4.

and for $\Delta t \rightarrow 0$ we have $\Delta\theta \rightarrow 0$, so that in the limit $\Delta\mathbf{r}$ and $\mathbf{e}$ are perpendicular to $\mathbf{r}$. Introducing the instantaneous angular velocity vector $\boldsymbol{\omega}$ as shown, we find

$$\frac{d\mathbf{r}}{dt} = \dot{\mathbf{r}} = r\omega\mathbf{e} = \boldsymbol{\omega} \times \mathbf{r} \tag{4.1}$$

It may be seen that this vector expression for the derivative $\dot{\mathbf{r}}$ of a constant length vector has the proper magnitude and direction.

It is important to notice in equation (4.1) that $\boldsymbol{\omega}$ is the *total* instantaneous angular velocity of the constant length vector $\mathbf{r}$. If the vector is located on a plate and rotating with instantaneous angular velocity $\boldsymbol{\omega}_r$ relative to the plate, and if the plate has the instantaneous angular velocity $\boldsymbol{\omega}$, then $\boldsymbol{\Omega} = \boldsymbol{\omega}_r + \boldsymbol{\omega}$ must be used instead of $\boldsymbol{\omega}$ in the formula (4.1), where the magnitudes of $\boldsymbol{\omega}_r$ and $\boldsymbol{\omega}$ may be added numerically with proper signs to give the total instantaneous angular velocity $\boldsymbol{\Omega}$ of $\mathbf{r}$.

Differentiating $\dot{\mathbf{r}}$ from (4.1), we find

$$\ddot{\mathbf{r}} = \boldsymbol{\omega} \times \dot{\mathbf{r}} + \dot{\boldsymbol{\omega}} \times \mathbf{r} = \boldsymbol{\omega} \times (\boldsymbol{\omega} \times \mathbf{r}) + \dot{\boldsymbol{\omega}} \times \mathbf{r} \tag{4.2}$$

The total angular velocity and acceleration $\boldsymbol{\Omega}$ and $\dot{\boldsymbol{\Omega}}$ must be used in (4.2).

4.3 VELOCITY IN CIRCULAR MOTION

Fig. 4.5 shows a point A in circular motion. The position vector from O to A is $\mathbf{r}$ where $|\mathbf{r}|$ = constant. By definition $\dot{\mathbf{r}} = \mathbf{V}_A$ is the velocity of point A. Using (4.1) we find $\mathbf{V}_A = \boldsymbol{\omega} \times \mathbf{r}$. The vector $\boldsymbol{\omega} \times \mathbf{r}$ is perpendicular to $\mathbf{r}$ as shown, and its magnitude is

$$|\boldsymbol{\omega}| \cdot |\mathbf{r}| \sin 90° = r\omega = |\mathbf{V}_A|$$

The direction and magnitude of $\mathbf{V}_A$ are in agreement with previous results.

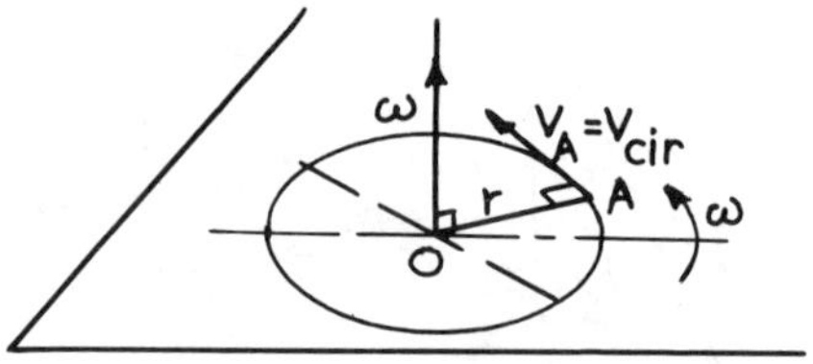

Fig. 4.5.

4.4 ACCELERATION IN CIRCULAR MOTION

The acceleration of point A in Fig. 4.5 is by definition $\ddot{\mathbf{r}}$, and from (4.2)

$$\ddot{\mathbf{r}} = \boldsymbol{\omega} \times (\boldsymbol{\omega} \times \mathbf{r}) + \dot{\boldsymbol{\omega}} \times \mathbf{r} = \mathbf{A}_A = \mathbf{A}_{cir}$$

Taking the first component of $\mathbf{A}_A$ as $\mathbf{A}_n = \boldsymbol{\omega} \times (\boldsymbol{\omega} \times \mathbf{r})$ we find the magnitude

$$|\mathbf{A}_n| = |\boldsymbol{\omega}| \cdot |\boldsymbol{\omega} \times \mathbf{r}| \sin 90° = |\boldsymbol{\omega}| \cdot |\boldsymbol{\omega}| \cdot |\mathbf{r}| \sin^2 90° = r\omega^2$$

This component is called the normal acceleration of point A; its direction is shown in Fig. 4.6; it is always directed towards the centre of rotation O.

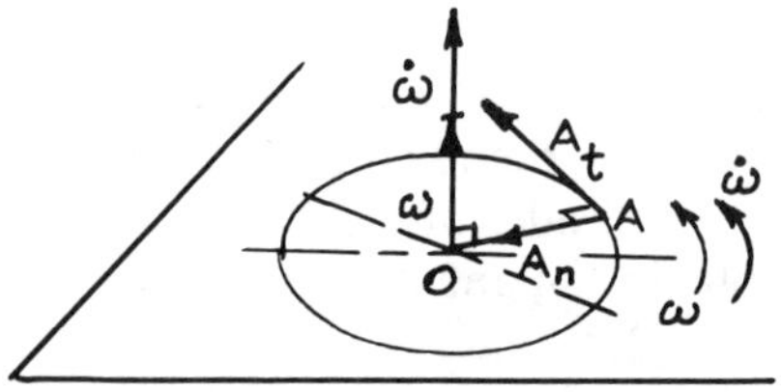

Fig. 4.6.

The second component of $\mathbf{A}_A$ is $\mathbf{A}_t = \dot{\boldsymbol{\omega}} \times \mathbf{r}$; its magnitude is $|\mathbf{A}_t| = |\dot{\boldsymbol{\omega}}||\mathbf{r}| \sin 90° = r\dot{\omega}$. This component is called the *tangential acceleration* of point A; its direction is shown in Fig. 4.6 along the tangent at point A; it is, therefore, perpendicular to the radius of the circle. We have now $\mathbf{A}_{cir} = \mathbf{A}_n + \mathbf{A}_t = \mathbf{A}_A$.

The components $\mathbf{A}_n$ and $\mathbf{A}_t$ of the acceleration of point A are in agreement with previous results.

4.5 MOTION OF A FIGURE IN A PLANE

The motion of an unchangeable figure in a plane may be determined by considering the motion of a line on the figure. Fig. 4.7 shows a line moving in a plane; the initial

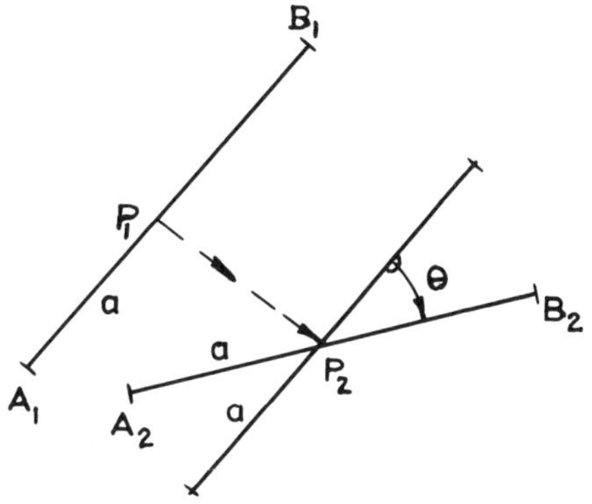

Fig. 4.7.

position is A_1B_1 and the final position A_2B_2. The line is moved to the final position by taking *any* point P_1 on the line and translating the line so that P_1 ends up in its final position P_2. The line may now be rotated through an angle θ as shown to bring it into its final position A_2B_2. The distance of the translation depends on the point P_1 chosen; for instance, the distance A_1A_2 is different from P_1P_2; however, the angle of rotation is always the same magnitude and direction.

Consider now an infinitesimal translation and rotation in the time Δt as shown in Fig. 4.8. Point O is a fixed point in the plane and the position vector of P is $\mathbf{r}$. The

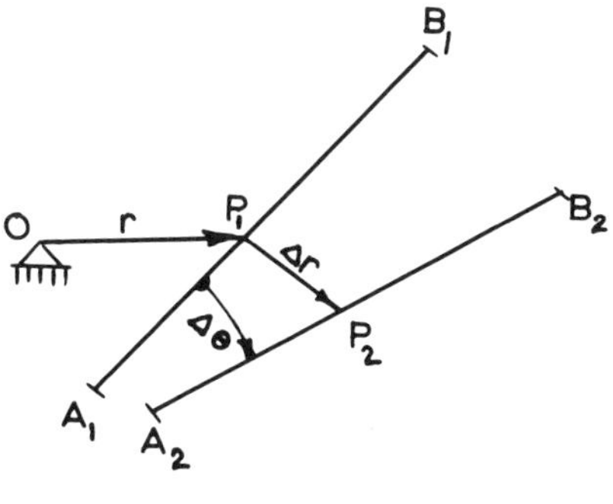

Fig. 4.8.

translation of the line is $\Delta\mathbf{r}$ and the rotation $\Delta\theta$, where the rotation is superimposed on the translation, with both happening at the same time. The instantaneous velocity of P is now defined as usual by

$$\mathbf{V}_\mathrm{P} = \lim_{\Delta t \to 0} \frac{\Delta \mathbf{r}}{\Delta t} = \frac{\mathrm{d}\mathbf{r}}{\mathrm{d}t} = \dot{\mathbf{r}}$$

and the instantaneous angular velocity is defined as usual by

$$\omega = \lim_{\Delta t \to 0} \frac{\Delta\theta}{\Delta t} = \frac{d\theta}{dt} = \dot{\theta}$$

where the angular velocity may be taken as a vector $\boldsymbol{\omega}$ perpendicular to the plane, and in this case inward directed following the right-hand screw rule. It must be emphasised that in general $\mathbf{V}_P$ depends on the point P chosen, while $\boldsymbol{\omega}$ is the same for any point chosen; that is, $\boldsymbol{\omega}$ has the same magnitude, direction and sense for any point P but we locate it at the point P chosen.

4.6 TIME DERIVATIVE OF A VECTOR IN A FIXED AND ROTATING COORDINATE SYSTEM

Fig. 4.9 shows a plate which is moving in the fixed or absolute coordinate system XY.

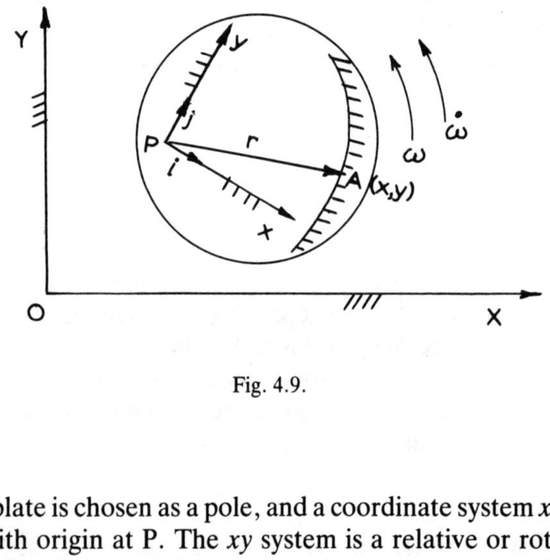

Fig. 4.9.

A point P on the plate is chosen as a pole, and a coordinate system xy is drawn on the plate as shown with origin at P. The xy system is a relative or rotating coordinate system, and the plate is assumed to have an instantaneous rotation about P with instantaneous angular velocity and acceleration ω and $\dot{\boldsymbol{\omega}}$ as shown. A point A(x,y) moves along a curve drawn on the plate as shown. The position vector from P to A is $\mathbf{r}$. In the following it is assumed that if we refer $\mathbf{r}$ to the rotating system xy it will be given the subscript r, with $\mathbf{r} = \mathbf{r}_r$. The time derivative of $\mathbf{r}$ will be denoted $\dot{\mathbf{r}}_r$ when it is taken in the rotating system xy, and $\dot{\mathbf{r}}$ without a subscript when it is taken in the absolute system XY.

The unit vectors $\mathbf{i}$ and $\mathbf{j}$ have a fixed length and direction in the xy system, and therefore

$$\left(\frac{d\mathbf{i}}{dt}\right)_r = \mathbf{0} \quad \text{and} \quad \left(\frac{d\mathbf{j}}{dt}\right)_r = \mathbf{0}$$

We have $\mathbf{r}_r = x\mathbf{j} + y\mathbf{j} = \mathbf{r}$ and

$$\left(\frac{d\mathbf{r}_r}{dt}\right)_r = \dot{\mathbf{r}}_r = \dot{x}\mathbf{i} + x\left(\frac{d\mathbf{i}}{dt}\right)_r + \dot{y}\mathbf{j} + y\left(\frac{d\mathbf{j}}{dt}\right)_r$$

or $\dot{\mathbf{r}}_r = \dot{x}\mathbf{i} + \dot{y}\mathbf{j}$. Similarly we find $\ddot{\mathbf{r}}_r = \ddot{x}\mathbf{i} + \ddot{y}\mathbf{j}$. Differentiating **i** and **j** in the absolute XY system, we use the formula (4.1), from which

$$\frac{d\mathbf{i}}{dt} = \boldsymbol{\omega} \times \mathbf{i} \quad \text{and} \quad \frac{d\mathbf{j}}{dt} = \boldsymbol{\omega} \times \mathbf{j}$$

Differentiating **r** in the absolute system XY gives the result

$$\begin{aligned}\frac{d\mathbf{r}}{dt} &= \dot{\mathbf{r}} = \dot{x}\mathbf{i} + x\frac{d\mathbf{i}}{dt} + \dot{y}\mathbf{j} + y\frac{d\mathbf{j}}{dt}\\ &= (\dot{x}\mathbf{i} + \dot{y}\mathbf{j}) + x\boldsymbol{\omega} \times \mathbf{i} + y\boldsymbol{\omega} \times \mathbf{j} = (\dot{x}\mathbf{i} + \dot{y}\mathbf{j}) + \boldsymbol{\omega} \times (x\mathbf{i} + y\mathbf{j})\end{aligned}$$

or

$$\dot{\mathbf{r}} = \dot{\mathbf{r}}_r + \boldsymbol{\omega} \times \mathbf{r} \tag{4.3}$$

If **r** is a vector of *constant length* and a *fixed direction* in the xy system, we have $\dot{\mathbf{r}}_r = \mathbf{0}$ and (4.3) gives the result $\mathbf{r} = \boldsymbol{\omega} \times \mathbf{r}$ where $\boldsymbol{\omega}$ is the instantaneous angular velocity of the xy system, that is of the plate in Fig. 4.9. The result is in agreement with the formula (4.1). If **r** is a vector of *constant length* but *changing direction* in the xy system, we obtain $\dot{\mathbf{r}}_r = \boldsymbol{\omega}_r \times \mathbf{r}$, where $\boldsymbol{\omega}_r$ is the instantaneous angular velocity of **r** relative to the xy system on the plate; if the xy system has the instantaneous angular velocity $\boldsymbol{\omega}$, the result is

$$\dot{\mathbf{r}} = \dot{\mathbf{r}}_r + \dot{\boldsymbol{\omega}} \times \mathbf{r} = \boldsymbol{\omega}_r \times \mathbf{r} + \boldsymbol{\omega} \times \mathbf{r} = (\boldsymbol{\omega}_r + \boldsymbol{\omega}) \times \mathbf{r} = \boldsymbol{\Omega} \times \mathbf{r}$$

where $\boldsymbol{\Omega}$ is the *total* instantaneous angular velocity of **r** in the XY system.

To determine the second time derivative of **r** we may take $\mathbf{r} = x\mathbf{i} + y\mathbf{j}$ and proceed to differentiate this expression twice in absolute terms; it is, however, simpler to use the formula (4.3). We find

$$\ddot{\mathbf{r}} = \frac{d}{dt}(\dot{\mathbf{r}}) = \frac{d}{dt}(\dot{\mathbf{r}}_r) + \dot{\boldsymbol{\omega}} \times \mathbf{r} + \boldsymbol{\omega} \times \dot{\mathbf{r}}$$

$$= (\ddot{\mathbf{r}}_r + \dot{\boldsymbol{\omega}} \times \dot{\mathbf{r}}_r) + \dot{\boldsymbol{\omega}} \times \mathbf{r} + \boldsymbol{\omega} \times (\dot{\mathbf{r}}_r + \boldsymbol{\omega} \times \mathbf{r})$$

so that

$$\ddot{\mathbf{r}} = \boldsymbol{\omega} \times (\boldsymbol{\omega} \times \mathbf{r}) + \dot{\boldsymbol{\omega}} \times \mathbf{r} + \ddot{\mathbf{r}}_r + 2\boldsymbol{\omega} \times \dot{\mathbf{r}}_r \tag{4.4}$$

This important formula a holds for all vector quantities.If $\mathbf{r}$ is a vector of *constant* length and *fixed* in the *xy* system, we have $\dot{\mathbf{r}}_r = \mathbf{0}$ and $\ddot{\mathbf{r}}_r = \mathbf{0}$, and (4.4) gives the result $\ddot{\mathbf{r}} = \boldsymbol{\omega} \times (\boldsymbol{\omega} \times \mathbf{r}) + \dot{\boldsymbol{\omega}} \times \mathbf{r}$ in agreement with (4.2).

If $\mathbf{r}$ is of *constant* length but changing direction with instantaneous relative velocity and acceleration $\boldsymbol{\omega}_r$ and $\dot{\boldsymbol{\omega}}_r$, the formula (4.4) may be shown to give the result

$$\ddot{\mathbf{r}} = \boldsymbol{\Omega} \times (\boldsymbol{\Omega} \times \mathbf{r}) + \dot{\boldsymbol{\Omega}} \times \mathbf{r}$$

where $\boldsymbol{\Omega} = \boldsymbol{\omega} + \boldsymbol{\omega}_r$ and $\dot{\boldsymbol{\Omega}} = \dot{\boldsymbol{\omega}} + \dot{\boldsymbol{\omega}}_r$.

4.7 VELOCITY AND ACCELERATION OF A POINT MOVING ON A ROTATING BODY

The body in question here is taken as a ship which is sailing forward and at the same time rotating as shown in Fig. 4.10. The deck of the ship is assumed to be in the

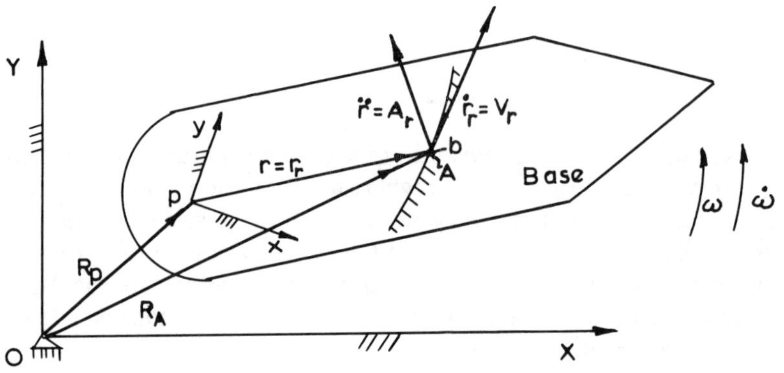

Fig. 4.10.

horizontal plane. Choosing an arbitrary point 'p' fixed on the deck, we consider p as a *pole* for the motion and the deck as a *base*. The coordinate system *xy* is drawn on the deck with origin at p and the instantaneous angular velocity and acceleration of the ship ω and $\dot{\omega}$ are taken to be rotation about p. A point A is shown moving across the deck on a curve drawn on the deck. The position vector from p to A is $\mathbf{r}$, and the time derivative of $\mathbf{r}$ in the *xy* system is shown as $\dot{\mathbf{r}}_r$ which is the *relative* velocity $\mathbf{V}_r$ of A along the curve, which is tangential to the curve; the second derivative is $\ddot{\mathbf{r}}_r$ which is the *relative* acceleration $\mathbf{A}_r$ of point A along the curve. We now introduce a fixed point O on the shore as shown in the plane of the deck. An absolute coordinate system *XY* is fixed on the shore with origin at O. A point 'b' is fixed on the curve on the deck, and we consider the situation at the instant when A passes through point b. Point b is called a *base point* for A. It must be kept in mind that although points A and b coincide at an instant, they have quite different motions.

We may now draw the absolute position vectors to point p and point A from the fixed point O on the shore; these are shown as $\mathbf{R}_p$ and $\mathbf{R}_A$ respectively.

From the basic vector triangle OpA, we have now $\mathbf{R}_A = \mathbf{R}_p + \mathbf{r}$; differentiating this in absolute terms we find $\dot{\mathbf{R}}_A = \dot{\mathbf{R}}_p + \dot{\mathbf{r}}$, and substituting $\dot{\mathbf{r}} = \dot{\mathbf{r}}_r + \boldsymbol{\omega} \times \mathbf{r}$ from equation (4.3) we find

$$\dot{\mathbf{R}}_A = [\dot{\mathbf{R}}_p + \boldsymbol{\omega} \times \mathbf{r}] + \dot{\mathbf{r}}_r \tag{4.5}$$

Equation (4.5) determines the absolute velocity of point A in terms of the absolute velocity of point p and the relative velocity of point A. The equation also holds in three-dimensional motion and will be discussed in detail in section 4.8. Differentiating $\dot{\mathbf{R}}_A = \dot{\mathbf{R}}_p + \dot{\mathbf{r}}$ we obtain $\ddot{\mathbf{R}}_A = \ddot{\mathbf{R}}_p + \ddot{\mathbf{r}}$, and substituting $\ddot{\mathbf{r}}$ from equation (4.4) there results

$$\ddot{\mathbf{R}}_A = [\ddot{\mathbf{R}}_p + \boldsymbol{\omega} \times (\boldsymbol{\omega} \times \mathbf{r}) + \dot{\boldsymbol{\omega}} \times \mathbf{r}] + \ddot{\mathbf{r}}_r + 2\boldsymbol{\omega} \times \dot{\mathbf{r}}_r \tag{4.6}$$

Equation (4.6) determines the absolute acceleration of point A in terms of various acceleration terms to be discussed in detail in section 4.10. The equation also holds in three-dimensional motion.

4.8 VELOCITY IN PLANE MOTION

In the formula (4.5) we may substitute the absolute velocity of point A as $\mathbf{V}_A = \dot{\mathbf{R}}_A$, the absolute pole velocity $\mathbf{V}_p = \dot{\mathbf{R}}_p$ and use the notation $\mathbf{V}_{cir} = \boldsymbol{\omega} \times \mathbf{r}$. Comparison with the velocity of a point in circular motion will show that $\mathbf{V}_{cir}$ is determined exactly as if point A was in circular motion about the pole with radius r and angular velocity ω, the magnitude of $|\mathbf{V}_{cir}| = r\omega$, it is perpendicular to $\mathbf{r}$ and with direction to give the angular velocity direction of rotation. The subscript 'cir' stands for circular motion. Finally the term $\dot{\mathbf{r}}_r$ is the relative velocity of A across the body, for which we use the notation $\mathbf{V}_r = \dot{\mathbf{r}}_r$. Substituting these notations in (4.5) leads to

$$\mathbf{V}_A = [\mathbf{V}_p + \mathbf{V}_{cir}] + \mathbf{V}_r \tag{4.7}$$

where it must be kept in mind that $\mathbf{V}_{cir} = \boldsymbol{\omega} \times \mathbf{r}$.

No restrictions were made in the development of (4.7); it may also be used for points fixed in the body. If (4.7) is used for the base point b, which has no relative velocity we find $\mathbf{V}_b = [\mathbf{V}_p + \mathbf{V}_{cir}]$. Substituting $\mathbf{V}_b$ in (4.7) gives the result

$$\mathbf{V}_A = \mathbf{V}_b + \mathbf{V}_r \tag{4.8}$$

Equation (4.8) is the shortest possible expression for the velocity of the moving point A on the rotating body.

The formula divides the problem into two stages: we first fix A on the body and determine its velocity $\mathbf{V}_b$ in this simpler case; we then fix the body and determine the velocity of A in this much simpler case where $\mathbf{V}_A = \mathbf{V}_r$. The vector sum of the two velocities $\mathbf{V}_b$ and $\mathbf{V}_r$ then gives the absolute velocity of point A in the plane of motion.

In the solution of problems in kinematics it is convenient to divide the problems into *two cases*. Those in case 1 are problems where all points are *fixed* in the *moving rigid body*; the term $\mathbf{V}_r$ does not appear in these problems and they are consequently the simpler types of problem. Those in *case* 2 are problems where at least one point is *moving across the body* with relative velocity $\mathbf{V}_r$ of sliding or rolling along the body. These problems are not as simple as problems of the case 1 type.

4.8.1 Summary of velocity in plane motion for case 1 problems

With *point A fixed on the base*, the point becomes a base point for itself.

With $\mathbf{V}_r = \mathbf{0}$, (4.7) gives

$$\mathbf{V}_A = \mathbf{V}_p + \mathbf{V}_{cir}$$

with $\mathbf{V}_{cir} = \boldsymbol{\omega} \times \mathbf{r}$. The formula may be illustrated by *taking out* the line connecting p and A as shown in Fig. 4.11, which shows the two components of $\mathbf{V}_A$. The vectors $\mathbf{V}_p$

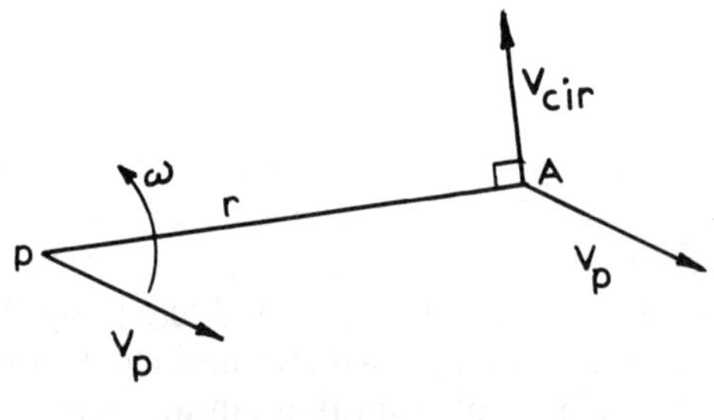

Fig. 4.11.

and $\mathbf{V}_{cir}$ must, of course, be added vectorially. All the vectors are in the plane of the paper.

Example 4.1

Fig. 4.12(a) shows an equilateral triangle of side length 0.61 m. The point A slides on the x-axis, while the point B slides on the y-axis. At the instant considered, the position is as shown and the velocity of point A is 1.22 m/s in the positive x direction.

Determine the instantaneous velocity of point B and the instantaneous angular velocity of the triangle.

Solution

Taking A as a pole for AB, equation (4.7) states that $\mathbf{V}_B = \mathbf{V}_A + \mathbf{V}_{cir}$, where $|\mathbf{V}_{cir}| = r\omega$ and perpendicular to AB; the formula is illustrated in Fig. 4.12(b). Drawing a

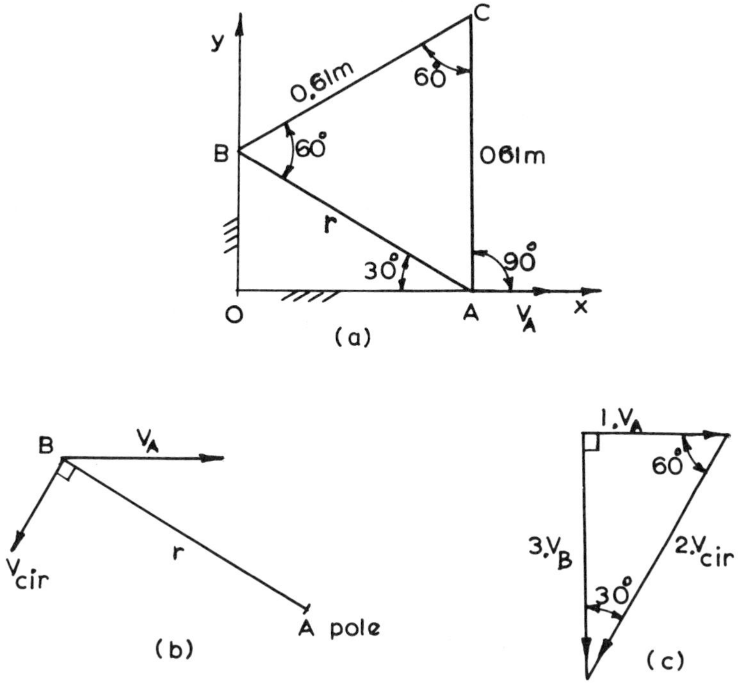

Fig. 4.12.

velocity diagram to a suitable scale, for instance 1 cm ~ 0.2 m/s, we start with the vector $\mathbf{V}_A$ as shown in Fig. 12(c). $\mathbf{V}_{cir}$ is at an angle of 60° to horizontal and the solution is determined by the fact that $\mathbf{V}_B$ is vertical, which enables us to finish the velocity diagram. From the diagram, $\mathbf{V}_B = 2.113$ m/s $\downarrow$ and $\mathbf{V}_{cir} = 2.44$ m/s$\swarrow$, from which $r\omega = 2.44$ m/s or $\omega = 4$ rad/s$\circlearrowleft$. That ω is anti-clockwise follows from the fact that $\mathbf{V}_{cir}$ is the circular velocity of point B about the pole A.

The solution may be compared to an analytical solution; taking $\angle 0BA = \theta$, we obtain the position of point A by $x = 0.61 \sin\theta$, from which $\dot{x} = 0.61 (\cos\theta)\dot{\theta}$, or $\dot{\theta} = \omega = \dot{x}/0.61 \cos\theta$. At the position considered, $\theta = 60°$ and $\dot{x} = V_A = 1.22$ m/s; substituting gives $\omega = 4$ rad/s$\circlearrowleft$. The distance OB $= y = 0.61 \cos\theta$, with $\dot{y} = V_B = -0.61(\sin\theta)\dot{\theta}$. At the position $\theta = 60°$, $\dot{\theta} = 4$ rad/s, and $V_B = -2.113$ m/s. The negative sign indicates that $\mathbf{V}_B$ is in the negative y direction as before.

Example 4.2

Fig. 4.13(a) shows a slider–crank mechanism. The radial arm OP of length r rotates in the plane about the fixed point O with angular velocity ω and angular acceleration $\dot{\omega}$ as shown; the crank arm OP is pinned to a connecting rod PC of length l at P, and the connecting rod is pinned to a piston at C. The piston is constrained to move in a horizontal cylinder. The angle θ determines the position of the system at the instant considered. In actual cases with the dimensions given, Fig. 4.13(a) is drawn to a certain scale, and is called a configuration diagram.

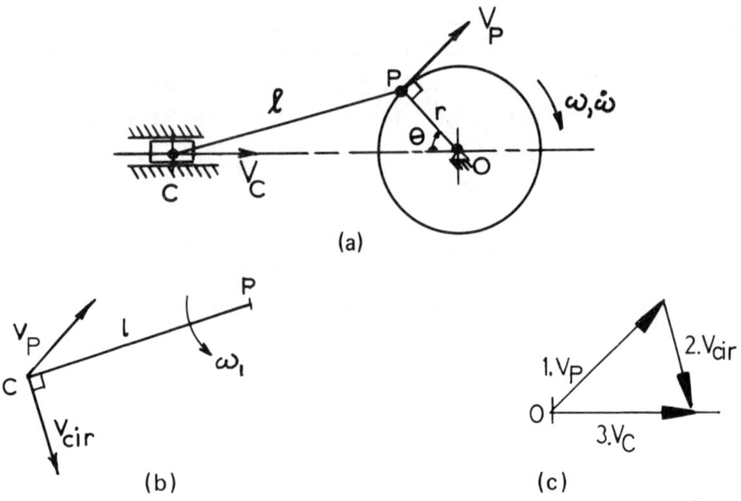

Fig. 4.13.

To determine the piston velocity $\mathbf{V}_C$ at this position, the speed of P may be calculated as $V_P = r\omega$, and the direction of $\mathbf{V}_P$ is as shown perpendicular to OP. Considering P as a point on the connecting rod, P may be used as a pole for the motion of the rod. Equation (4.7) now gives $\mathbf{V}_C = \mathbf{V}_P + \mathbf{V}_{cir}$. The equation is illustrated in Fig. 4.13(b) where $|\mathbf{V}_{cir}| = l\omega_1$, with ω_1 the instanteous angular velocity of PC, and $\mathbf{V}_{cir}$ perpendicular to PC.

The velocity diagram, which represents the vector equation $\mathbf{V}_C = \mathbf{V}_P + \mathbf{V}_{cir}$, may now be drawn to a suitable scale. This is shown in Fig. 4.13(c), where the vectors in the diagram have been numbered to indicate the order of the construction. The piston velocity $\mathbf{V}_C$ may now be measured from the diagram, which also gives the magnitude and direction of $\mathbf{V}_{cir}$. The instantaneous angular velocity ω_1 of PC is determined from $V_{cir} = l\omega_1$, and the direction of ω_1 is anti-clockwise; this is determined by the direction of $\mathbf{V}_{cir}$ in the diagram, which may be taken as circular motion of C about the pole P.

If the velocity diagram is drawn with starting point P, as shown in Fig. 4.14, it may

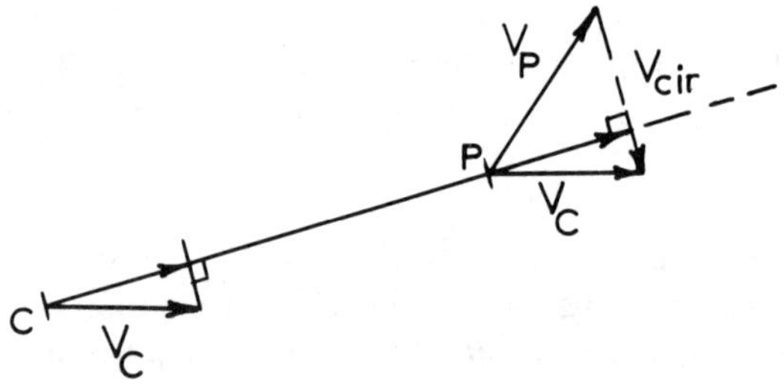

Fig. 4.14.

be seen that the components of $\mathbf{V}_P$ and $\mathbf{V}_C$ along the bar are the same magnitude and direction. This must be the case, since PC is assumed rigid, so the distance PC must be constant at all times; the construction in Fig. 4.14 gives a simple way of determining the velocity of one point in a rigid body with known direction of velocity, if the magnitude and direction of the velocity of one other point in the body are known.

4.8.2 Summary of velocity in plane motion for case 2 problems

The velocity formula for these cases is (4.7): $\mathbf{V}_A = [\mathbf{V}_P + \mathbf{V}_{cir}] + \mathbf{V}_r$, or (4.8): $\mathbf{V}_A = \mathbf{V}_b + \mathbf{V}_r$, where $\mathbf{V}_b = \mathbf{V}_P + \mathbf{V}_{cir}$.

The formula may be illustrated by the vectors in Fig. 4.15 where $\mathbf{V}_{cir} = \boldsymbol{\omega} \times \mathbf{r}$, with

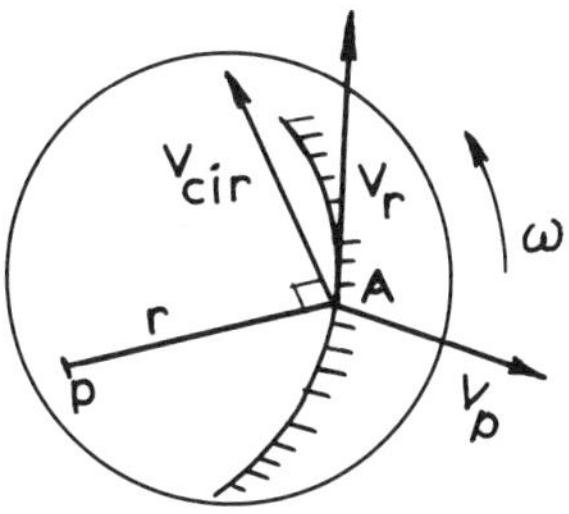

Fig. 4.15.

$|\mathbf{V}_{cir}| = r\omega$ and perpendicular to $\mathbf{r}$ to give the rotational direction ω about the pole p. The vectors must be added by vector addition to give the resultant vector $\mathbf{V}_A$.

Example 4.3

Fig. 4.16(a) shows a horizontal disc of radius r, which is rotating about a fixed vertical

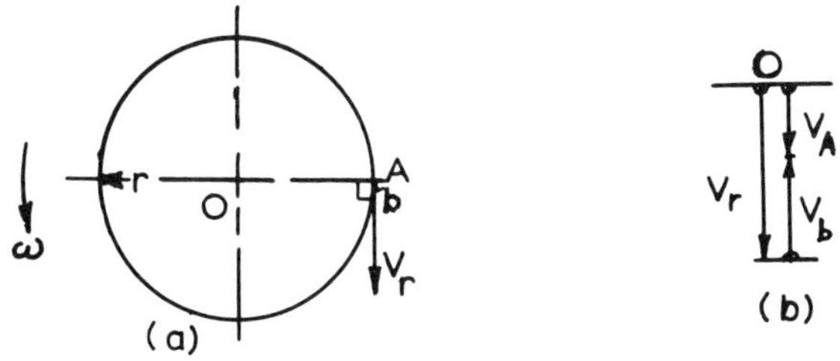

Fig. 4.16.

axis through O. The constant angular velocity of the disc is ω as shown. A point A is moving along the rim of the disc, the relative velocity of A being of constant magnitude and equal to $\mathbf{V}_r$. The direction of $\mathbf{V}_r$, is shown on the figure, and $V_r > r\omega$. Determine the absolute velocity of A at the position shown, when A is passing through the base point b.

Solution

Equation (4.8) gives $\mathbf{V}_A = \mathbf{V}_b + \mathbf{V}_r$, where $\mathbf{V}_b$ is the absolute velocity of a base point fixed in the disc at the instantaneous position of A, so that $V_b = r\omega$ acting upwards. The velocity diagram (Fig. 4.16(b)) gives the result $V_A = V_r - r\omega$ which acts downwards, since $V_r > r\omega$.

If $V_r = r\omega$, $V_A = 0$, so that point A is stationary in space. If $V_r < r\omega$, $V_A = (r\omega - V_r)$ acting upwards.

Example 4.4

Fig. 4.17 shows a plane mechanism. The crank arm OB rotates about the fixed point

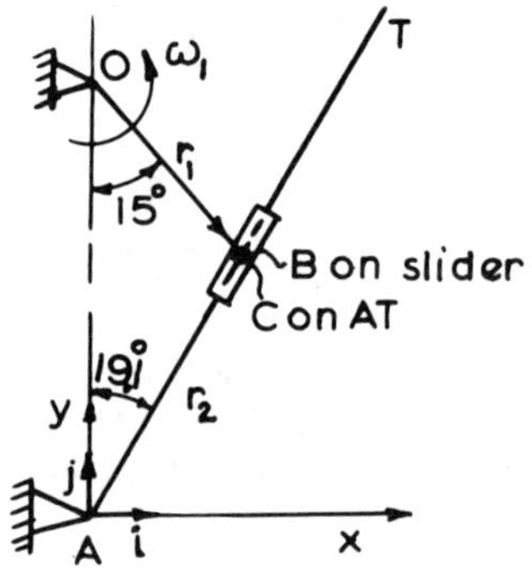

Fig. 4.17.

O with a constant angular velocity $\omega_1 = 10$ rpm anti-clockwise. The slider at B moves along the bar AT, and AT rotates about the point A. The length OB $= r_1 = 17.8$ cm. At the instant considered, the length AB $= r_2 = 14.1$ cm. This is a *case* 2 problem. Determine the instantaneous angular velocity ω_2 of AT, the sliding velocity of the slider along AT and the instantaneous velocity of the point C on AT which coincides with point B at the instant considered.

Solution

$$\omega_1 = 2\pi \frac{10}{60} = 1.047 \text{ rad/s} \circlearrowleft \text{ constant.}$$

$$\mathbf{V}_B = r_1\omega_1 \nearrow = 0.178 \times 1.047 = 0.1864 \text{ m/s } \angle 15^\circ.$$

From the formula (4.8) we have $\mathbf{V}_B = \mathbf{V}_b + \mathbf{V}_r$. Taking the point C on AT as base point for point B, we obtain

$$\mathbf{V}_B = \mathbf{V}_C + \mathbf{V}_r$$

$$\pm\ \mathbf{V}_C = r_2\omega_2 \updownarrow = 0.141\ \omega_2$$

perpendicular to AT, and $\mathbf{V}_r$ 

is along AT, so that $\mathbf{V}_C \perp \mathbf{V}_r$.

We may now draw a velocity diagram as shown in Fig. 4.18. From the diagram we

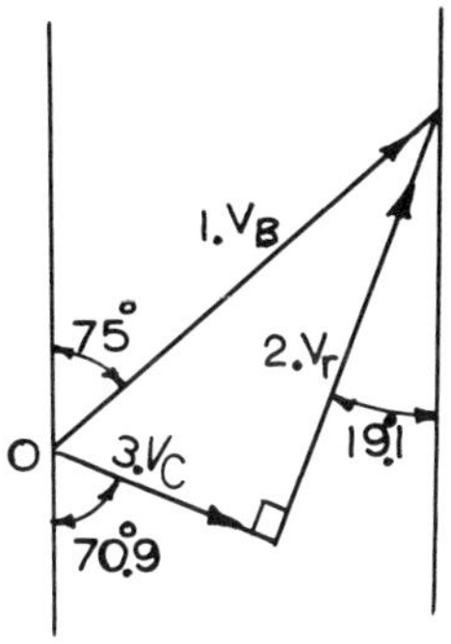

Fig. 4.18.

obtain $\mathbf{V}_C = 0.1864 \sin 55.°9 \searrow\ = 0.1544$ m/s $\searrow$.

Since $\mathbf{V}_C = r_2\omega_2 \searrow$, we find $\omega_2 = 0.1544/0.141 = 1.095$ rad/s $\circlearrowright$. $\mathbf{V}_r \uparrow\ = 0.1864 \cos 55°9 \uparrow\ = 0.1045$ m/s $\uparrow$; this is the instantaneous velocity of sliding of the slider along AT.

4.9 VELOCITY BY THE INSTANTANEOUS CENTRE METHOD

Fig. 4.19 shows a body in plane motion. The instantaneous velocities of any point of

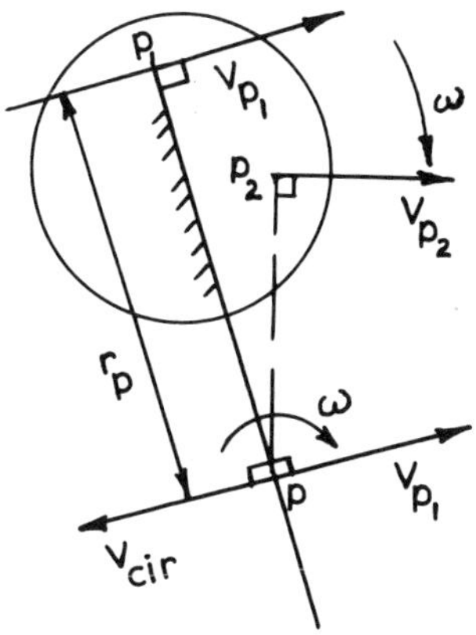

Fig. 4.19.

the body may be determined from the given velocity $\mathbf{V}_{p_1}$ of a point p_1 fixed in the body, together with the instantaneous angular velocity ω. If we now draw a line on the body perpendicular to $\mathbf{V}_{p_1}$ as shown and consider this, if necessary, extended beyond the body on a rigid extension of the body, we may choose a new pole p for the motion on this line a distance r_p from p_1. The velocity of point p from formula (4.7) is $\mathbf{V}_p = \mathbf{V}_{p_1} + \mathbf{V}_{cir}$, where $\mathbf{V}_{cir} = \boldsymbol{\omega} \times \mathbf{r}_p$. If we choose $\mathbf{r}_p$ so that $\omega\, r_p = |\mathbf{V}_{p_1}|$, or $r_p = |\mathbf{V}_{p_1}/\omega|$, it may be seen from the figure that $\mathbf{V}_p = \mathbf{0}$. The point p is the most useful pole for the motion, since the instantaneous velocity of any other point is now $\boldsymbol{\omega} \times \mathbf{r}$ where $|\mathbf{r}|$ is the distance from the point to the pole p. The point p is called *the instantaneous centre of velocity*, and the motion of the body may be considered a pure rotation about p for the purpose of determining velocities; for instance, the point p_2 has the velocity $\mathbf{V}_{p_2}$ as shown, where

$$|\mathbf{V}_{p_2}| = r_{p_2}\omega$$

If ω is small, the distance $|\mathbf{r}_p| = |\mathbf{V}_{p_1}/\omega|$ becomes large and the method becomes impracticable, specifically if the body is translating, $\omega = 0$ and the instantaneous centre is at infinity. The instantaneous centre is not in general a fixed point; although its instantaneous velocity is zero, it generally has an acceleration, and a moment later its position in the plane has changed. The instantaneous centre may *not* therefore be applied to accelerations.

The use of the instantaneous centre to determine velocities is often very convenient. The method is best shown by some examples.

Example 4.5

Solve Example 4.1 by the instantaneous centre method.

Solution

The instantaneous centre is determined by drawing lines perpendicular to the velocities of points A and B as shown in Fig. 4.20. The intersection of the lines gives

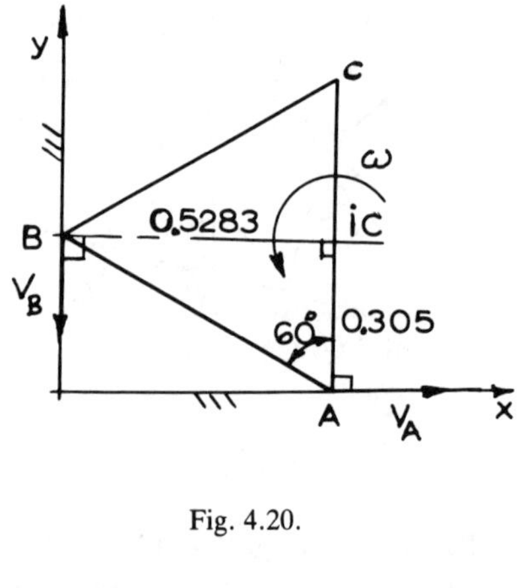

Fig. 4.20.

the instantaneous centre.

Since $V_A = 1.22$ m/s $= 0.305\ \omega$, we find $\omega = 4$ rad/s$\curvearrowleft$, and $\mathbf{V}_B = 0.5283\ \omega \downarrow$ = 2.113 m/s $\downarrow$; for point C we have $\mathbf{V}_C = 0.305\ \omega = 1.22$ m/s$\leftarrow$.

It is important to notice that the two velocities used to determine the instantaneous centre *must be on the same rigid body.*

Example 4.6

Determine the instantaneous centre for the connecting rod PC in Fig. 4.21.

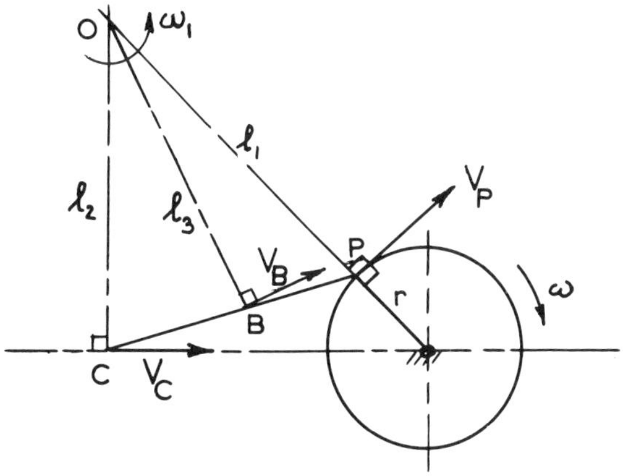

Fig. 4.21.

Solution

Drawing perpendicular lines to the velocity vectors of points P and C, the instantaneous centre is at O as shown on the figure. If the magnitude of ω and the length r are given, we have $V_P = r\omega = l_1\omega_1$. By measuring l_1 the instantaneous angular velocity ω_1 of the connecting rod may be determined. By measuring the lengths l_2 and l_3 as shown, the velocities of points C and B may be determined as $V_C = l_2\omega_1$ and $V_B = l_3\omega_1$ with directions as shown.

Example 4.7

Fig. 4.22 shows a circular disc of radius r which *rolls without slipping* along the line as shown. Determine the instantaneous centre.

Solution

Taking the distance from a fixed point in the plane to the centre of the disc as x and the angle of rotation as θ, a length of circular arc $r\theta$ steps on the straight line during a rotation θ; since there is no slip we have $x = r\theta$, from which $\dot{x} = V_C = r\dot{\theta} = r\omega$, where V_C is the instantaneous velocity of the centre C of the disc. Point C moves on a straight line a distance r from the base line, and the direction of $\mathbf{V}_C$ is as shown. The

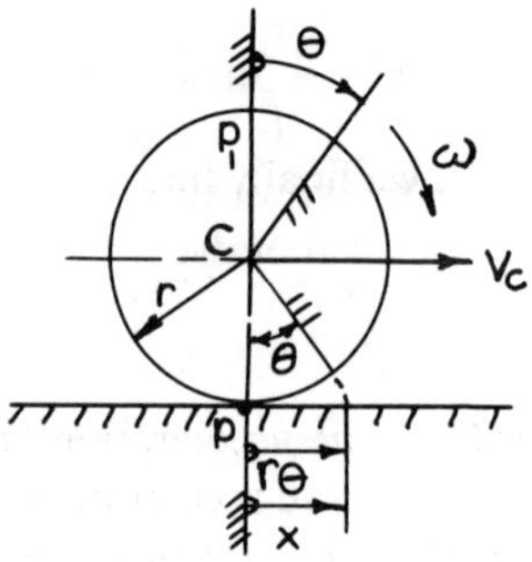

Fig. 4.22.

relationship $V_C = r\omega$ holds only for the no-slip condition. Taking the centre C as pole, the velocity of the contact point p is $\mathbf{V}_p = \mathbf{V}_C + \boldsymbol{\omega} \times \mathbf{r} = r\overrightarrow{\omega} + r\overleftarrow{\omega} = \mathbf{0}$.

The contact point p is then the instantaneous centre. The instantaneous velocity of the top point p_1 of the disc is $\mathbf{V}_{p_1} = 2r\overrightarrow{\omega}$, or twice the velocity of point C.

4.10 ACCELERATION IN PLANE MOTION

In the acceleration formula (4.6):

$$\ddot{\mathbf{R}}_A = [\ddot{\mathbf{R}}_p + \boldsymbol{\omega} \times (\boldsymbol{\omega} \times \mathbf{r}) + \dot{\boldsymbol{\omega}} \times \mathbf{r}] + \ddot{\mathbf{r}}_r + 2\boldsymbol{\omega} \times \dot{\mathbf{r}}_r$$

we now introduce new notations. The absolute acceleration of point A is $\ddot{\mathbf{R}}_A = \mathbf{A}_A$; the absolute acceleration of point p is $\ddot{\mathbf{R}}_p = \mathbf{A}_p$. The components $\boldsymbol{\omega} \times (\boldsymbol{\omega} \times \mathbf{r})$ and $\dot{\boldsymbol{\omega}} \times \mathbf{r}$ are recognised as the *normal* and *tangential* components of acceleration for the point A in *circular motion* about the pole p; consequently, we use the notations

$$\boldsymbol{\omega} \times (\boldsymbol{\omega} \times \mathbf{r}) = \mathbf{A}_n \quad \text{and} \quad \dot{\boldsymbol{\omega}} \times \mathbf{r} = \mathbf{A}_t$$

with $\mathbf{A}_{cir} = \mathbf{A}_n + \mathbf{A}_t$. The component $\ddot{\mathbf{r}}_r$ is the *relative acceleration* of point A in its motion across the base, so we take $\ddot{\mathbf{r}}_r = \mathbf{A}_r$. Finally in the component $2\boldsymbol{\omega} \times \dot{\mathbf{r}}_r$, we first introduce $\dot{\mathbf{r}}_r = \mathbf{V}_r$, where $\mathbf{V}_r$ is the *relative velocity* of point A in its motion across the base, and we then introduce the notation $\mathbf{A}_{cor} = 2\boldsymbol{\omega} \times \mathbf{V}_r$; this component is called *Coriolis Acceleration* after the French mathematician who was the first to discuss this acceleration component about 1830. With these new notations the formula (4.6) takes the form

$$\mathbf{A}_A = [\mathbf{A}_p + \mathbf{A}_{cir}] + \mathbf{A}_r + \mathbf{A}_{cor} \tag{4.9}$$

Applying the formula (4.9) to the base point b for which $\mathbf{V}_r = \mathbf{0}$ and $\mathbf{A}_r = \mathbf{0}$ and consequently also $\mathbf{A}_{cor} = \mathbf{0}$, we find

$$\mathbf{A}_b = [\mathbf{A}_p + \mathbf{A}_{cir}] \tag{4.10}$$

Substituting this in equation (4.9), we finally find

$$\mathbf{A}_A = \mathbf{A}_b + \mathbf{A}_r + \mathbf{A}_{cor} \tag{4.11}$$

This is the shortest possible form for the acceleration and compares to the velocity formula (4.8), $\mathbf{V}_A = \mathbf{V}_b + \mathbf{V}_r$. The absolute acceleration $\mathbf{A}_A$ of a point A may then be determined by adding vectorially the acceleration of the *base* point $\mathbf{A}_b$ to the *relative* acceleration $\mathbf{A}_r$ and the *Coriolis* acceleration $\mathbf{A}_{cor}$. The formula indicates that we may find the absolute acceleration of a point A moving on a rotating body by first determining the acceleration of the point *as if it were fixed on the body* at the instant considered; this gives $\mathbf{A}_b$.

We may then determine the acceleration of the point by *fixing the body*, which gives the acceleration $\mathbf{A}_r$ of the point moving across the body. We must finally add the special component $\mathbf{A}_{cor} = 2\boldsymbol{\omega} \times \mathbf{V}_r$, where ω is the *instantaneous angular velocity of the body or base*, and $\mathbf{V}_r$ is the instantaneous velocity of the point in its motion across the body, that is its *relative velocity*.

The Coriolis acceleration component $\mathbf{A}_{cor} = 2\boldsymbol{\omega} \times \mathbf{V}_r$ vanishes in the two cases where $\boldsymbol{\omega} = \mathbf{0}$, that is the body is translating, or when $\mathbf{V}_r = \mathbf{0}$ which means that the point is at rest or fixed in the body at the instant considered. In three-dimensional motion, $\mathbf{A}_{cor} = \mathbf{0}$ also if $\boldsymbol{\omega}$ and $\mathbf{V}_r$ *are parallel*, but this case does not exist in plane motion. It is convenient to divide the problem into two cases, as was done in the use of velocity. *Case* 1 are problems where point A is fixed in a body, and *case* 2 are problems where point A is moving across a rotating body.

4.10.1 Summary for acceleration in plane motion. Case 1

We have here $\mathbf{V}_r = \mathbf{0}$ and $\mathbf{A}_r = \mathbf{0}$ so that $\mathbf{A}_{cor} = \mathbf{0}$. The acceleration formula is (4.11), $\mathbf{A}_A = \mathbf{A}_b$, so that from (4.10), $\mathbf{A}_A = \mathbf{A}_p + \mathbf{A}_n + \mathbf{A}_t = \mathbf{A}_p + \mathbf{A}_{cir}$, where $\mathbf{A}_p$ is the absolute acceleration of the pole and $\mathbf{A}_n = \boldsymbol{\omega} \times (\boldsymbol{\omega} \times \mathbf{r})$, $\mathbf{A}_t = \dot{\boldsymbol{\omega}} \times \mathbf{r}$ with magnitudes $|\mathbf{A}_n| = r\omega^2$ and $|\mathbf{A}_t| = r\dot{\omega}$ determined in the usual way, assuming that point A is in circular motion about the pole p. The formula may be visualised by *taking out* the line connecting points A and p as shown in Fig. 4.23. The next three examples are all case 1 problems.

Example 4.8

Assuming that point A in Example 4.1 has an instantaneous acceleration $\mathbf{A}_A = 1.53$ m/s$^2 \leftarrow$, determine the instantaneous acceleration of point B and the instantaneous angular acceleration of the triangle.

Solution

From the velocity solution, $\omega = 4$ rad/s ↺. Using point A as a pole for point B, we have the formula (4.10), $\mathbf{A}_B = \mathbf{A}_A + \mathbf{A}_n + \mathbf{A}_t$, in which $\mathbf{A}_A$ is given.

$$\mathbf{A}_n = r\omega^2 \searrow = 0.61 \cdot 4^2 \searrow = 9.76 \text{ m/s} \;\; \searrow 30^\circ$$

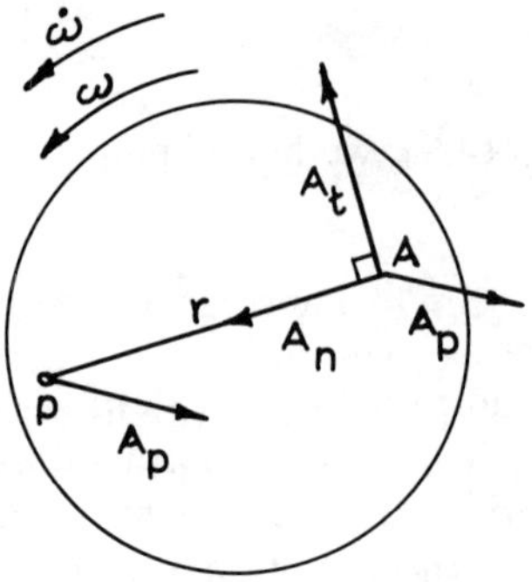

Fig. 4.23.

$$\mathbf{A}_t = r\dot{\omega} \;⤢\; = 0.61\dot{\omega}$$

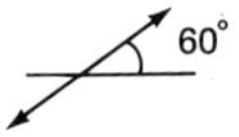

and $\mathbf{A}_B$ is vertical. The acceleration diagram may now be drawn to a suitable scale as shown in Fig. 4.24. From the diagram,

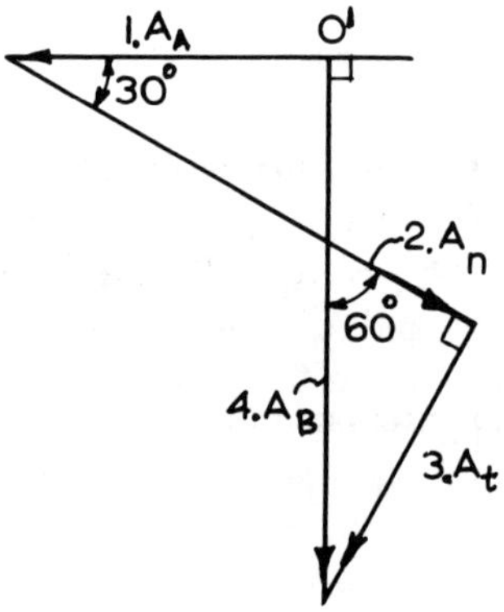

Fig. 4.24.

$$\mathbf{A}_B = 16.9 \text{ m/s}^2 \downarrow$$

and

$$\mathbf{A}_t = 13.85 \text{ m/s}^2 \swarrow$$

so that

$$\dot{\omega} = 13.85/0.61 = 22.7 \text{ rad/s}^2 \circlearrowleft$$

That $\dot{\omega}$ is anti-clockwise is seen by considering point B in circular motion about the pole A with the direction of $\mathbf{A}_t$ as found in the acceleration diagram.

For comparison, an analytical solution of this example may be determined. From Example 4.1 we have the expression $\dot{x} = 0.61\dot{\theta}\cos\theta$, from which

$$\ddot{x} = 0.61\ddot{\theta}\cos\theta - 0.61\dot{\theta}^2\sin\theta, \quad \text{or} \quad \ddot{\theta} = \dot{\omega} = \frac{\ddot{x} + 0.16\dot{\theta}^2\sin\theta}{0.61\cos\theta}$$

At the instant considered,

$$\theta = 60°, \quad \dot{\theta} = \omega = 4 \text{ rad/s}, \quad \ddot{x} = A_A = -1.53 \text{ m/s}^2$$

Substituting these values gives $\dot{\omega} = 22.69$ rad/s^2. The positive sign indicates that $\dot{\omega}$ is anti-clockwise. We also have $\dot{y} = -0.61\dot{\theta}\sin\theta$, from which $\ddot{y} = -0.61\ddot{\theta}\sin\theta - 0.16\dot{\theta}^2\cos\theta$. Substituting the numerical values gives $\ddot{y} = A_B = -16.87$ m/s^2 as before.

Example 4.9
Sketch the acceleration diagram for the point C on the piston in Example 4.2.

Solution
Taking p as a pole for C, we have from equation (4.10), $\mathbf{A}_C = \mathbf{A}_p + \mathbf{A}_{Cn} + \mathbf{A}_{Ct}$. $\mathbf{A}_p = \mathbf{A}_{pn} + \mathbf{A}_{pt} = r\omega^2$  $+ r\dot{\omega}$  ; these are the usual two perpendicular components for a point in circular motion. $\mathbf{A}_{Cn} = l\omega_1^2$  , and $\mathbf{A}_{Ct} = l\dot{\omega}_1$ 

$\mathbf{A}_{Ct}$ is of unknown magnitude and sense since $\dot{\omega}_1$ is unknown, but its line of action is perpendicular to $\mathbf{A}_{cn}$. For given numerical values of r, l and ω, an acceleration diagram may now be drawn to a suitable scale, using the fact that $\mathbf{A}_C$ is horizontal. A sketch of the diagram is shown in Fig. 4.25. The magnitude of $\mathbf{A}_{Ct}$ may be taken from the diagram to give the instantaneous angular acceleration of the connecting rod, since $|\mathbf{A}_{Ct}| = l\dot{\omega}_1$. From the direction of $\mathbf{A}_{Ct}$, the direction of $\dot{\omega}_1$ is clockwise.

Example 4.10
Determine the acceleration of the contact point for the rolling wheel in Example 4.7. This point was shown to be the instantaneous centre.

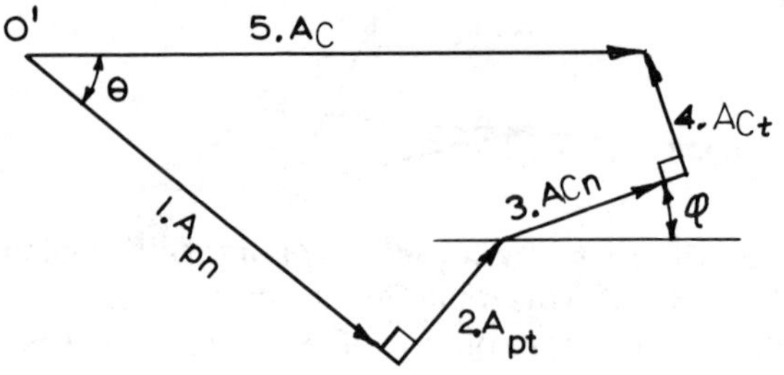

Fig. 4.25.

Solution

$x = r\theta$, $\dot{x} = r\dot{\theta}$ and $\ddot{x} = r\ddot{\theta} = r\dot{\omega}$, so that $\mathbf{A}_C = r\dot{\omega}\rightarrow$. Taking C as a pole, we find from formula (4.10) for the contact point, $\mathbf{A} = \mathbf{A}_C + \mathbf{A}_n + \mathbf{A}_t = r\dot{\omega}\rightarrow + r\omega^2\uparrow + r\dot{\omega}\leftarrow = r\omega^2\uparrow$.

The instantaneous centre has an acceleration directed towards the centre of the wheel and of magnitude $r\omega^2$.

4.10.2 Summary for acceleration in plane motion. Case 2

The acceleration formula is (4.11): $\mathbf{A}_A = \mathbf{A}_b + \mathbf{A}_r + \mathbf{A}_{cor}$, where $\mathbf{A}_b = \mathbf{A}_p + \mathbf{A}_{cir}$ from (4.10). $\mathbf{A}_p$ is the acceleration of the pole and $\mathbf{A}_{cir} = \mathbf{A}_n + \mathbf{A}_t$, where $\mathbf{A}_n = \boldsymbol{\omega} \times (\boldsymbol{\omega} \times \mathbf{r})$. $\mathbf{A}_t = \dot{\boldsymbol{\omega}} \times \mathbf{r}$, with magnitudes $|\mathbf{A}_n| = r\omega^2$ and $|\mathbf{A}_t| = r\dot{\omega}$, determined as usual by assuming that point A is in circular motion about the pole p. The relative acceleration component $\mathbf{A}_r$ is determined by *fixing the body* or base and considering point A in its motion across the body; $\mathbf{A}_r$ is then determined as usual for a point moving in a fixed plane.

The Coriolis component of acceleration $\mathbf{A}_{cor} = 2\boldsymbol{\omega} \times \mathbf{V}_r$, where $\boldsymbol{\omega}$ is the angular velocity of the body and $\mathbf{V}_r$ is the relative velocity of point A across the body; both are instantaneous values at the instant considered. Since $\boldsymbol{\omega}$ is perpendicular to the plane of motion, the magnitude of $\mathbf{A}_{cor}$ is $|\mathbf{A}_{cor}| = 2\omega V_r \sin 90° = 2\omega V_r$.

The sense and direction of $\mathbf{A}_{cor}$ may be determined directly from the vector cross-product $\boldsymbol{\omega} \times \mathbf{V}_r$, but in the case of plane motion considered here it may also be determined from the *following simple rule*: rotate the vector $\mathbf{V}_r$ in the plane of motion 90° in the direction of rotation ω; this gives the direction and sense of $\mathbf{A}_{cor}$.

The formulae (4.10) and (4.11) may be visualised by *taking out* the line connecting the points A and p as shown in Fig. 4.26. The next two examples are case 2 problems.

Example 4.11

Determine the instantaneous acceleration of point A in Example 4.3.

Solution

From formula (4.11) the acceleration of point A is $\mathbf{A}_A = \mathbf{A}_b + \mathbf{A}_r + \mathbf{A}_{cor}$.

The base point b is a point in circular motion with constant angular velocity ω, so that $\mathbf{A}_b = r\omega^2\leftarrow$. The relative acceleration is determined by stopping the disc; point

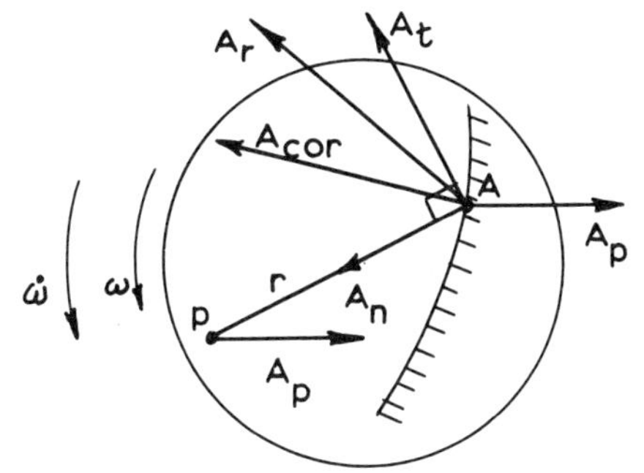

Fig. 4.26.

A is then in circular motion with constant velocity $|\mathbf{V}_r|$, and we have $\mathbf{A}_r = V_r^2/r \leftarrow$. $\mathbf{A}_{cor} = 2\boldsymbol{\omega} \times \mathbf{V}_r$, with magnitude $|\mathbf{A}_{cor}| = 2\omega V_r$.

The direction of $\mathbf{A}_{cor}$ is determined by rotating the vector $\mathbf{V}_r$ in the plane 90° in the direction of rotation ω; this gives $\mathbf{A}_{cor} \rightarrow$. We have now $\mathbf{A}_A = r\omega^2 \leftarrow + V_r^2/r \leftarrow + 2\omega V_r \rightarrow$.

Taking the direction towards the centre of the disc as positive, we obtain

$$|\mathbf{A}_A| = r\omega^2 + \frac{V_r^2}{r} - 2\omega V_r = \frac{1}{r}(r^2\omega^2 + V_r^2 - 2\omega V_r r) = \frac{1}{r}(V_r - r\omega)^2$$

this is always positive so that $\mathbf{A}_A$ is always directed towards the centre of the disc. In the special case where $|\mathbf{V}_r| = r\omega$, we have found that $\mathbf{V}_A = \mathbf{0}$, and the acceleration in this case is $\mathbf{A}_A = \mathbf{0}$ so that the point A is stationary in the plane as it should be. In the present simple case the acceleration of point A may also be determined without consideration of Coriolis acceleration.

Since point A is in circular motion with constant magnitude of velocity $|\mathbf{V}_A| = V_r - r\omega$, the acceleration is the usual acceleration towards the centre of the disc, and the magnitude is $|\mathbf{A}_A| = V_A^2/r = \frac{1}{r}(V_r - r\omega)^2$, as determined above.

Example 4.12

In Example 4.4, determine the instantaneous angular acceleration of the bar AT, the relative acceleration of sliding of the slider along AT and the instantaneous acceleration of point C on AT.

Solution

The point B is in circular motion, with acceleration $\mathbf{A}_B = \mathbf{A}_{Bn} + \mathbf{A}_{Bt}$.

$$\mathbf{A}_{Bn} = r_1\omega^2 \nwarrow = 0.178 \cdot 1.047^2 \nwarrow = 0.1951 \text{ m/s}^2$$

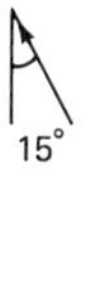

$$\mathbf{A}_{Bt} = \mathbf{0}$$

Using AT as base for B with base point C fixed on AT at the instantaneous position of B, we have from formula (4.10), $\mathbf{A}_B = \mathbf{A}_C + \mathbf{A}_r + \mathbf{A}_{cor}$.

Point C is in circular motion, with $\mathbf{A}_C = \mathbf{A}_{Cn} + \mathbf{A}_{Ct}$; $\mathbf{A}_{Cn} = r_2\omega_2^2 \swarrow = 0.1691$ m/s^2 

, and $\mathbf{A}_{Ct} = r^2\dot{\omega}_2 | = 0.141\dot{\omega}_2 \updownarrow$ perpendicular to $\mathbf{A}_{Cn}$ · $\mathbf{A}_r$ is the relative acceleration of sliding along AT 

. $\mathbf{A}_{cor} = 2\boldsymbol{\omega}_2 \times \mathbf{V}_r$; $|\mathbf{A}_{cor}| = 2\omega_2 V_r = 2 \cdot 1.095 \cdot 0.1045 = 0.2289$ m/s^2; rotating $\mathbf{V}_r$ 90° in the plane in direction ω_2 gives the direction of $\mathbf{A}_{cor}$ perpendicular to AT $\searrow$.

We may now draw an acceleration diagram to a suitable scale or as a sketch not to scale. A sketch of the diagram is shown in Fig. 4.27. From the diagram we find

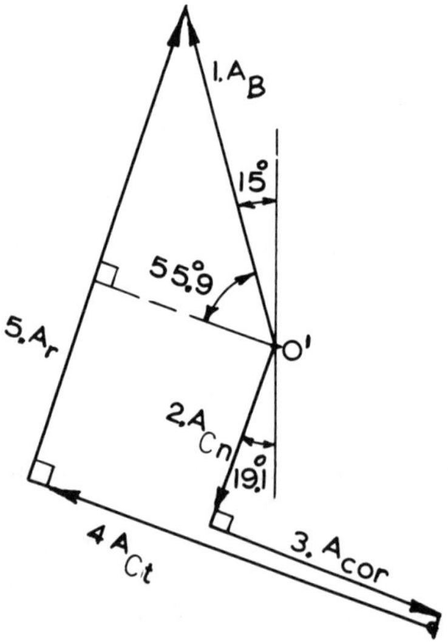

Fig. 4.27.

$$|\mathbf{A}_{Ct}| - |\mathbf{A}_{cor}| = 0.1951 \cos 55.°9 = 0.1094$$

$$|\mathbf{A}_{Ct}| = 0.1094 + 0.2289 = 0.3383 \text{ m/s}^2 = 0.141\dot{\omega}_2$$

from which $\underline{\dot{\omega}_2 = 2.40 \text{rad/s}^2 \circlearrowleft}$. The counterclockwise direction of $\dot{\omega}_2$ is determined from the direction of $\mathbf{A}_{Ct}$ and considering C in its circular motion about point A. The acceleration $\mathbf{A}_r$ is determined as $|\mathbf{A}_r| = |\mathbf{A}_{Cn}| + |\mathbf{A}_B| \sin 55.°9 = \underline{0.3307}$ m/s^2 directed as shown in the diagram.

$$|\mathbf{A}_C| = \sqrt{|\mathbf{A}_{Cn}|^2 + |\mathbf{A}_{Ct}|^2} = \sqrt{0.1691^2 + 0.3383^2} = 0.3782 \text{ m/s}^2.$$

4.11 ANALYTICAL DETERMINATION OF VELOCITIES

In some cases the displacement of a point on a mechanism may be expressed as a function of time, and we may determine the velocity and acceleration of the point as continuous functions of time by differentiation, an example of which is the engine mechanism in Chapter 1. The procedure is usually only simple if the point is in rectilinear motion or in circular motion. In more complicated cases we use the graphical method described for instantaneous solutions at a given position; the solution may of course be repeated for any number of positions of the mechanism. The vector formulae developed may also be used for instantaneous analytical solutions at a given position. The procedure is best shown by some examples.

Example 4.13

Fig. 4.28 shows a plane four-bar mechanism, consisting of bars pinned together at the

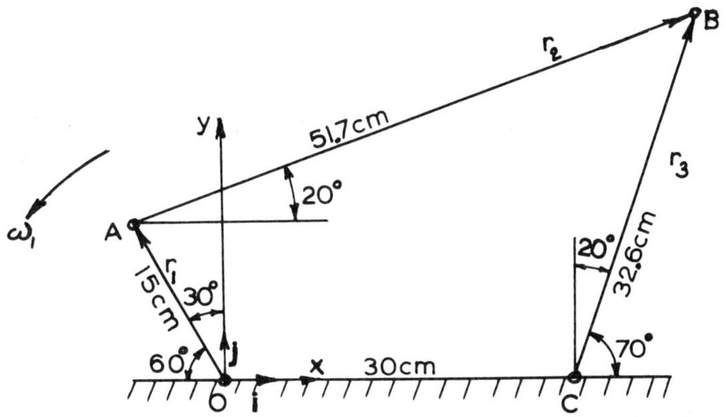

Fig. 4.28.

ends; the fixed ground OC is counted as one bar. The mechanism is able to move in the plane of the mechanism. Bar OA is rotated by an exteral torque about O at a constant angular velocity $\omega_1 = 12$ rad/s counterclockwise. Determine the instantaneous velocity of points A and B and the angular velocities of AB and BC at the given configuration. This is a case 1 problem.

Solution

A coordinate system (x,y) is introduced at O and instantaneous position vectors $\mathbf{r}_1$, $\mathbf{r}_2$ and $\mathbf{r}_3$ drawn along the bars as shown. From the given geometry

$$\mathbf{r}_1 = -0.075\mathbf{i} + 0.13\mathbf{j} \text{ m}$$

$$\mathbf{r}_2 = 0.486\mathbf{i} + 0.177\mathbf{j} \text{ m}$$

$$\mathbf{r}_3 = 0.1113\mathbf{i} + 0.306\mathbf{j} \text{ m}$$

The angular velocities are

$$\boldsymbol{\omega}_1 = 12\mathbf{k}\ \text{rad/s} \qquad \boldsymbol{\omega}_2 = \omega_2\mathbf{k}\ \text{rad/s} \quad \text{and} \quad \boldsymbol{\omega}_3 = \omega_3\mathbf{k}\ \text{rad/s}$$

The velocities may now be calculated from formula (4.7):

$$\mathbf{V}_A = \boldsymbol{\omega}_1 \times \mathbf{r}_1 = -1.56\mathbf{i}–0.90\mathbf{j}\ \text{m/s} \qquad V_A = 1.80\ \text{m/s}$$

$$\mathbf{V}_B = \mathbf{V}_A + \mathbf{V}_{cir}$$

$$\mathbf{V}_{cir} = \boldsymbol{\omega}_2 \times \mathbf{r}_2 = -0.177\ \omega_2\mathbf{i} + 0.486\omega_2\mathbf{j}$$

Substituting in $\mathbf{V}_B$ gives

$$\mathbf{V}_B = -(1.56 + 0.177\omega_2)\mathbf{i} - (0.90–0.486\omega_2)\mathbf{j}$$

$\mathbf{V}_B$ may also be found from

$$\mathbf{V}_B = \boldsymbol{\omega}_3 \times \mathbf{r}_3 = 0.306\omega_3\mathbf{i} + 0.1113\omega_3\mathbf{j}$$

Equating the **i** terms and the **j** terms for the two expressions for $\mathbf{V}_B$ gives two equations to determine ω_2 and ω_3; the solution is

$$\omega_2 = 3.48\ \text{rad/s} \qquad \text{and} \qquad \omega_3 = 7.10\ \text{rad/s}$$

both anti-clockwise. Substituting back gives

$$\mathbf{V}_B = -2.17\mathbf{i} + 0.791\mathbf{j}\ \text{m/s} \qquad V_B = 2.13\ \text{m/s}$$

$$\mathbf{V}_{cir} = -0.616\mathbf{i} + 1.69\mathbf{j}\ \text{m/s} \qquad V_{cir} = 1.80\ \text{m/s}$$

Example 4.14
As an example of a *case 2* problem, calculate the instantaneous velocities in Example 4.4 (Fig. 4.17).

Solution
From the geometry, the instantaneous position vectors are

$$\mathbf{r}_1 = 4.6070\ \mathbf{i} - 17.1935\ \mathbf{j}\ \text{cm}$$

$$\mathbf{r}_2 = 4.6138\ \mathbf{i} + 13.3238\ \mathbf{j}\ \text{cm}$$

The angular velocities are $\boldsymbol{\omega}_1 = 1.0472\mathbf{k}$ rad/s and $\boldsymbol{\omega}_2 = \omega_2\mathbf{k}$ rad/s.

$$\mathbf{V}_B = \boldsymbol{\omega}_1 \times \mathbf{r}_1 = 18.0050\ \mathbf{i} + 4.8245\ \mathbf{j}\ \text{cm/s}$$

$$|\mathbf{V}_B| = 18.6402 \text{ cm/s}$$

Taking point C on AT as base point for point B,

$$\mathbf{V}_B = \mathbf{V}_C + \mathbf{V}_r$$

$\mathbf{V}_C = \boldsymbol{\omega}_2 \times \mathbf{r}_2 = -13.3238\omega_2\mathbf{i} + 4.6138\,\omega_2\mathbf{j}$ cm/s $\mathbf{V}_r = V_r \sin 19.°1\,\mathbf{i} + V_r \cos 19.°1\,\mathbf{j} = 0.3272\,V_r\mathbf{i} + 0.9449\,V_r\mathbf{j}$. Substituting $\mathbf{V}_C$ and $\mathbf{V}_r$ gives $\mathbf{V}_B = (0.3272\,V_r - 13.3238\,\omega_2)\,\mathbf{i} + (0.9449\,V_r + 4.6138\,\omega_2)\mathbf{j}$. Equating the **i** and **j** terms in the two expressions for $\mathbf{V}_B$ gives two algebraic equations in V_r and ω_2. The solution is $V_r = 10.4505$ cm/s and $\omega_2 = -1.0947$ rad/s. The result is $\omega_2 = 1.0947$ rad/s↶, $\mathbf{V}_r = 3.4196\,\mathbf{i} + 9.8752\,\mathbf{j}$ cm/s, and $\mathbf{V}_C = 14.5856\mathbf{i} - 5.0507\mathbf{j}$ cm/s; $|\mathbf{V}_C| = 15.4353$ cm/s.

4.12 ANALYTICAL DETERMINATION OF ACCELERATIONS

The analytical determination of accelerations is best shown by an example.

Example 4.15

For the mechanism in Fig. 4.28, determine the instantaneous accelerations of points A and B, and the instantaneous angular accelerations $\dot{\omega}_2$ and $\dot{\omega}_3$ of AB and BC respectively.

This is a case 1 problem

Solution

The linear and angular velocities were determined in Example 4.13. The accelerations are determined from formula (4.7). The angular accelerations are $\dot{\boldsymbol{\omega}}_1 = \mathbf{0}$, $\dot{\boldsymbol{\omega}}_2 = \dot{\omega}_2 k$ and $\dot{\boldsymbol{\omega}}_3 = \dot{\omega}_3\mathbf{k}$.

$$\mathbf{A}_A = \boldsymbol{\omega}_1 \times (\boldsymbol{\omega}_1 \times \mathbf{r}_1) + \dot{\boldsymbol{\omega}}_1 \times \mathbf{r}_1 = \boldsymbol{\omega}_1 \times \mathbf{V}_A = 10.8\mathbf{i} - 18.7\mathbf{j} \text{ m/s}^2$$

$$A_A = 21.6 \text{ m/s}^2$$

$$\mathbf{A}_B = \mathbf{A}_A + \boldsymbol{\omega}_2 \times (\boldsymbol{\omega}_2 \times \mathbf{r}_2) + \dot{\boldsymbol{\omega}}_2 \times \mathbf{r}_2$$

$$\boldsymbol{\omega}_2 \times (\boldsymbol{\omega}_2 \times \mathbf{r}_2) = \boldsymbol{\omega}_2 \times \mathbf{V}_{cir} = -5.88\mathbf{i} - 2.14\mathbf{j} \text{ m/s}^2$$

$$\dot{\boldsymbol{\omega}}_2 \times \mathbf{r}_2 = -0.1777\dot{\omega}_2\mathbf{i} + 0.486\dot{\omega}_2\mathbf{j} \text{ m/s}^2$$

We have now

$$\mathbf{A}_B = (4.92 - 0.177\dot{\omega}_2)\mathbf{i} - (20.84 - 0.486\dot{\omega}_2)\mathbf{j} \text{ m/s}^2$$

We also have

$$\mathbf{A}_B = \boldsymbol{\omega}_3 \times (\boldsymbol{\omega}_3 \times \mathbf{r}_3) + \dot{\boldsymbol{\omega}}_3 \times \mathbf{r}_3$$

$$\boldsymbol{\omega}_3 \times (\boldsymbol{\omega}_3 \times \mathbf{r}_3) = \boldsymbol{\omega}_3 \times \mathbf{V}_B = -5.61\mathbf{i} - 15.4\mathbf{j} \text{ m/s}^2$$

$$\dot{\boldsymbol{\omega}}_3 \times \mathbf{r}_3 = -0.306\dot{\omega}_3\mathbf{i} + 0.1113\dot{\omega}_3\mathbf{j} \text{ m/s}^2$$

so that

$$\mathbf{A}_B = -(5.61 + 0.306\dot{\omega}_3)\mathbf{i} - (15.4 - 0.1113\dot{\omega}_3)\mathbf{j} \text{ m/s}^2$$

Equating the **i** and **j** terms from the two expressions for $\mathbf{A}_B$ gives two equations in $\dot{\omega}_2$ and $\dot{\omega}_3$; the solution is $\dot{\omega}_2 = 4.0$ rad/s^2 (anti-clockwise) and $\dot{\omega}_3 = -32.1$ rad/s^2 (clockwise). Substituting back in the previous expression for $\mathbf{A}_B$ gives $\mathbf{A}_B = 4.21\mathbf{i} - 18.9\mathbf{j}$ m/s^2, so that

$$A_B = 19.35 \text{ m/s}^2$$

The direction of $\mathbf{A}_B$ is given by the vector expression.

Example 4.16
Determine the acceleration in Example 4.12 analytically. This is a *case 2* problem.

Solution
$\mathbf{A}_B = \boldsymbol{\omega}_1 \times (\boldsymbol{\omega}_1 \times \mathbf{r}_1) + \dot{\boldsymbol{\omega}}_1 \times \mathbf{r}_1$. $\boldsymbol{\omega}_1 \times \mathbf{r}_1 = \mathbf{V}_B$ and $\dot{\boldsymbol{\omega}}_1 = \mathbf{0}$, so that

$$\mathbf{A}_B = \boldsymbol{\omega}_1 \times \mathbf{V}_B = -5.0522\mathbf{i} + 18.8548\mathbf{j} \text{ cm/s}^2$$

$$|\mathbf{A}_b| = 19.5199 \text{ cm/s}^2$$

Taking point C on AT as a base point for point B, we have also $\mathbf{A}_B = \mathbf{A}_C + \mathbf{A}_r + \mathbf{A}_{cor}$. With $\dot{\boldsymbol{\omega}}_2 = \dot{\omega}_2\mathbf{k}$, the acceleration of point C is $\mathbf{A}_C = \boldsymbol{\omega}_2 \times (\boldsymbol{\omega}_2 \times \mathbf{r}_2) + \dot{\boldsymbol{\omega}}_2 \times \mathbf{r}_2 = \boldsymbol{\omega}_2 \times \mathbf{V}_C + \dot{\boldsymbol{\omega}}_2 \times \mathbf{r}_2 = -(5.5290 + 13.3238\,\dot{\omega}_2)\mathbf{i} - (15.9669 - 4.6138\,\dot{\omega}_2)\mathbf{j}$ cm/s^2. The relative acceleration $\mathbf{A}_r$ is along AT, so that $\mathbf{A}_r = A_r\,(\sin 19°.1)\mathbf{i} + A_r(\cos 19°.1)\mathbf{j} = 0.3272\,A_r\mathbf{i} + 0.9449\,A_r\mathbf{j}$ cm/s^2.

$$\mathbf{A}_{cor} = 2\boldsymbol{\omega}_2 \times \mathbf{V}_r = 2(-1.0947\mathbf{k}) \times (3.4196\mathbf{i} + 9.8752\mathbf{j}) =$$

$$21.6208\mathbf{i} - 7.4869\mathbf{j} \text{ cm/s}^2.\ |\mathbf{A}_{cor}| = 22.8804 \text{ cm/s}^2.$$

Accumulating **i** and **j** terms, this gives

$$\mathbf{A}_B = (16.0918 - 13.3238\,\dot{\omega}_2 + 0.3272\,A_r)\mathbf{i} + (-23.4538 + 4.6138\,\dot{\omega}_2 + 0.9449\,A_r)\mathbf{j}$$

Equating the **i** and **j** terms in the two expressions for $\mathbf{A}_B$ gives two algebraic equations in $\dot{\omega}_2$ and A_r, and the solution is $A_r = 33.0649$ cm/s^2 and $\dot{\omega}_2 = 2.3983$ rad/s^2).

The relative acceleration is now $\mathbf{A}_r = 10.8188\mathbf{i} + 31.2430\mathbf{j}$ cm/s^2.

Substituting back in the expression for $\mathbf{A}_C$ gives the result $\mathbf{A}_C = -37.4835\mathbf{i} - 4.9016\mathbf{j}$ cm/s^2, with $|\mathbf{A}_C| = 37.8026$ cm/s^2.

PROBLEMS

4.1 Fig. 4.29 shows a mechanism consisting of a circular disc with centre C and radius

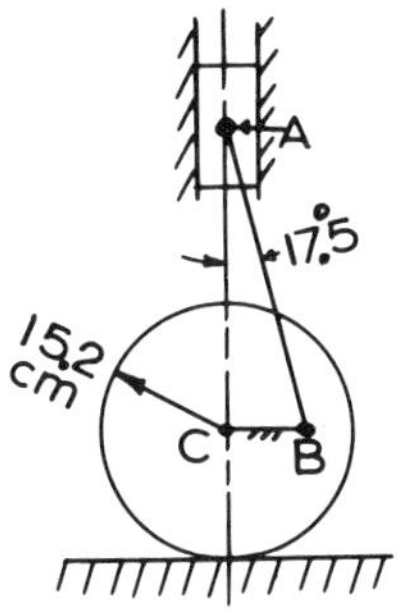

Fig. 4.29.

15.2 cm. The disc rolls on the horizontal plane without slipping; a link AB is attached to the disc at B and rotates freely around B; the other end of the link A is constrained to move in the vertical direction only.

For the position shown, the velocity of C is 0.61 m/s to the right. Find the magnitude and direction of the velocity of A and the angular velocity of AB for the given position, given that CB = 7.62 cm, AB = 25.4 cm.

4.2 Fig. 4.30 shows an articulated connecting rod used on a two-cylinder V-engine.

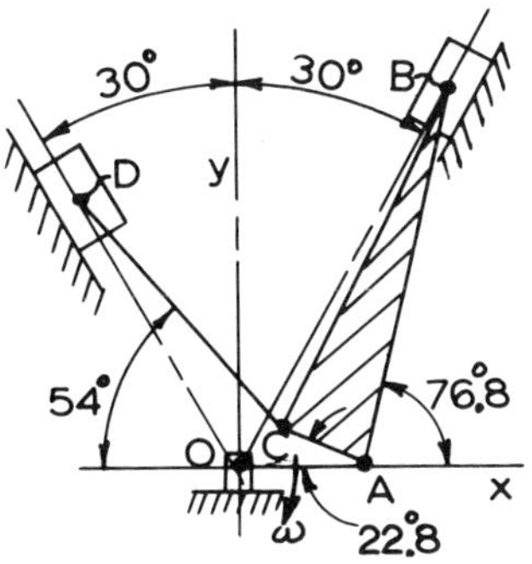

Fig. 4.30.

The crank OA rotates at a constant 2000 rev/min in the clockwise direction. The dimensions are OA = AC = 5.08 cm, AB = BC = 15.25 cm, and CD = 12.7 cm.

Find the velocity of the pistons B and D for the given position in which OA is horizontal.

4.3 The four-bar plane-motion mechanism (Fig. 4.31) has a crank angular velocity

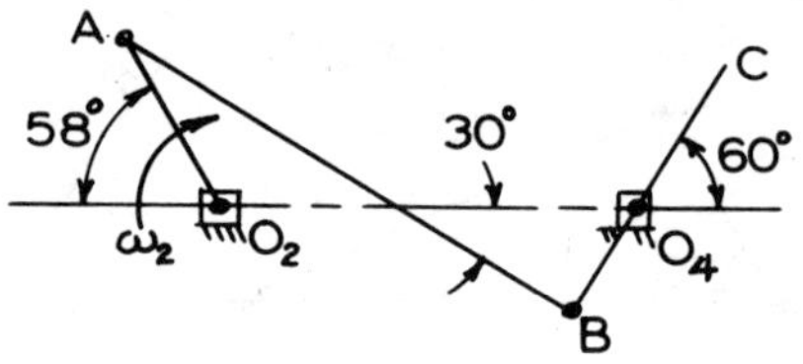

Fig. 4.31.

$\omega_2 = 5$ rad/s. Find the velocity of point C and the angular velocity of link 4, given that $O_2A = 7.61$ cm, $0_4C = 10.15$ cm, AB = 25.4 cm, BC = 17.8 cm, $O_20_4 =$ 21.6 cm.

4.4 An indicator reducing gear shown in Fig. 4.32 consists of four bars: AB =

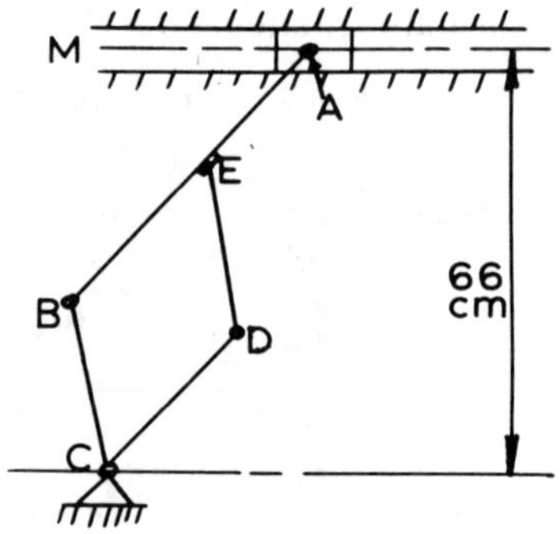

Fig. 4.32.

55.9 cm, BC = ED = 20.3 cm, DC = EB = 15.25 cm, pinned together to form a pantograph.

Point A is attached to the crosshead of an engine and C is a fixed point on the engine frame. When angle MAE = 60°, A is moving to the left with a velocity of 3.05 m/s and an acceleration of 6.10 m/s^2.

In the given position find the velocity and acceleration of point E and the angular velocity and acceleration of BC.

4.5 In the plane mechanism shown in Fig. 4.33, CA = 10 cm and AB = 20 cm. The

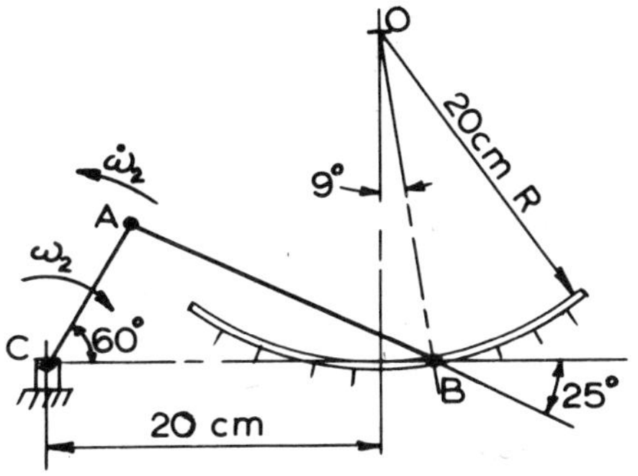

Fig. 4.33.

arm CA is rotating about C, and has, at the instant considered, an angular velocity $\omega_2 = 30$ rad/s and angular acceleration $\dot{\omega}_2 = 240$ rad/s^2, directed as shown on the figure. The end B of AB is sliding along a fixed circular guide of 20 cm radius.

Find the velocity and acceleration of point B and the angular velocity and angular acceleration of AB for the position shown.

4.6 In Fig. 4.34 a horizontal turntable rotates about O with constant angular velocity

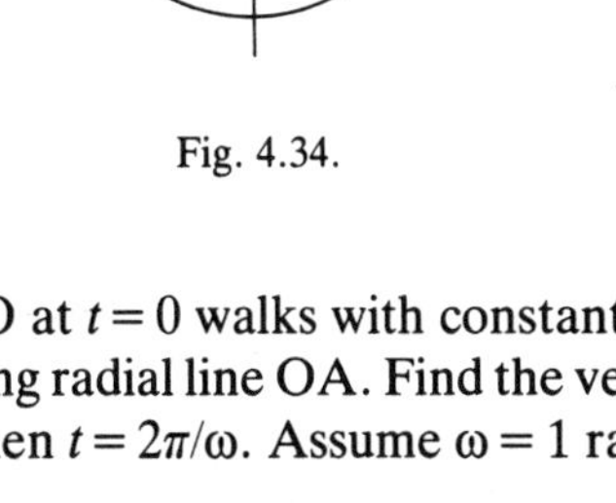

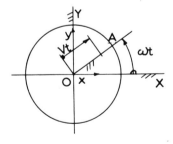

Fig. 4.34.

ω. A man starting from O at $t = 0$ walks with constant relative speed $V = 3$ m/s outwards along the rotating radial line OA. Find the velocity and acceleration of the man at the instant when $t = 2\pi/\omega$. Assume $\omega = 1$ rad/s.

4.7 A point P_1 (Fig. 4.35) moves with constant $V_r = 3.66$ m/s along AB while the disc

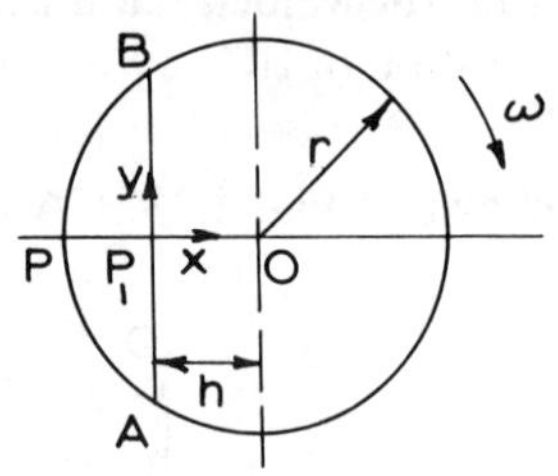

Fig. 4.35.

rotates with constant $\omega = 62.9$ rad/s. Find V and A of P_1 for the position shown, if $r = 0.305$ m and $h = 0.153$ m. Assume that P_1 at this instant is moving towards B.

4.8 A gun fires a shell with muzzle velocity $V_r = 3000$ m/s ($A_r \simeq 0$), while rotating in a horizontal plane about a vertical axis (Fig. 4.36) at $\omega = 0.1$ rev/s. If the gun barrel

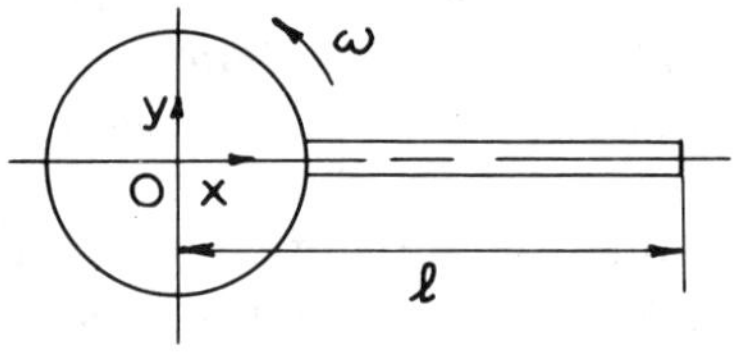

Fig. 4.36.

is of length $l = 15$ m, find the velocity and acceleration of the shell just as it leaves the barrel.

4.9 In the plane mechanism shown in Fig. 4.37 the crank arm AB of length 5.08 cm

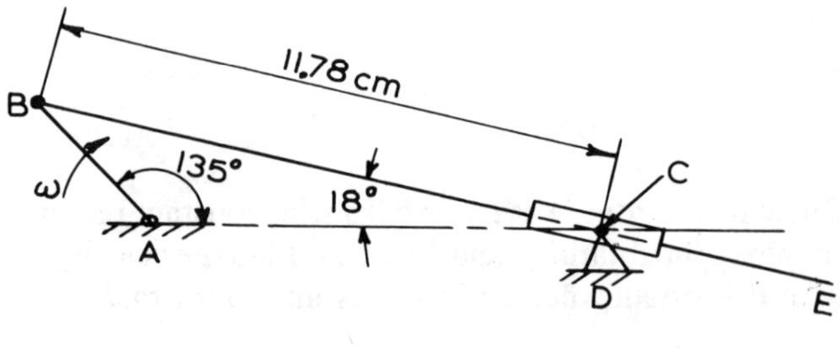

Fig. 4.37.

rotates about A in the clockwise direction with a constant angular velocity $\omega = 12$ rad/s.

The link BE is pinned to AB at B and slides freely in the guide at D; BE is of length 17.8 cm. The guide at D rotates freely in the plane about D, and AD is of length 7.62 cm.

At the position where angle BAD = 135°, find the velocity and acceleration of point C on BE at the centre of the guide D, and the velocity and acceleration of point E.

Find also the angular velocity and angular acceleration of BE.

5

Moments of area and moments of inertia

5.1 MOMENTS OF AREA OF A PLANE FIGURE

5.1.1 First moment of area. Centre of area

The *first moment of area* of a plane figure in Fig. 5.1 with respect to the x-axis is

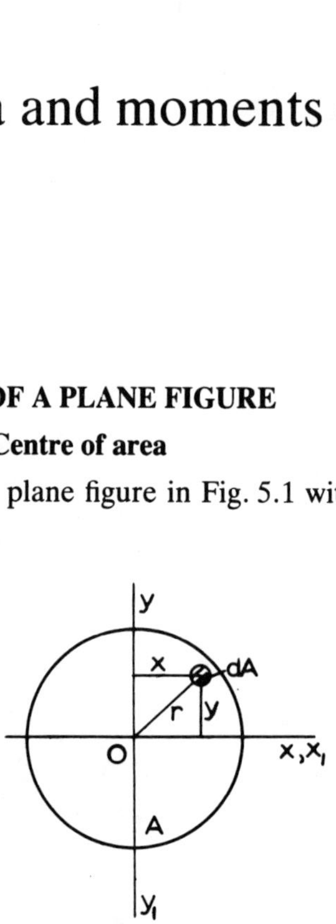

Fig. 5.1.

defined by the integral $S_x = \int_A y \, dA$, and for the y-axis, $S_y = \int_A x \, dA$; the units are m^3, and S_x and S_y may be positive, negative or zero.

By writing $S_x = y_C A$, $S_y = x_C A$, where A is the total area of the figure, we define a point C (x_C, y_C) called the *centre of area or centroid*; the centre of area is then determined by the expressions (both in metres)

$$x_C = \frac{\int x \, dA}{A} \qquad y_C = \frac{\int y \, dA}{A} \tag{5.1}$$

For an axis through C, $x_C = y_C = 0$, so for centroidal axes

$$\int x\,dA = \int y\,dA = 0 \tag{5.2}$$

If the y-axis is an axis of symmetry, we obtain contributions $x dA$ and $-x dA$ from symmetrical elements; hence $\int x dA = 0$ or $x_C = 0$, which means that the centre of area is on the axis of symmetry. For two axes of symmetry the centre of area is at the intersection of the axes.

For the rectangular area shown in Fig. 5.2, $dA = b\,dy$. Hence $S_x = \int_A y\,dA =$

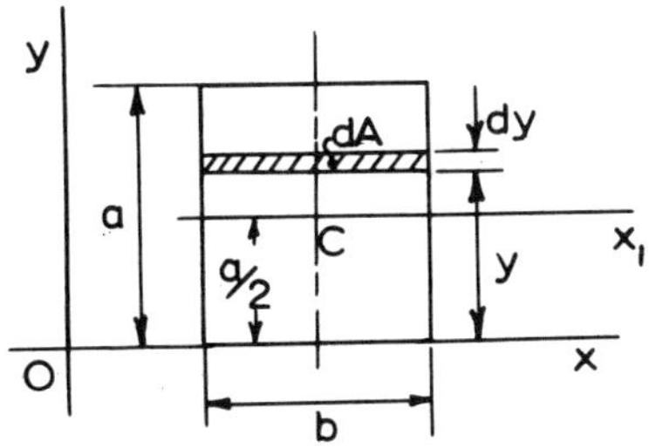

Fig. 5.2.

$\int_0^a by\,dy = \frac{1}{2}ba^2 = Ay_C = Aa/2$.

For an area under a given curve $y = f(x)$ (Fig. 5.3)

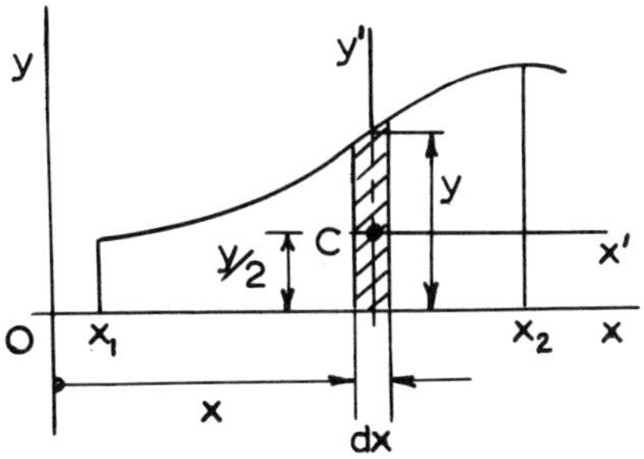

Fig. 5.3.

$$dS_x = (y\mathrm{d}x)\,y/2 \qquad S_x = \tfrac{1}{2}\int_{x_1}^{x_2} y^2\,\mathrm{d}x = y_C A$$

$$dS_y = (y\mathrm{d}x)\,x \qquad S_y = \int_{x_1}^{x_2} xy\,\mathrm{d}x = y_C A$$

These two formulae determine $C(x_C, y_C)$ for the area.

5.1.2 Second moments of area of a plane area

The *second moments of area* are defined by the expressions

$$I_x = \int_A y^2\,\mathrm{d}A \qquad I_y = \int_A x^2\,\mathrm{d}A$$

These are always positive. The units for second moments of area are m^4. Analytical calculation may be done by dividing the area (Fig. 5.2) into rectangles:

$$I_{x_1} = 2\int_0^{a/2} y^2\,b\,\mathrm{d}y = \frac{2}{3}b\,(a/2)^3 = \frac{1}{12}ba^3$$

$$I_x = \int_0^a (b\mathrm{d}y)y^2 = \frac{1}{3}ba^3$$

Writing $I_x = Ar_x^2$, we call r_x the *radius of gyration* with respect to the x-axis, and r_x is defined and calculated from this expression; similarly $I_y = Ar_y^2$.

Second moment of area with respect to an axis perpendicular to the area

This second moment is called the *polar moment of area* J_0, and is defined by the expression $J_0 = \int_A r^2\,\mathrm{d}A$ (Fig. 5.1), so

$$J_0 = \int_A (x^2+y^2)\mathrm{d}A = \int_A x^2\,\mathrm{d}A + \int_A y^2\,\mathrm{d}A = I_y + I_x \qquad (5.3)$$

The polar moment of area J_0 is equal to the sum of I_x and I_y *for any* set of perpendicular axes in the area through O. Hence $I_x + I_y$ = constant for any set of perpendicular axes in the plane of the area through the same point; this is called *the perpendicular-axis theorem*.

Example 5.1

For a circular area (Fig. 5.4), J_0 for an axis throught the centre O is, determined by dividing the area into elements as shown; the contribution of one element $\mathrm{d}A = 2\pi r\,\mathrm{d}r$ *is* $\mathrm{d}J_0 = \mathrm{d}A(r^2) = (2\pi r\,\mathrm{d}r)r^2$, so

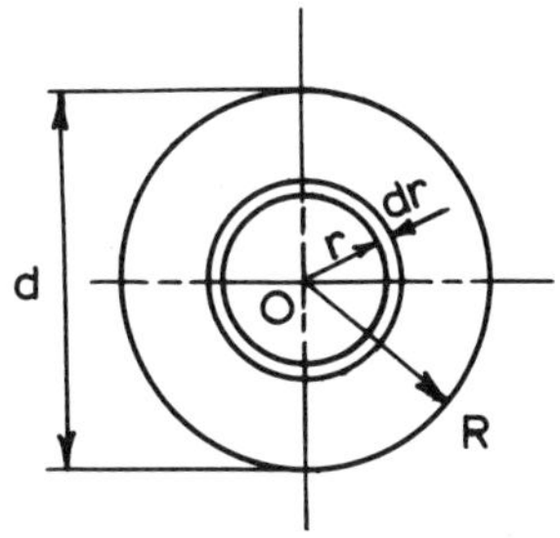

Fig. 5.4.

$$J_0 = \int_0^R (2\pi r\,\mathrm{d}r)r^2 = \int_0^R 2\pi r^3\,\mathrm{d}r = \frac{\pi}{2}[r^4]_0^R = \frac{\pi}{2}R^4$$

Since $I_x + I_y = J_0$, and $I_y = I_x$, we find $I_x = j_0/2 = (\pi/4)R^4 = (\pi/64)d^4$; this is much simpler than integrating directly for I_x by rectangular division of the area.

Parallel-axis theorem

Fig. 5.5 shows an area with centre of area C and a set of coordinate axes (x, y)

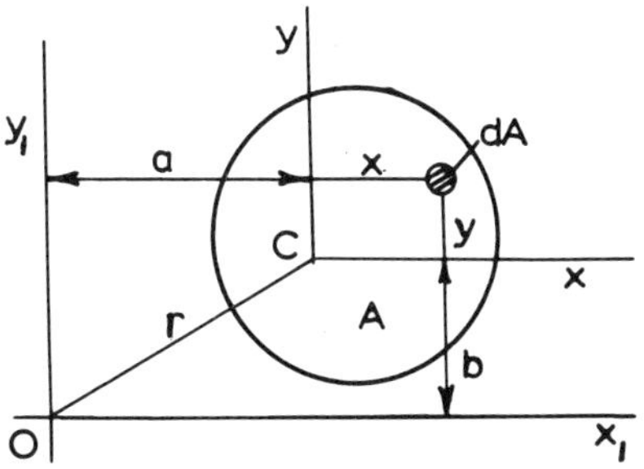

Fig. 5.5.

through C; suppose that $I_x = \int_A y^2 \mathrm{d}A$ is known for the x-axis; for a parallel axis x_1 we find

$$I_{x_1} = \int_A y_1^2\,\mathrm{d}A = \int_A (y+b)^2\,\mathrm{d}A = \int_A y^2\,\mathrm{d}A + 2b\int_A y\,\mathrm{d}A + b^2\int \mathrm{d}A$$

$$= I_x + b^2 A \tag{5.4}$$

In a similar manner

$$I_{y_1} = I_y + a^2 A \qquad (5.5)$$

These formulae constitute the *parallel-axis theorem*; the x- and y-axes *must* be through the centre of area. The formulae are often used in the form

$$I_x = I_{x1} - b^2 A \quad \text{and} \quad I_y = I_{y1} - a^2 A$$

Adding the expressions (5.4) and (5.5) gives

$$I_{x_1} + I_{y_1} = I_x + I_y + (a^2 + b^2)A = I_x + I_y + r^2 A$$

Hence

$$J_0 = J_C + r^2 A \qquad (5.6)$$

The parallel-axis theorem holds also for *polar moments of area*. The theorem is particularly useful for calculations of second moments of area of composite areas.

Products of area

The *product of an area A* (Fig. 5.1) with respect to the x- and y-axes is defined by the summation: $I_{xy} = \int_A xy\,\mathrm{d}A$; the units are m^4. It follows at once that $I_{xy} = I_{yx}$; I_{xy} may be positive, negative or vanish; that $I_{xy} = 0$ for a certain position of xy may be seen by changing the y-axis to y_1; we find $x_1 = x$, $y_1 = -y$, so

$$I_{x_1 y_1} = \int_A x_1 y_1\,\mathrm{d}A = \int_A x(-y)\mathrm{d}A = -I_{xy}$$

Since I_{xy} changes to $-I_{xy}$ as we rotate the axes about O, it may be shown that at some position $I_{xy} = 0$; *these particular axes are called principal axes* for the area at point O.

If a figure has an axis of symmetry (Fig. 5.6) the contribution from dA to I_{xy} is $dI_{xy} = xy\,dA + (-x)y\,dA = 0$, so that $I_{xy} = 0$; the x-axis may be positioned anywhere perpendicular to the axis of symmetry; such a set of axes are principal axes.

Parallel-axis theorem for products of area

Suppose $I_{xy} = \int_A xy\,dA$ is known for a set of axes xy *through the centre of area C* (Fig. 5.5); for a parallel set of axes (x_1, y_1), we find

$$I_{x_1y_1} = \int_A x_1y_1 dA = \int_A (x+a)(y+b)dA$$

$$= \int_A xy\,dA + b\int_A x\,dA + a\int_A y\,dA + ab\int_A dA$$

Therefore

$$I_{x_1y_1} = I_{xy} + abA \tag{5.7}$$

a and b are coordinates of C in the system (x_1, y_1) (positive or negative), *or* coordinates of O in the system (x, y).

For an area under a curve $y = f(x)$ (Fig. 5.3) we find for the rectangular strip with centre C

$$I_{xys} = I_{x'y's} + A_s(x)\,(y/2)$$

$$I_{x'y's} = 0 \qquad \text{from symmetry}$$

Hence $I_{xys} = A_s xy/2 = (y\,dx)\,xy/2$, and therefore

$$I_{xy} = \frac{1}{2}\int_{x_1}^{x_2} xy^2\,dx$$

5.1.3 Determination of principal axes for a point of a plane area

Assume that for a given area A in Fig. 5.7, we know $I_x = \int_A y^2\,dA$, $I_y = \int_A x^2\,dA$ and $I_{xy} = \int_A xy\,dA$. The formulae connecting the two coordinate systems are

$$\left.\begin{aligned} x_1 &= x\cos\theta + y\sin\theta \\ y_1 &= y\cos\theta - x\sin\theta \end{aligned}\right\} \tag{5.8}$$

Substituting (5.8) in $I_{x_1} = \int_A y_1^2\,dA$ and $I_{y_1} = \int_A x_1^2\,dA$ leads to

$$I_{x_1} = I_x\cos^2\theta + I_y\sin^2\theta - I_{xy}\sin 2\theta$$

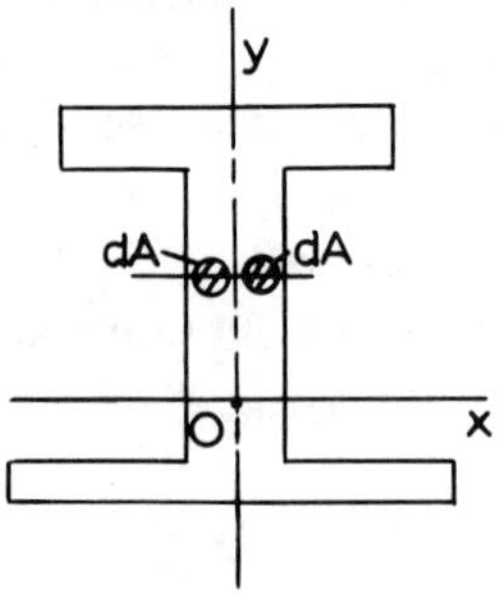

Fig. 5.6.

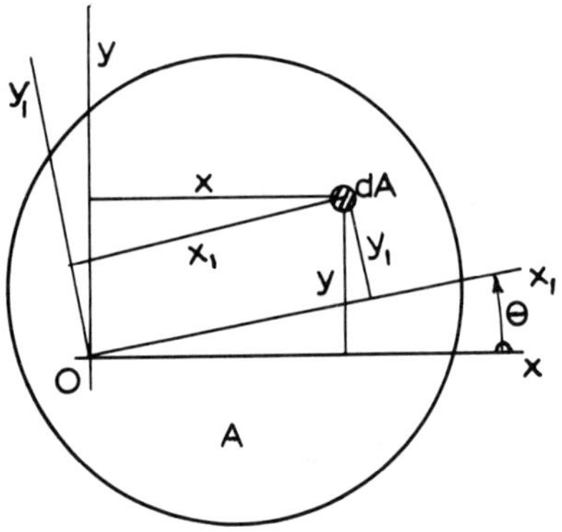

Fig. 5.7.

$$I_{y_1} = I_x \sin^2\theta + I_y \cos^2\theta + I_{xy} \sin 2\theta$$

The sum and difference of these expressions give

$$\left.\begin{aligned} I_{x_1} + I_{y_1} &= I_x + I_y \\ I_{x_1} - I_{y_1} &= (I_x - I_y)\cos 2\theta - 2I_{xy}\sin 2\theta \end{aligned}\right\} \tag{5.9}$$

Equations (5.9) give the simplest numerical calculation of I_{x_1} and I_{y_1}. In the same manner we find

$$I_{x_1y_1} = \int_A x_1 y_1 \, \mathrm{d}A = \tfrac{1}{2}(I_x - I_y)\sin 2\theta + I_{xy}\cos 2\theta \tag{5.10}$$

The principal axes are found for $I_{x_1y_1} = 0$:

$$\tfrac{1}{2}(I_x - I_y)\sin 2\theta + I_{xy}\cos 2\theta = 0$$

that is at the position determined by

$$\tan 2\theta = \frac{2I_{xy}}{I_y - I_x} \tag{5.11}$$

If we sum the expressions (5.9), the result is

$$I_{x_1} = \frac{I_x + I_y}{2} + \frac{(I_x - I_y)}{2}\cos 2\theta - I_{xy}\sin 2\theta$$

Taking $dI_{x_1}/d\theta = 0$ gives $\tan 2\theta = 2I_{xy}/(I_y - I_x)$, which shows that I_{x_1} is a maximum or minimum where $I_{x_1y_1} = 0$ (the same result is found for I_{y_1}). Therefore, the *principal axes* also determine the maximum and minimum second moments of area for any set of orthogonal axes through O. The actual maximum or minimum axis is determined from the numerical result for I_{x_1} and I_{y_1} for this value pf θ.

The formulae relate to the axes as shown in Fig. 5.7. If one of the axes is in the opposite direction, the sign of I_{xy} and $I_{x_1y_1}$ must be changed in the formulae; if the rotation is clockwise, $-\theta$ must be substituted for θ in the formulae.

5.2 MOMENTS OF INERTIA OF A LAMINA

Consider a lamina, as shown in Fig. 5.8, of homogeneous material of density ρ and

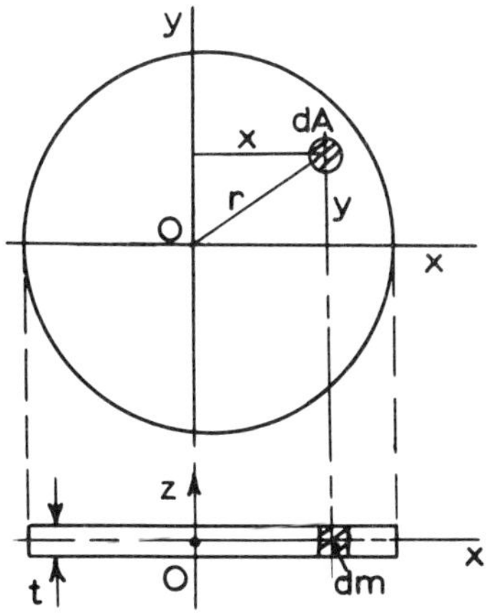

Fig. 5.8.

uniform thickness t; a coordinate system (x, y, z) is fixed at O in the middle plane of the lamina. We assume that t is small compared to the other dimensions.

By *definition* the *moment of inertia* of the lamina with respect to the z-axis is

$$I_z = \int_V (x^2 + y^2)\,\mathrm{d}m = \int_V r^2\,\mathrm{d}m$$

where $\mathrm{d}m = (\mathrm{d}A)t\rho$ and V denotes the volume. Hence

$$I_z = \int_A r^2 \rho t\,\mathrm{d}A = (\rho t)\int_A r^2\,\mathrm{d}A = \rho t J_0 \tag{5.12}$$

where J_0 is the *polar moment of area* with respect to the z-axis. Now

$$I_x = \int_V (y^2 + z^2)\,\mathrm{d}m$$

and with $|z| \leqslant t/2$ we have $|z| \ll |y|$. Neglecting z^2, we find

$$I_x \cong \int_V y^2\,\mathrm{d}m = (\rho t)\int_A y^2\,\mathrm{d}A = (\rho t)I_{xA} \tag{5.13}$$

where I_{xA} is the *second moment of area* with respect to the x-axis; similarly

$$I_y = \int_V (x^2 + z^2)\,\mathrm{d}m \cong \int_V x^2\,\mathrm{d}m = \rho t\int_A x^2\,\mathrm{d}A = (\rho t)I_{yA}$$

which means that *second moments of area* may be used directly to find *moments of inertia* of a lamina by multiplication by the constant factor (ρt).

Perpendicular-axis theorem

For second moments of area, Fig. 5.8, equation (5.3), we have $I_{xA} + I_{yA} = J_0 = I_{zA}$. Multiplying by ρt gives

$$I_x + I_y = I_z \tag{5.14}$$

Hence the perpendicular-axis theorem holds also for *moments of inertia* of a lamina with xy in the middle plane. (The theorem does *not* hold for three-dimensional bodies in general.)

Parallel-axis theorem

For the area in Fig. 5.5 we have formula (5.4).

$$I_{x_1A} = I_{xA} + Ab^2$$

Multiplying by ρt gives

$$I_{x_1} = I_x + Mb^2 \tag{5.15}$$

In (5.15), M is the total mass of the lamina.

It is sometimes convenient to use expressions of the form $I_x = Mr_x^2$, where r_x is called the *radius of gyration* with respect to the x-axis; r_x is defined and calculated from this expression, which is particularly useful when it is possible to cancel the mass M in an equation.

5.3 MOMENTS OF INERTIA OF THREE-DIMENSIONAL BODIES

5.3.1 Definition of moments and products of inertia

The moment of inertia, with respect to the x-axis, of the body in Fig. 5.9 is *defined* as

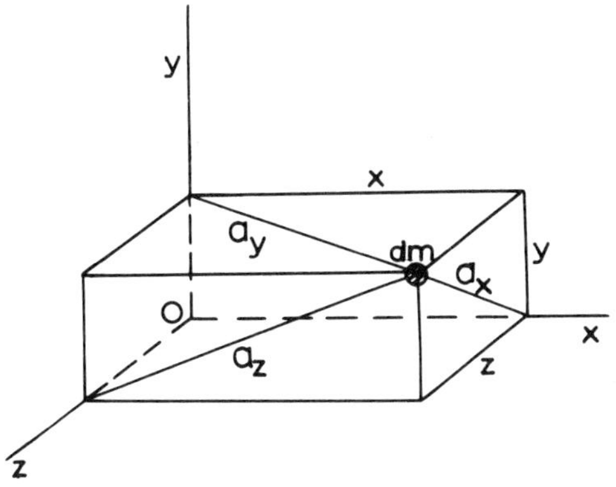

Fig. 5.9.

follows:

$$I_x = \int_V (y^2 + z^2)\,\mathrm{d}m$$

that is the sum of each mass element multiplied by the square of its distance from the x-axis:

$$I_x = \int_V a_x^2\,\mathrm{d}m$$

In the same way

$$I_y = \int_V (x^2 + z^2)\,\mathrm{d}m = \int_V a_y^2\,\mathrm{d}m$$

$$I_z = \int_V (x^2 + y^2)\,\mathrm{d}m = \int_V a_z^2\,\mathrm{d}m$$

These expressions are always positive.

It is sometimes convenient to write $I_x = Mr_x^2$, in the same way as for a lamina, where M is the total mass of the body and r_x the *radius of gyration*; similarly $I_y = Mr_y^2$ and $I_z = Mr_x^2$.

The *product of inertia* with respect to the xy-plane is *defined* by the expression

$$I_{xy} = \int_V xy \, \mathrm{d}m = \int_V yx \, \mathrm{d}m = I_{yx}$$

The product of inertia is the sum of each mass element multiplied by the product of its x and y coordinates, that is by the product of its distances from the yz- and xz-planes; similarly

$$I_{xz} = I_{zx} = \int_V xz \, \mathrm{d}m \qquad I_{yz} = I_{zy} = \int_V yz \, \mathrm{d}m$$

Products of inertia may be positive, negative or vanish, as will be shown in the following.

5.3.2 Centre of mass of a rigid body

Consider a series of particles m_i, as shown in Fig. 5.10, with position vectors $\mathbf{r}_i$; the

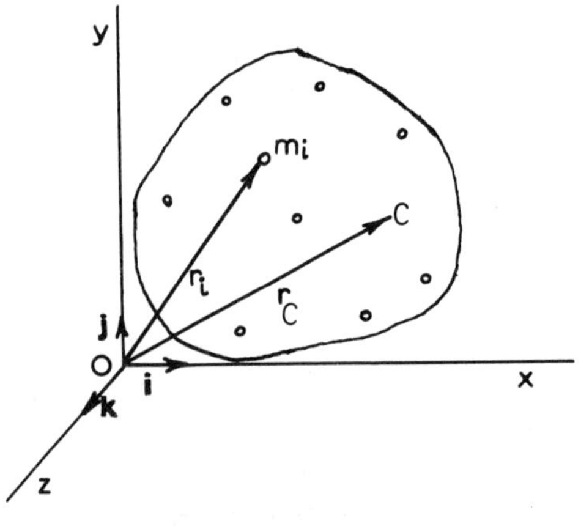

Fig. 5.10.

centre of mass C of the particles is *defined* by the position vector

$$\mathbf{r}_\mathrm{C} = \frac{\Sigma m_i \mathbf{r}_i}{\Sigma m_i} \tag{5.16}$$

With $\mathbf{r}_\mathrm{C} = x_\mathrm{C}\mathbf{i} + y_\mathrm{C}\mathbf{j} + x_\mathrm{C}\mathbf{k}$ and $\mathbf{r}_i = x_i\mathbf{i} + y_i\mathbf{j} + z_i\mathbf{k}$, we find by substitution that $x_\mathrm{C} = (\Sigma m_i x_i)/\Sigma m_i$, with similar expressions for y_C and z_C.

For a *rigid body* these expressions take the form

$$x_\mathrm{C} = \frac{\int x \, \mathrm{d}m}{M} \qquad y_\mathrm{C} = \frac{\int y \, \mathrm{d}m}{M} \qquad z_\mathrm{C} = \frac{\int z \, \mathrm{d}m}{M}$$

where M is the total mass of the body.

If the origin O of the coordinate system is taken at C, then

$$\int x\,\mathrm{d}m \;=\; \int y\,\mathrm{d}m \;=\; \int z\,\mathrm{d}m \;=\; 0 \tag{5.17}$$

This holds for axes through the centre of mass; since C is also determined by (5.16), we have $\mathbf{r}_\mathrm{C} = \mathbf{0}$. Therefore

$$\Sigma m_i\,\mathbf{r}_i \;=\; \mathbf{0} \tag{5.18}$$

for position vectors $\mathbf{r}_i$ taken from C. We also find

$$\Sigma m_i\dot{\mathbf{r}}_i \;=\; \Sigma m_i\mathrm{d}\mathbf{r}_i/\mathrm{d}t \;=\; \Sigma\mathrm{d}(m_i\mathbf{r}_i)/\mathrm{d}t \;=\; \mathrm{d}(\Sigma m_i\mathbf{r}_i)/\mathrm{d}t \;=\; \mathbf{0} \tag{5.19}$$

These two vector formulae will be found useful in future developments of dynamics.

5.3.3 Parallel-axis theorem for a rigid body

Fig. 5.11 shows the centre of mass C of a rigid body of total mass M; a coordinate

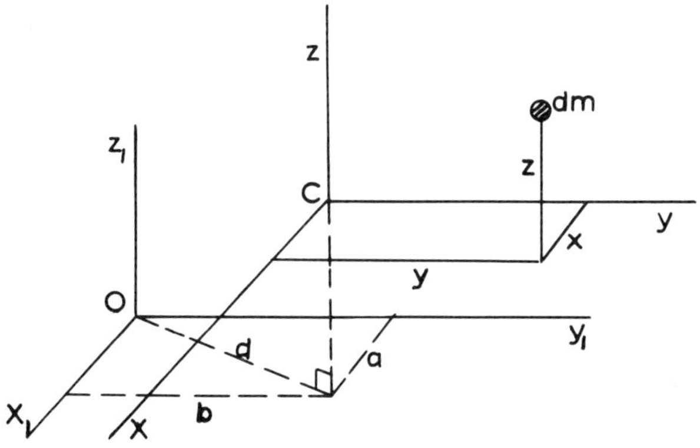

Fig. 5.11.

system (x, y, z) has been introduced with origin at C and a particular mass element dm of the body is shown at position (x, y, z); a coordinate system (x_1, y_1, z_1) with origin O has been introduced with axes *parallel* to (x, y, z).

By definition

$$I_{z_1} \;=\; \int_V (x_1^2 + y2_1)\,\mathrm{d}m \;=\; \int_V[(a+x)^2 + (b+y)^2]\,\mathrm{d}m$$

$$\;=\; \int_V (x^2+y^2)\,\mathrm{d}m + 2a\int_V x\,\mathrm{d}m + 2b\int_V y\,\mathrm{d}m + (a^2+b^2)\int_V \mathrm{d}m$$

$$= I_z + d^2M \tag{5.20}$$

The formula (5.20) states the parallel-axis theorem for a rigid body; it is essential to remember that the z axis, in this case, *must be through the centre of mass*; the formula is often used in the form $I_z = I_{z_1} - d^2M$.

The product of inertia $I_{x_1y_1}$ is by definition

$$I_{x_1y_1} = \int_V x_1y_1\, \mathrm{d}m = \int_V (x+a)(y+b)\, \mathrm{d}m$$

$$= \int_V xy\, \mathrm{d}m + b\int_V x\, \mathrm{d}m + a\int_V y\, \mathrm{d}m + ab\int_V \mathrm{d}m$$

$$= I_{xy} + abM \tag{5.21}$$

In (5.21), a and b are coordinates, with proper signs, of C in the system (x_1, y_1), *or* the coordinates of O in the system (x, y); the formula is opften used in the form $I_{xy} = I_{x_1y_1} - abM$; the x- and y-axes *must be through the centre of mass C*.

The formulae (5.20) and (5.21) constitute the parallel-axis theorem for a rigid body; the *perpendicular-axis theorem* developed for areas and laminae does *not* hold in general for a three-dimensional body.

Example 5.2

Fig. 5.12 shows a right circular homogeneous cylinder of density ρ and total mass $M =$

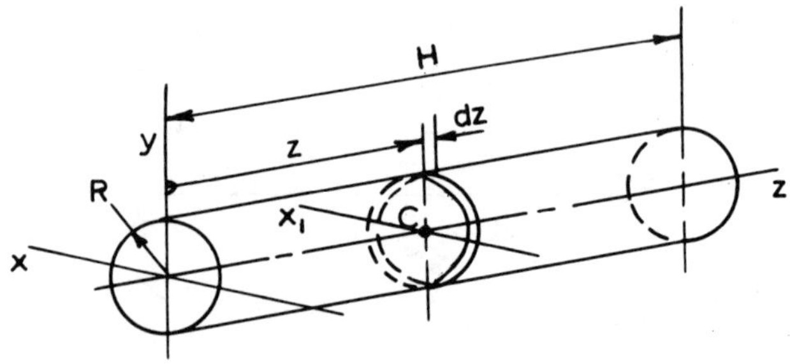

Fig. 5.12.

$\pi R^2 H\rho$. The determination of the moment of inertia I_x with respect to a diameter in the base is done most simply in the following manner: for the lamina shown parallel to the base, the second moment of area of the circular surface with respect to the x_1-axis parallel to the x-axis is $I_{x_1A} = (\pi/4)R^4$; the moment of inertia is then $I_{x_1} = I_{x_1A}(\rho t) = (\pi/4)R^4\rho\, \mathrm{d}z$; the mass of the lamina is $m = \pi R^2\, \mathrm{d}z\, \rho$; the contribution of the lamina to I_x is from (5.20):

$$\mathrm{d}I_x = I_{x_1} + mz^2 = (\pi/4)\rho R^4\, \mathrm{d}x + \pi R^2\, \mathrm{d}z\, z^2$$

Therefore

$$I_x = (\pi/4)\rho R^4 \int_0^H \mathrm{d}z + \pi R^2 \rho \int_0^H z^2\,\mathrm{d}z$$

$$= (\pi R^2 H)\,\rho\left(\frac{1}{4}R^2 + \frac{1}{3}H^2\right) = M\left[\frac{3R^2 + 4H^2}{12}\right] = Mr_x^2$$

5.3.4 Principal axes for a rigid body

Fig. 5.13 shows a coordinate system (x, y, z) and a mass element dm of a rigid body;

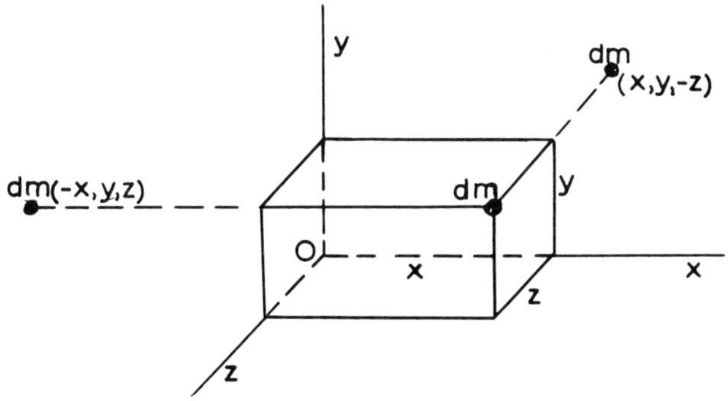

Fig. 5.13.

dm is located at position (x, y, z); suppose that the body is *homogeneous* and has a *plane of symmetry yz*; corresponding to the element dm at (x, y, z), there will always be an element dm at $(-x, y, z)$; the contribution to I_{xy} from these elements is then d$m\,(x)\,(y) +$ d$m\,(-x)\,(y) = 0$. Therefore

$$I_{xy} = \int_V xy\,\mathrm{d}m = 0$$

for the body; similarly

$$I_{xz} = \int_V xz \, dm = 0$$

If the xy-plane is also a plane of symmetry, we find that $I_{xz} = \int_V yz \, dm = 0$, so for a homogeneous body with two perpendicular planes of symmetry, the line of intersection between the planes (here the y-axis), together with two perpendicular axes, one in each plane of symmetry (here the x-and z-axes), will determine a set of axes for which $I_{xy} = I_{xz} = I_{yz} = 0$; such axes are called *principal axes for the body* at O, and the moments of inertia I_x, I_y and I_z are called *principal moments of inertia* for the body at point 0.

Principal axes may be determined by inspection for regular homogeneous bodies with two planes of symmetry; for instance the geometrical centre line of a right circular cyclinder or cone, any diameter in a sphere, and the centre line of a right rectangular prism are all principal axes.

Some useful principal moments of inertia

These are shown in Fig. 5.14–5.19 and their captions. All bodies are assumed to be homogeneous and of total mass M.

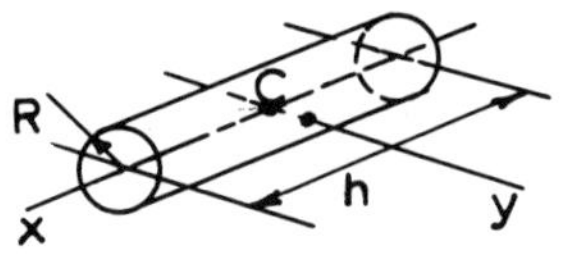

Fig. 5.14 — Solid right circular cylinder or disc: $I_x = MR^2/2$, $I_y = M(h^2 + 3R^2)/12$.

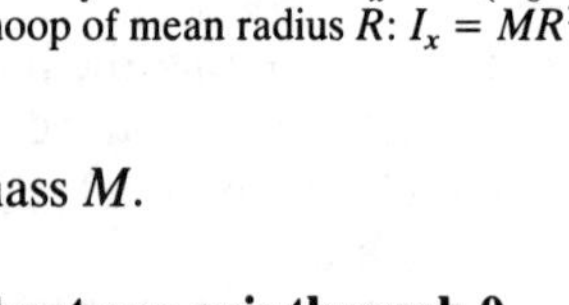

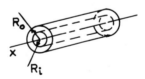

Fig. 5.15 — Hollow right circular cylinder or disc: $I_x = M(R_o^2 + R_i^2)/2$; thin cylindrical shell or hoop of mean radius R: $I_x = MR^2$.

5.3.5 Moments of inertia about any axis through 0

Fig. 5.20 shows a coordinate system (x_1, y_1, z_1) and a mass element dm of a rigid body; dm is located at position (x_1, y_1, z_1) with position vector **r**; a second set of axes (x, y) is shown rotated to a different position from (x_1, y_1) about O.

Suppose that I_{x_1}, I_{y_1}, I_{z_1} and $I_{x_1z_1}$, $I_{y_1z_1}$ are known for the body, and that the

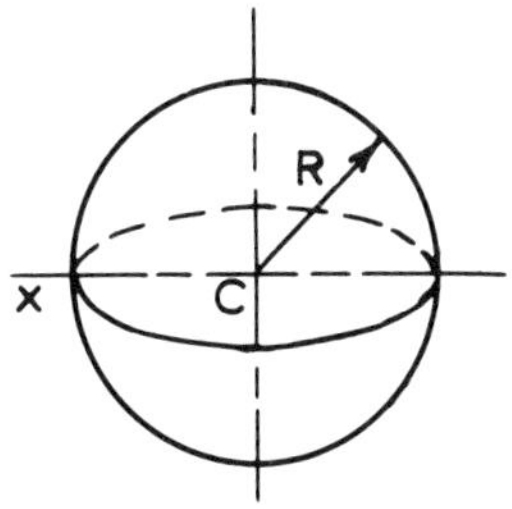

Fig. 5.16 — Solid sphere: $I_x = \frac{2}{5}MR^2$; hollow sphere with inner and outer radii R_i and R_o: $I_x = \frac{2}{5}M(R_o^5 - R_i^5)/(R_o^3 - R_i^3)$; thin spherical shell with mean radius R: $I_x = \frac{2}{3}MR^2$.

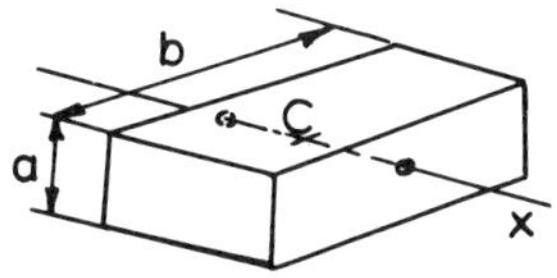

Fig. 5.17 — Right regular rectangular prism: $I_x = M(a^2 + b^2)/12$.

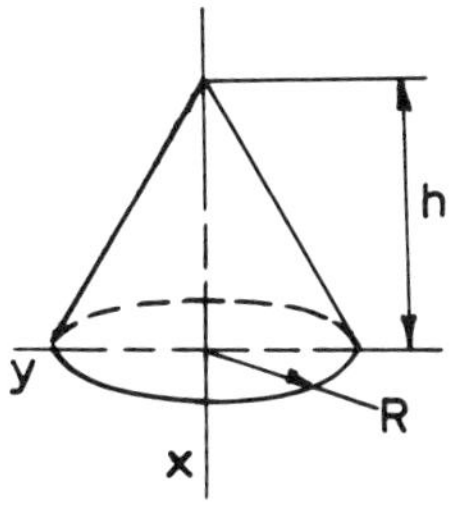

Fig. 5.18 — Right circular cone: $I_x = \frac{3}{10}MR^2$, $I_y = M(3R^2 + 2h^2)/20$.

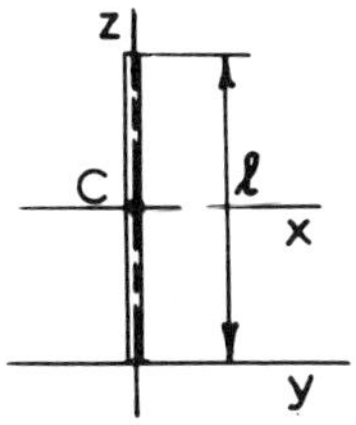

Fig. 5.19 — Slender uniform prismatic bar: $I_x = Ml^2/12$, $I_y = Ml^2/3$, $I_z = 0$.

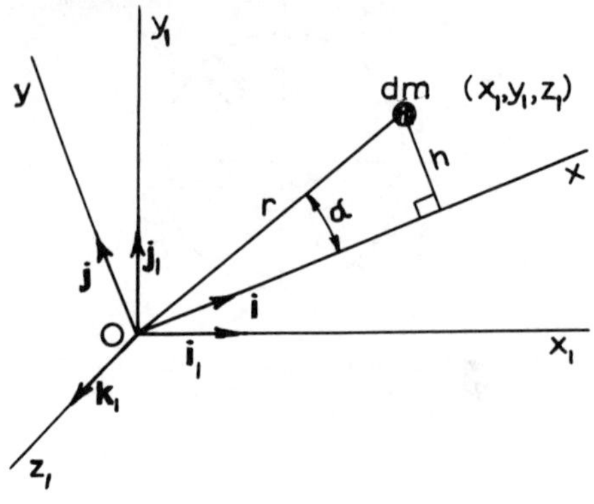

Fig. 5.20.

moment of inertia I_x is required. The x-axis is determined by its direction cosines: $\cos \angle (x, x_1) = l$, $\cos \angle (x, y_1) = m$ and $\cos \angle (x, z_1) = n$; for the y-axis, the direction cosines are l', m' and n'.

The unit vectors **i** and **j** are determined by

$$\mathbf{i} = l\mathbf{i}_1 + m\mathbf{j}_1 + n\mathbf{k}_1 \qquad \mathbf{j} = l'\mathbf{i}_1 + m'\mathbf{j}_1 + n'\mathbf{k}_1$$

The position vector **r** is $\mathbf{r} = x_1\mathbf{i}_1 + y_1\mathbf{j}_1 + z_{1\mathbf{K}1}$, and by definition $I_x = \int_V h^2\,\mathrm{d}m$, where

$$h = |\mathbf{r}| \sin \alpha = |\mathbf{r} \times \mathbf{i}|$$

$$\mathbf{r} \times \mathbf{i} = (y_1 n - z_1 l)\,\mathbf{i}_1 + (z_1 l - x_1 n)\,\mathbf{j}_1 + (x_1 m - y_1 l)\,\mathbf{k}_1$$

$$h^2 = |\mathbf{r} \times \mathbf{i}|^2 = (y_1 n - z_1 l)^2 + (z_1 l - x_1 n)^2 + (x_1 m - y_1 l)^2$$

$$= (x_1^2 + y_1^2)n^2 + (x_1^2 + z_1^2)m^2 + (y_1^2 + z_1^2)l^2$$
$$- 2(x_1 y_1 lm + x_1 z_1 ln + y_1 z_1 mn)$$

Therefore

$$I_x = l^2 I_{x_1} + m^2 I_{y_1} + n^2 I_{z_1} - 2(lm I_{x_1y_1} + ln I_{x_1z_1} + mn I_{y_1z_1}) \tag{5.22}$$

Similar expressions may be developed for I_y and I_z. The use of (5.22) to determine I_x is much simpler than a complicated integration for an inclined axis.

If the x_1-, y_1-, z_1-axes are *principal axes*, $I_{x_1y_1} = I_{x_1z_1} = I_{y_1z_1} = 0$, and (5.22) takes the simpler form

$$I_x = l^2 I_{x_1} + m^2 I_{y_1} + n^2 I_{z_1} \tag{5.23}$$

For the *product of inertia* I_{xy} we have by definition $I_{xy} = \int_V xy \, dm$; introducing

$$x = \mathbf{r}\cdot\mathbf{i} = x_1 l + y_1 m + z_1 n \qquad \text{and} \qquad y = \mathbf{r}\cdot\mathbf{j} = x_1 l' + y_1 m' + z_1 n'$$

we find

$$xy = x_1^2 ll' + y_1^2 mm' + z_1^2 nn' + x_1 y_1 (lm' + ml') + x_1 z_1 (ln' + nl') + y_1 z_1 (mn' + nm')$$

Subtracting from xy the expression

$$(\mathbf{r}\cdot\mathbf{r})(\mathbf{i}\cdot\mathbf{j}) = (x_1^2 + y_1^2 + z_1^2)(ll' + mm' + nn') = 0$$

the first three terms in xy change to

$$-(y_1^2 + z_1^2)ll' - (x_1^2 + z_1^2)mm' - (x_1^2 + y_1^2)nn'$$

therefore

$$I_{xy} = -(I_{x_1} ll' + I_{y_1} mm' + I_{z_1} nn') + [I_{x_1y_1}(lm' + ml') + I_{x_1z_1}(ln' + nl') + I_{y_1z_1}(mn' + nm')] \tag{5.24}$$

Similar expressions may be developed for I_{xz} and I_{yz}. If the x_1-, y_1- and z_1-axes are principal axes, (5.24) reduces to the simpler form

$$I_{xy} = -(I_{x_1} ll' + I_{y_1} mm' + I_{z_1} nn') \tag{5.25}$$

Example 5.3
Fig. 5.21 shows a thin circular disc of radius r and mass M; determine the moment of inertia about the inclined axis x and the product of inertia with respect to the axes xy. The z- and z_1-axes are perpendicular to the paper.

Solution
From symmetry,

$$I_{x_1y_1} = I_{x_1z_1} = I_{y_1z_1} = 0$$

so x_1, y_1 and z_1 are principal axes.

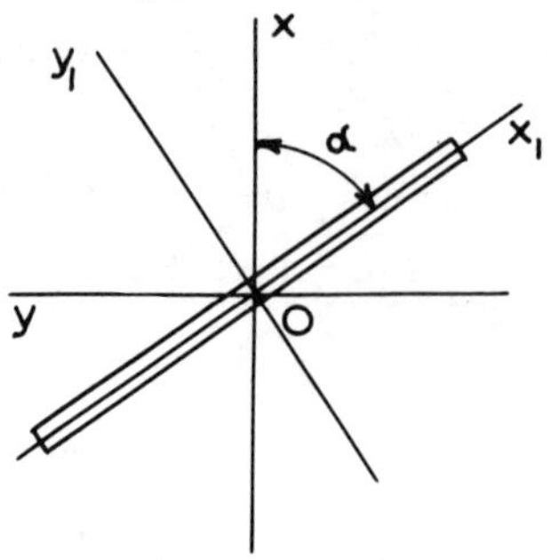

Fig. 5.21.

$$I_{x_1} = I_{z_1} = Mr^2/4 \qquad I_{y_1} = Mr^2/2$$
$$l = \cos \angle (x, x_1) = \cos \alpha$$
$$m = \cos \angle (x, y_1) = \cos (90° - \alpha) = \sin \alpha$$
$$n = \cos \angle (x, z_1) = \cos 90° = 0$$
$$l' = \cos \angle (y, x_1) = \cos (270° - \alpha) = -\sin \alpha$$
$$m' = \cos \angle (y, y_1) = \cos \alpha$$
$$n' = \cos \angle (y, z_1) = \cos 90° = 0$$

Substituting in (5.23) gives

$$I_x = (\cos^2 \alpha)\frac{Mr^2}{4} + (\sin^2 \alpha)\frac{\mathrm{Mr}^2}{2} = \frac{Mr^2(1 + \sin^2 \alpha)}{4}$$

Equation (5.25) gives

$$I_{xy} = -\left[\frac{Mr^2}{4}\cos \alpha\,(-\sin \alpha) + \frac{Mr^2}{2}\sin \alpha \cos \alpha\right]$$
$$= -\frac{Mr^2}{4}\sin \alpha \cos \alpha = -\frac{Mr^2}{8}\sin 2\alpha$$

5.3.6 Moment of momentum of a rigid body

Fig. 5.22 shows a homogeneous, rigid body which is rotating about a fixed axis.

The instantaneous angular velocity of the body is **ω** and the **ω** vector has been placed on the axis of rotation as shown. A coordinate system (x, y, z) is fixed in the body and therefore rotating with it; the origin of the system is a point O on the axis of rotation, ploint O is then a *fixed* point. A particular of the body of mass dm is shown in position (x, y, z), and with position vector **r** from point O. The instantaneous velocity of the particle is **V** as shown.

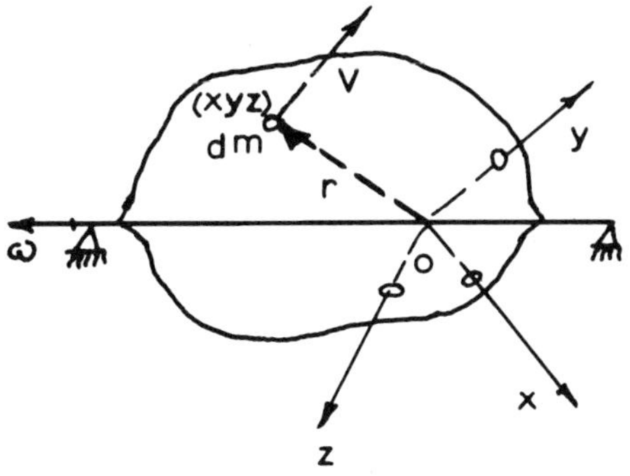

Fig. 5.22.

The *momentum* of the particle is *defined* as the vector $(dm)\mathbf{V}$, and we now define the *moment of momentum* $\mathbf{h}$ of the particle as the moment of the momentum vector in point O, that is $\mathbf{h} = \mathbf{r} \times (dm)\mathbf{V}$.

The *moment of momentum* $\mathbf{H}$ of the rigid body in point O is now *defined* as the sum of the moment of momentum vectors in point O of all the particles in the body, so that $\mathbf{H} = \int \mathbf{r} \times \mathbf{V}\, dm$, where the summation is extended over the volume of the body.

Taking point O as a pole, we have $\mathbf{V}_0 = \mathbf{0}$ and $\mathbf{V} = \mathbf{V}_0 + \boldsymbol{\omega} \times \mathbf{r} = \boldsymbol{\omega} \times \mathbf{r}$.

Substituting this in the expression for $\mathbf{H}$ leads to

$$\mathbf{H} = \int \mathbf{r} \times (\boldsymbol{\omega} \times \mathbf{r}) dm \qquad (5.26)$$

To express $\mathbf{H}$ in coordinates, we introduce

$$\mathbf{r} = x\mathbf{i} + y\mathbf{j} + z\mathbf{k}$$

and

$$\boldsymbol{\omega} = \omega_x \mathrm{i} + \omega_y \mathbf{j} + \omega_z \mathbf{k}$$

Using the formula $\mathbf{a} \times (\mathbf{b} \times \mathbf{c}) = (\mathbf{a.c})\mathbf{b} - (\mathbf{a.b})\mathbf{c}$ for the double vector cross-product, we obtain

$$\begin{aligned}
\mathbf{r} \times (\boldsymbol{\omega} \times \mathbf{r}) &= (\mathbf{r.r})\boldsymbol{\omega} - (\mathbf{r}.\boldsymbol{\omega})\mathbf{r} \\
&= (x^2 + y^2 + z^2)(\omega_x \mathbf{i} + \omega_y \mathbf{j} + \omega_z \mathbf{k}) - (x\omega_x + y\omega_y + z\omega_z)(x\mathbf{i} + y\mathbf{j} + z\mathbf{k}) \\
&= [(y^2 + z^2)\omega_x - xy\omega_y - xz\omega_z]\mathbf{i} + [(x^2 + z^2)\omega_y - xy\omega_x - yz\omega_z]\mathbf{j} \\
&+ [(x^2 + y^2)\omega_z - xz\omega_x - yz\omega_y]\mathbf{k}
\end{aligned}$$

Multiplying by dm and integrating over the volume of the body, we find the components of $\mathbf{H}$.

$$\left.\begin{aligned} H_x &= I_x\,\omega_x - I_{xy}\,\omega_y - I_{xz}\,\omega_z \\ H_y &= -I_{xy}\,\omega_x + I_y\,\omega_y + I_y\,\omega_y - I_{yz}\,\omega_z \\ H_z &= -I_{xz}\,\omega_x - I_{yz}\,\omega_y + I_z\,\omega_z \end{aligned}\right\} \tag{5.27}$$

5.3.7 Determination of principal axes

Fig. 5.23 shows a rigid body rotating about a fixed *principal* axis, which has been

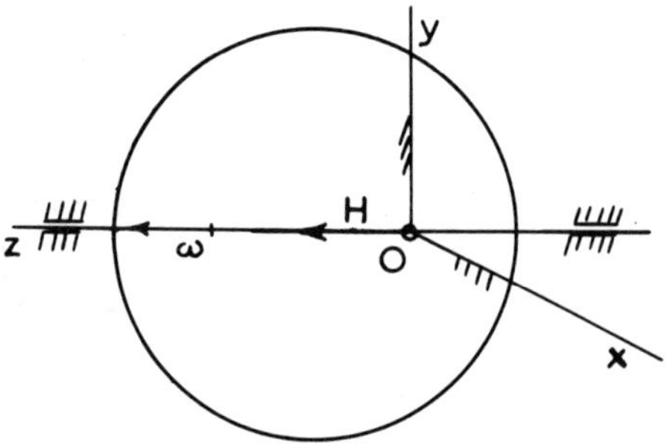

Fig. 5.23.

taken as the z-axis; this means that $I_{xz} = I_{yz} = 0$ and the $\boldsymbol{\omega}$ vector is on the axis of rotation, so that in the rotating system (x, y, z) which is fixed in the body we have $\omega_x = \omega_y = 0$, and $\omega_z = \omega$. The formula (5.27) for the moment of momentum vector **H** now gives $H_x = 0$, $H_y = 0$ and $H_z = I_z\,\omega_z$, so for rotation about a principal axis, the moment of momentum vector **H** is along the axis and equal to $\mathbf{H} = I\boldsymbol{\omega}$; this may in fact be taken as a definition of a principal axis.

Suppose the body in Fig. 5.24 is rotating with angular velocity $\boldsymbol{\omega}$ about a *fixed*

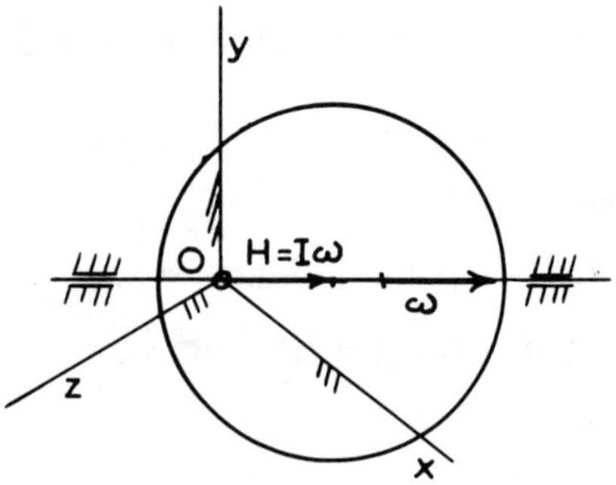

Fig. 5.24.

principal axis, so that $\mathbf{H} = I\boldsymbol{\omega}$, where I is the principal moment of inertia for the axis of rotation; for the system (x, y, y) fixed in the body with origin O on the axis of rotation, we find from formula (5.27)

$$\begin{aligned} H_x &= I_x\omega_x - I_{xy}\,\omega_y - I_{xz}\omega_z = I\omega_x \\ H_y &= -I_{xy}\omega_x + I_y\omega_y - I_{yz}\omega_z = I\omega_y \\ H_z &= -I_{xz}\,\omega_x - I_{yz}\,\omega_y + I_z\,\omega_z = I\omega_z \end{aligned}$$

or

$$\left.\begin{aligned} (I_x - I)\omega_x - I_{xy}\omega_y - I_{xz}\omega_z &= 0 \\ -I_{xy}\omega_x + (I_y - I)\omega_y - I_{yz}\omega_z &= 0 \\ -I_{xz}\omega_x - I_{yz}\omega_y + (I_z - I)\omega_z &= 0 \end{aligned}\right\} \tag{5.28}$$

Since we have real solutions for ω_x, ω_y, ω_z, the determinant of the coefficients of (5.28) must vanish:

$$\begin{vmatrix} I_x - I & -I_{xy} & -I_{xz} \\ -I_{xy} & I_y - I & -I_{yz} \\ -I_{xz} & -I_{yz} & I_z - I \end{vmatrix} = 0$$

Expanding the determinant gives a cubic equation of the form $I^3 + bI^2 + cI + d = 0$; it may be shown that this equation always has three real roots, which also follows from the physical significance of the principal moments of inertia. The coefficients of the equation are

$$\left.\begin{aligned} b &= -(I_x + I_y + I_z) \\ c &= I_xI_y + I_xI_z + I_yI_z - I_{xy}^2 - I_{xz}^2 - I_{yz}^2 \\ d &= I_xI_{yz}^2 + I_yI_{xz}^2 + I_zI_{xy}^2 - I_xI_yI_z + 2I_{xy}I_{xz}I_{yz} \end{aligned}\right\} \tag{5.29}$$

By substituting numerical values for the inertia terms, we may find the three roots of the equation as follows: by calculating

$$P = \frac{c}{3} - \left(\frac{b}{3}\right)^2 \qquad Q = \frac{d}{2} - \frac{3}{2}\left(\frac{b}{3}\right)\left(\frac{c}{3}\right) + \left(\frac{b}{3}\right)^3 \qquad R = P^3 + Q^2$$

it follows from Vieta's geometrical solution of the cubic equation that there are two possible cases:

1. If $R = 0$, the roots are

$$I_1 = (2\sqrt[3]{-Q}) - \frac{b}{3}\; I_2 = I_3 = (-\sqrt[3]{-Q}) - \frac{b}{3}\,.$$

2. If $R < 0$, $I_1 = (2\sqrt{(-P)})\cos(\theta/3) - b/3$, where $\cos\theta = Q/P\sqrt{(-P)}$, and $0 < 0 < 180°$.

The other two roots may be found by dividing out I_1 from the cubic equation and solving the remaining quadratic equation; alternatively they may be calculated from

$$I_2 = (-2\sqrt{(-P)})\sin\left(\frac{\theta}{3} + 30°\right) - \frac{b}{3}$$

$$I_3 = (-2\sqrt{(-P)})\cos\left(\frac{\theta}{3} + 60°\right) - \frac{b}{3}$$

If the x-axis is a principal axis, $I_{xy} = I_{xz} = 0$, the determinant shows at once that $I_1 = I_x$, and the other two roots may be found from the remaining second-order determinant, which gives the equation $(I_y - I)(I_z - I) - I_{yz}^2 = 0$; in a similar manner if the y-axis is principal, $I_1 = I_y$ and I_2 and I_3 may be found from $(I_x - I)(I_z - I) - I_{xz}^2 = 0$; for the z-axis principal, $I_1 = I_z$ and $(I_x - I)(I_y - I) - I_{xy}^2 = 0$ for the remaining roots.

By substituting $I = I_1$ in equation (5.28), the *ratios* of ω_x, ω_y and ω_z may be calculated. The direction of the principal axis for I_1 i then given by the vector $\boldsymbol{\omega} = \omega_x\mathbf{i} + \omega_y\mathbf{j} + \omega_z\mathbf{k}$, and the direction of the other two principal axes may be determined in a similar manner. The procedure is best described by an example.

Example 5.4

Fig. 5.25 shows a homogeneous right rectangular prism. A coordinate system (x, y, z) has been introduced with origin O at one corner of the prism; the dimensions are as shown and the density $\rho = 7760\ \text{kg/m}^3$.

Determine the principal moments of inertia at O and the direction of the principal axes.

Solution

Depending on the figures, the calculation must be carried to five or more figures after the point.

The mass of the prism is $M = 0.12 \times 0.16 \times 0.20 \times 7760 = 29.79840$ kg.

The centre of mass C is located at (0.08, 0.06, 0.10) m. Introducing parallel axes

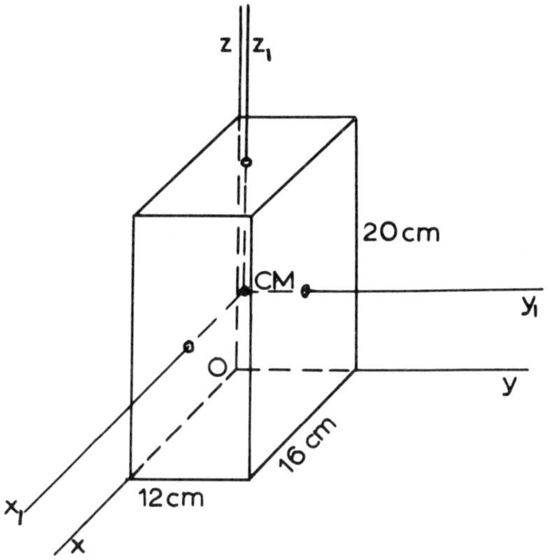

Fig. 5.25.

(x_1, y_1, z_1) at C, we find from the formula (Fig. 5.17). $I = M(a^2 + b^2)/12$ that

$$I_{x_1} = 2.48320\,(0.12^2 + 0.20^2) = 0.13509\text{ kg m}^2$$
$$I_{y_1} = 2.48320\,(0.16^2 + 0.20^2) = 0.16290\text{ kg m}^2$$
$$I_{z_1} = 2.48320\,(0.16^2 + 0.12^2) = 0.09933\text{ kg m}^2$$

From symmetry, $I_{x_1y_1} = I_{x_1y_1} = I_{y_1z_1} = 0$. Using the parallel-axis theorem ((5.20) and (5.21)), we find

$$I_x = 0.13509 + 29.79840 \times 0.0136 = 0.54034\text{ kg m}^2$$

$$I_y = 0.16290 + 29.79840 \times 0.0164 = 0.65159\text{ kg m}^2$$

$$I_z = 0.09933 + 29.79840 \times 0.0100 = 0.39731\text{ kg m}^2$$

$$I_{xy} = 29.79840 \times 0.08 \times 0.06 = 0.14303\text{ kg m}^2$$

$$I_{xz} = 29.79840 \times 0.08 \times 0.10 = 0.23839\text{ kg m}^2$$

$$I_{yz} = 29.79840 \times 0.06 \times 0.10 = 0.17879\text{ kg m}^2$$

Using these values in (5.29) we find

$$b = -1.58925 \qquad c = 0.71640 \qquad d = -0.06527$$

$$P = -0.04183 \qquad Q = 0.00846 \qquad R = -0.0000017 < 0$$

$$\cos\theta = \frac{Q}{P\surd(-P)} = -0.98855 \qquad \theta = 171.32°$$

Hence

$$I_1 = 0.75188 \text{ kg m}^2 \qquad I_2 = 0.12121 \text{ kg m}^2 \qquad I_3 = 0.71616 \text{ kg m}^2$$

These are the three principal moments of inertia at O.

Equation (5.28) may now be used to find the direction of the principal axes. The determinant of the coefficients is known to vanish for I equal to any one of the three principal moments of inertia; this means that one of ω_x, ω_y or ω_z may be taken arbitrarily, so that only two of the equations need to be used.

Substituting $I_1 = 0.75188$ in the first two equations of (5.28) and solving for ω_x and ω_y gives

$$\omega_x = 2.19933\omega_z \qquad \omega_y = -4.91938\omega_z$$

The direction of the principal axis for the maximum principal moment of inertia I_1 is now determined by the vector

$$\boldsymbol{\omega} = \omega_x\mathbf{i} + \omega_y\mathbf{j} + \omega_z\mathbf{k} = \omega_z(2.19933\mathbf{i} - 4.91938\mathbf{j} + \mathbf{k})$$

Therefore the direction is given by the vector $2.19933\mathbf{i} - 4.91938\mathbf{j} + \mathbf{k}$, or by the unit vector $0.40129\mathbf{i} - 0.89759\mathbf{j} + 0.18246\mathbf{k}$, so that the direction cosines are $l = 0.40129$, $m = -0.89759$, $n = 0.18246$.

In a similar manner the direction of the axis for the minimum principal moment I_2 is found to be $\boldsymbol{\omega} = \omega_z(0.75310\mathbf{i} + 0.54019\mathbf{j} + \mathbf{k})$, which is in the direction of the unit vector $0.55235\mathbf{i} + 0.39620\mathbf{j} + 0.73344\mathbf{k}$, which directly gives the direction cosines.

The direction of the intermediate axis I_3 may be found from the fact that it is perpendicular to the other two axes, or by a similar calculation, which gives $\boldsymbol{\omega} = \omega_z(-1.11799\mathbf{i} - 0.29244\mathbf{j} + \mathbf{k})$; the corresponding unit vector is $-0.73157\mathbf{i} - 0.19136\mathbf{j} + 0.65436\mathbf{k}$, which directly gives the direction cosines.

The three principal axes are shown in Fig. 5.26; the I_1- and I_3-axes only touch the body at O; the minimum axis I_2 goes through the body from O to a point on the top surface of the body as shown.

5.3.8 Maximum and minimum moments of inertia

Let us assume that we have determined *the principal axes* x_1, y_1, z_1 at a point O, and the principal moments of inertia I_{x_1}, I_{y_1} and I_{z_1} for a particular rigid body. We shall also assume that the notation of the axes has been arranged so that $I_{x_1} > I_{y_1} > I_{z_1}$.

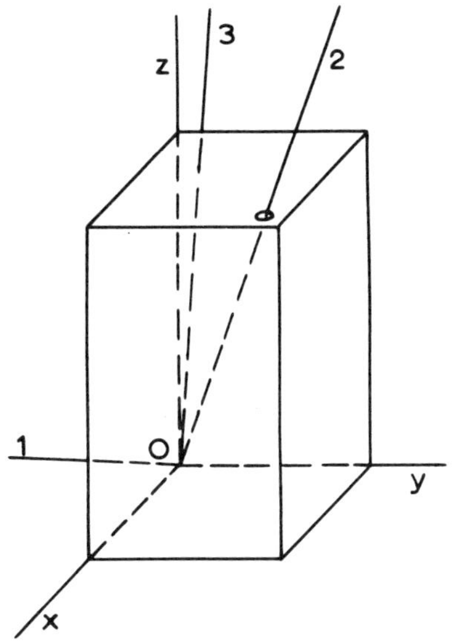

Fig. 5.26.

The moment of inertia of the body about some other axis x through O is now given by the formula (5.23):

$$I_x = l^2 I_{x_1} + m^2 I_{y1} + n^2 I_{z_1}$$

Introducing $l^2 + m^2 + n^2 = 1$, or $m^2 = 1 - l^2 - n^2$, leads to

$$I_x = l^2 I_{x_1} + (1 - l^2 - n^3) I_{y_1} + n^2 I_{z_1}$$

$$= [(I_{x_1} - I_{y_1}) l^2] - [(I_{y_1} - I_{z_1}) n^2] + [I_{y_1}]$$

Each of the brackets inside the square brackets is positive, so that I_x must have its maximum value when l is a maximum, $l = 1$, and n is a minimum, $n = 0$, so that

$$I_{x\,\max} = (I_{x_1} - I_{y_1}) + I_{y_1} = I_{x_1}$$

This shows that the largest principal moment of inertia is also the largest possible moment of inertia that can be obtained for any axis through O.

In a similar way, I_x is a minimum if $l = 0$ and $n = 1$, or

$$I_{x\,\min} = -(I_{y_1} - I_{z_1}) + I_{y_1} = I_{z_1}$$

The smallest possible moment of inertia for any axis through O is the smallest principal moment of inertia at O. These results correspond to the results previously obtained for principal axes of an area or a lamina.

It follows from the parallel-axis theorem, which states that $I_z = I_{z_1} + Md^2$, where I_{z_1} is the minimum principal moment of inertia at the centre of mass, that I_{z_1} is also the absolute minimum moment of inertia possible for any axis in space for the body in question.

5.3.9 Invariance of the sum of the moments of inertia at a point

Consider the rigid body shown in Fig. 5.27. The sum of the moments of inertia with

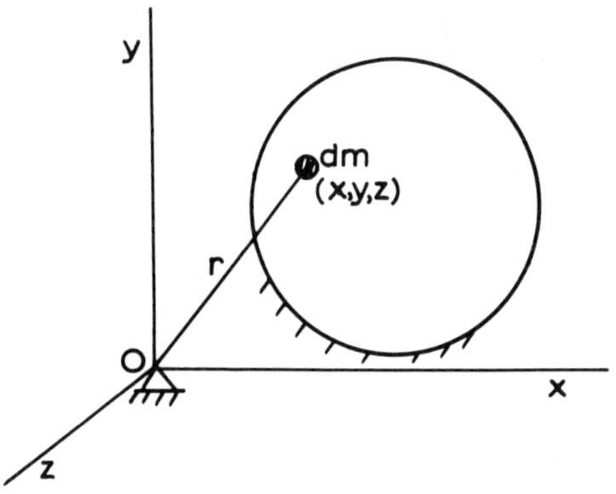

Fig. 5.27.

respect to the x-, y-and z-axes at point O is

$$I_x + I_y + I_z = \int(y^2 + z^2)\mathrm{d}m + \int(x^2 + z^2)\mathrm{d}m + \int(x^2 + y^2)\mathrm{d}m$$

$$= 2\int(x^2 + y^2 + z^2)\mathrm{d}m = 2\int r^2\,\mathrm{d}m = \text{constant}$$

The sum $(I_x + I_y + I_z)$ depend only on the position of point O relative to the body, independently of the *orientation* of the x-, y- and z-axes. It is therefore the same for any set of perpendicular axes at point O.

Example 5.5

Determine the moment of inertia about a diameter for a homogeneous solid sphere of radius R and mass M.

Solution

Introducing a coordinate system (x, y, z) at the centre of the sphere, we have $I_x + I_y + I_z = 3I_x = 2\int r^2\,\mathrm{d}m$. Dividing the sphere into thin spherical shells of thickness $\mathrm{d}r$ and surface $4\pi r^2$, and taking the density as ρ, we have $\mathrm{d}m = (4\pi r^2)dr\,\rho$, so that

$$I_x = \frac{2}{3}\int_0^R 4\pi r^4 \rho dr = \frac{8}{3}\frac{\pi\rho}{5}R^5$$

The volume of a sphere is $4/3\pi R^3$, and the total mass is $M = 4/3\pi R^3\rho$; introducing this gives $I_x = 2/5MR^2$.

Example 5.6
Determine the principal moment of inertia I_3 in Example 5.4, from the fact that $I_x + I_y + I_z$ = constant at a point.

Solution
From Example 5.4., we have

$$I_x + I_y + I_z = 1.58924 = I_1 + I_2 + I_3$$

so that

$$I_3 = 1.58924 - (I_1 + I_2) = 1.58924 - 0.87309 = 0.71615 \text{ kg m}^2$$

in agreement with the previous result in Example 5.4. The direction of this principal axis may be determined by the fact that it is perpendicular to the other two principal axes.

Consider now a set of perpendicular axes (x, y, z) in a rigid body at a point O. If the z-axis is *fixed* in the body, we have I_z = constant, and since $I_x + I_y + I_z$ = constant, we find also that $I_x + I_y$ = constant, even if the x- and y-axes are *rotated about the z-axes*.

This corresponds to the perpendicular-axis theorem for plane areas or laminae, the important difference being that for bodies in general, the sum $I_x + I_y$ is *not* equal to I_z. If the point O is moved along the z-axis, the sum $I_{x_1} + I_{y_1}$, for axes x_1 and y_1 at the new position of point O, will be different from $I_x + I_y$.

Example 5.7
Determine the moment of inertia I_y in Example 5.3, Fig. 5.21.

Solution
Since the z-axis is fixed in the body, we have

$$I_x + I_y = \text{constant} = I_{x_1} + I_{y_1} = \frac{Mr^2}{4} + \frac{Mr^2}{2} = \frac{3}{4}Mr^2$$

I_x was determined in Example 5.3 as

$$I_x = \frac{Mr^2}{4}(1 + \sin^2\alpha)$$

We have now

$$I_y = \frac{3}{4}Mr^2 - I_x = \frac{3}{4}Mr^2 - \frac{Mr^2}{4}(1+\sin^2\alpha) = \frac{Mr^2}{4}(2-\sin^2\alpha)$$

$$= \frac{Mr^2}{4}(1+\cos^2\alpha)$$

In this particular example, where I_x is determined as a function of the angle α, we may also find I_y by substituting $\alpha + 90°$ for α in the expression for I_x.

5.3.10 A principal axis through the centre of mass

Fig. 5.28 shows a rigid body with a set of *principal axes* (x_1, y_1, z_1) at a point O. A

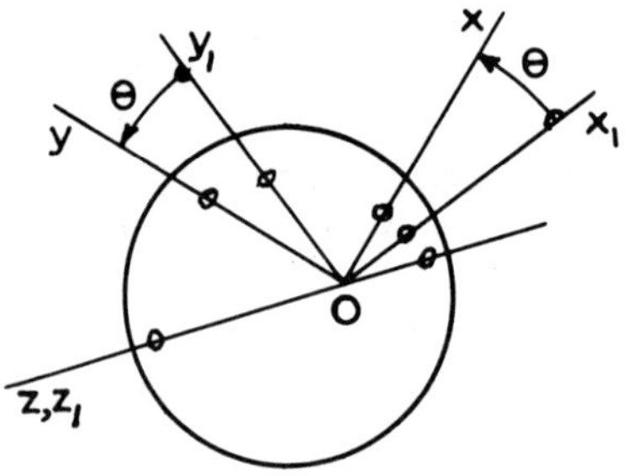

Fig. 5.28.

second coordinate system (x, y, z) is also shown rotated through an angle θ about the z_1-axis.

We have now, from (5.25),

$$I_{xz} = -(I_{x_1}ll'' + I_{y_1}mm'' + I_{z_1}nn'')$$

and

$$I_{yz} = -(I_{x_1}l'l'' + I_{y_1}m'm'' + I_{z_1}n'n'')$$

The direction cosines are (l, m, n) for the x-axis, (l', m', n') for the y-axis, and (l'', m'', n'') for the z-axis.

It may be seen from the figure that $l'' = \cos 90° = 0$; $m'' = \cos 90° = 0$ and $n'' = \cos 0° = 1$, and also $n = \cos 90° = 0$ and $n' = \cos 90° = 0$; substituting this gives $I_{xz} = 0$

and $I_{yz} = 0$. This means that if the *z-axis is a principal axis at* O, the product moments I_{xz} and I_{yz} are zero for *all positions* of the x- and y-axes.

If it is known that $I_{xz} = I_{yz} = 0$ for a set of axes at O, we find from the equation for principal moments of inertia in section 5.3.7 that

$$\begin{vmatrix} I_x - I & -I_{xy} & 0 \\ -I_{xy} & I_y - I & 0 \\ 0 & 0 & I_z - I \end{vmatrix} = 0$$

Expanding the determinant gives $(I_z - I)[(I_x - I)(I_y - I) - I_{xy}^2] = 0$, which is known to have three real positive roots, one of which is $I = I_z$. Hence the z-axis is a principal axis.

Fig. 5.29 shows a rigid body mass M with a coordinate system (x, y, z) with origin

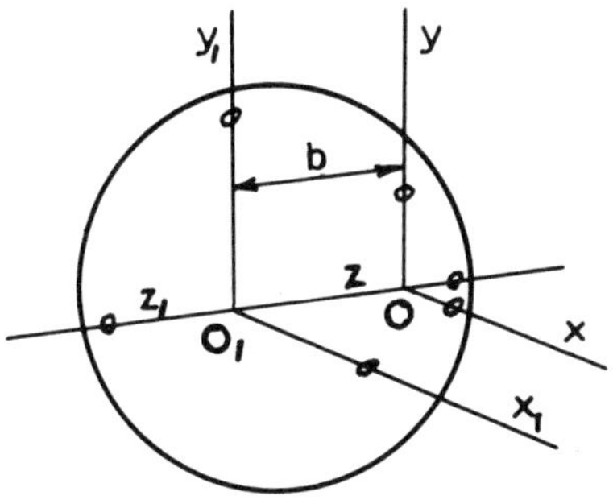

Fig. 5.29.

O and with axes taken along the principal axes at O. A parallel coordinate system (x_1, y_1, z_1) is shown with origin at a point O_1 on the z-axis, a distance b from O.

We have $I_{x_1z_1} = \int x_1 z_1 \, dm$ and $I_{y_1z_1} = \int y_1 z_1 \, dm$. Introducing $x_1 = x$, $y_1 = y$ and $z_1 = z - b$, we find

$$I_{x_1z_1} = \int x(z-b)\mathrm{d}m = \int xz\,\mathrm{d}m - b\int x\mathrm{d}m = I_{xz} - b\int x\,\mathrm{d}m$$

$$I_{y_1z_1} = \int y(z-b)\mathrm{d}m = \int yz\mathrm{d}m - b\int y\mathrm{d}m = I_{yz} - b\int y\,\mathrm{d}m$$

Since the z-axis is principal at O, we have $I_{xz} = I_{yz} = 0$. The centre of mass of the body is determined by the expressions

$$x_C M = \int x\,\mathrm{d}m \qquad y_C M = \int y\,\mathrm{d}m \qquad z_C M = \int z\mathrm{d}m$$

Assuming now that the centre of mass is *on the z-axis*, we have $x_C = y_C = 0$, or $\int x\,\mathrm{d}m$

$= \int y\,dm = 0$; under these conditions, then, $I_{x_1z_1} = I_{y_1z_1} = 0$, and the z_1-axis is a principal axis at O_1. Since O_1 was taken an arbitrary distance b from O, this holds for any point O_1 on the z_1-axis. We have hereby shown that if the z-axis is principal axis for a point on the axis, it will be a principal axis for any point on the axis, provided that the *centre of mass* is on the axis.

PROBLEMS

5.1 The section of a 40 mm by 5 mm angle iron may be regarded as a square of side 40 mm from one corner of which a square of side 35 mm has been removed (Fig. 5.30). Determine the position of the centroid of the figure and find the

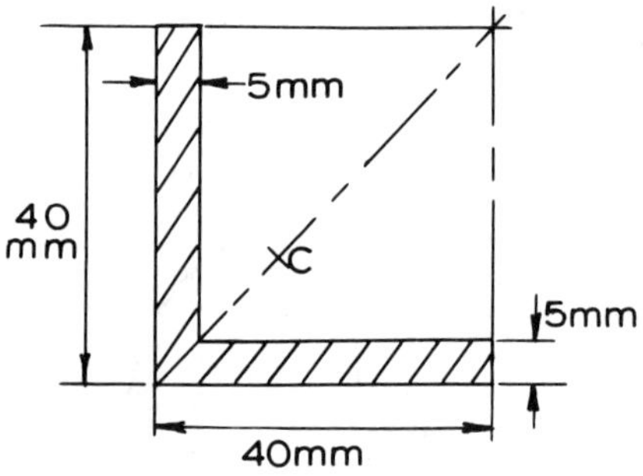

Fig. 5.30.

second moment of area about (a) the common diagonal of the two squares, (b) an axis in the plane of the section passing through the centroid and arranged at right angles to the common diagonal.

5.2 For the triangular area in Fig. 5.31 with centre of area C, find the products of

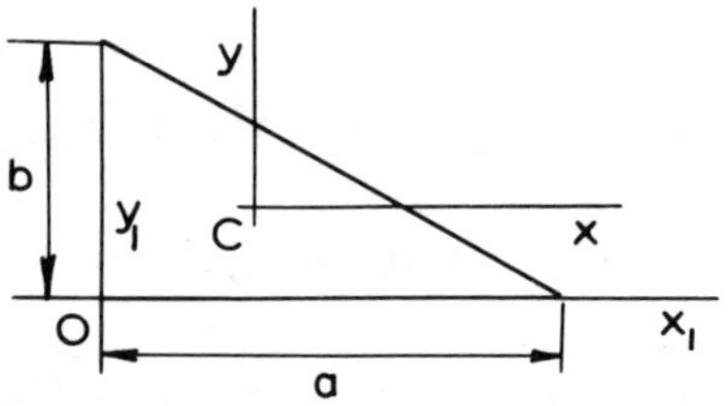

Fig. 5.31.

area I_{xy} and $I_{x_1y_1}$.

5.3 For the Z-section shown in Fig. 5.32, the dimensions are $a = 160\,mm$, $a_1 =$

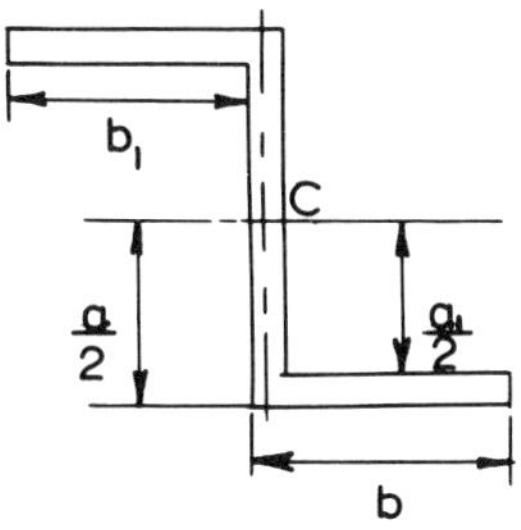

Fig. 5.32.

138 mm, $b = 70$ mm, $b_1 = 59$ mm. Find the location of the centroidal principal axis and the principal second moments of area.

5.4 Calculate the moment of inertia of a thin square plate of side length $2b$ and mass M with respect to an axis perpendicular to the plate through the mid-point of one side. Calculate the moment of inertia with respect to an axis perpendicular to the plane of a semi-circular thin plate of radius r and mass M, through a corner.

5.5 Calculate the moment of inertia of a homogeneous rectangular body with respect to an axis through one edge (Fig. 5.33).

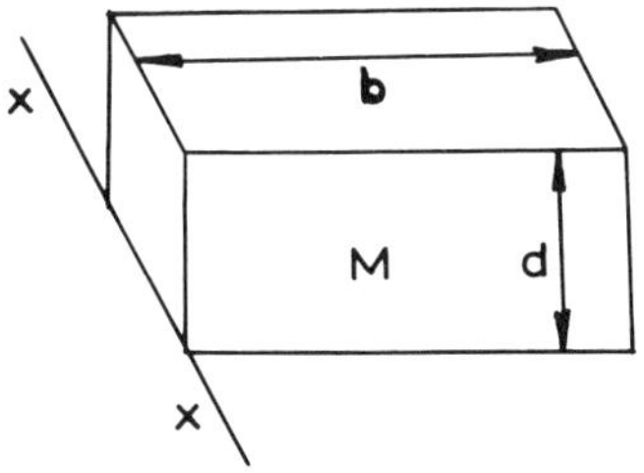

Fig. 5.33.

5.6 Show that the moments of inertia of a solid cone of mass M and semi-vertical angle α at its vertex are

$$I_x = \frac{3Mh^2 \tan^2 \alpha}{10}, \qquad I_y = I_z = \frac{3Mh^2}{20}(4 + \tan^2 \alpha)$$

where the axis of the cone is chosen as the axis OX with O at the vertex.

5.7 The moment of inertia of a solid sphere of mass M about a diameter is $2Ma^2/5$ and the mass centre of a solid hemisphere is at a distance $3a/8$ from its base, a being the radius in each case. Show from these facts that the moment of inertia of a solid hemisphere about a tangent at its pole is $(13/20)ma^2$, where m is the mass of the hemisphere.

6

Rotation of a rigid body about a fixed axis

Rotation about a fixed axis is one of the most common and important types of motion encountered in engineering. Before this motion can be considered it is necessary to discuss the motion of the centre of mass of a rigid body.

6.1 CENTRE OF PARALLEL FORCES. CENTRE OF GRAVITY

Fig. 6.1 shows a system of parallel forces acting on a *rigid* body; the resultant is a

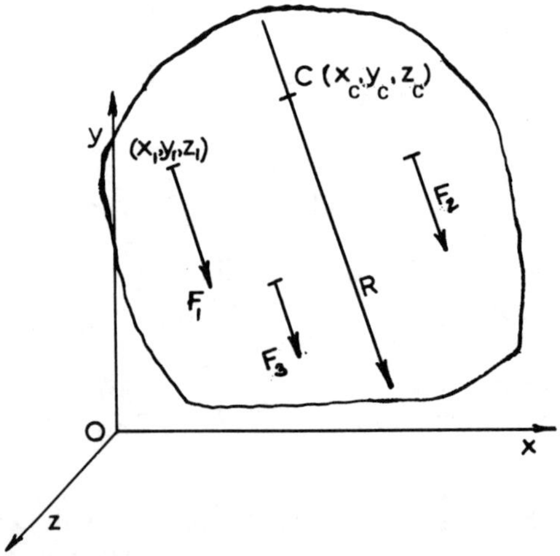

Fig. 6.1.

single force **R**, whose magnitude is found by adding the force magnitudes algebraically.

The point (x_C, y_C, z_C) on the line of action of **R** may be determined as follows: resolving each force into rectangular components, we have $F_{ix} = F_i \cos\theta_x$, $F_{iy} = F_i$ cos

θ_y, $F_{iz} = F_i \cos\theta_z$, $i = 1, 2, 3, \ldots, n$. The cosines are direction cosines for the forces, and in this case they are the same for each of the forces; $R_x = \Sigma F_{ix} = \Sigma F_i \cdot \cos\theta_x = \cos\theta_x\, \Sigma F_i$, $R_y = \cos\theta_y\, \Sigma F_i$, $R_z = \cos\theta_z\, \Sigma F_i$.

Taking moments of the x components about the z-axis gives

$$R_x y_C = \Sigma F_{ix} y_i = \Sigma(F_i \cos\theta_x) y_i = \cos\theta_x\, \Sigma F_i y_i$$

$$y_C = \frac{\cos\theta_x\, \Sigma F_i y_i}{\cos\theta_x\, \Sigma F_i} = \frac{\Sigma F_i y_i}{\Sigma F_i}$$

Similarly

$$x_C = \frac{\Sigma F_i x_i}{\Sigma F_i} \quad \text{and} \quad z_C = \frac{\Sigma F_i z_i}{\Sigma F_i}$$

Since the direction cosines cancel out, the same result is found if all the forces are rotated through the same angle to remain parallel.

If the acting forces are *gravity forces* $F_i = m_i g$ on a body consisting of a series of *rigidly connected particles* m_i; the centre C of the gravity forces is located at

$$x_C = \frac{\Sigma m_i g x_i}{\Sigma m_i g} = \frac{\Sigma m_i x_i}{\Sigma m_i} \quad y_C = \frac{\Sigma m_i y_i}{\Sigma m_i} \quad z_C \frac{\Sigma m_i z_i}{\Sigma m_i}$$

This point C is called the *centre of gravity of* the particles or body; its location is independent of the orientation of the body in space, since the same point is found if all the forces are rotated through the same angle, or if the body is rotated in any way about C.

For a homogeneous *rigid body* consisting of infinitely many particles, the centre of gravity is determined by

$$x_C = \frac{\int x\, dm}{M} \quad y_C = \frac{\int y\, dm}{M} \quad z_C = \frac{\int z\, dm}{M}$$

where M is the total mass of the body and the summation is extended over the volume of the body. Since the expression for the coordinates of the *centre of gravity* is the same as the expressions in Chapter 5 for the *centre of mass*, it follows that the centre of gravity and the centre of mass are the *same point* for a rigid body in a *uniform parallel gravitational field*. The concept of the centre of gravity enables us to deal with the distributed gravitational forces as one resultant force in rigid-body dynamics.

6.2 NEWTON'S SECOND LAW FOR A RIGID BODY; MOTION OF THE CENTRE OF MASS

For a typical mass particle m_i in a rigid body, the equation of motion, from Newton's second law, is $\mathbf{F}_{ei} + \mathbf{F}_{ii} = m_i\ddot{\mathbf{r}}_i$, where $\mathbf{F}_{ei}$ is the *resultant external force* on m_i, and $\mathbf{F}_{ii}$ is the *resultant internal force* on m_i from the other particles of the body.

Writing the above equation for all particles of the body and adding all the equations gives

$$\Sigma\mathbf{F}_{ei} + \Sigma\mathbf{F}_{ii} = \Sigma m_i\ddot{\mathbf{r}}_i$$

However, for a rigid body, the internal forces between the particles are always in equal and opposite pairs from Newton's third law; this means that $\Sigma\mathbf{F}_{ii} = \mathbf{0}$; taking the resultant of the external forces $\Sigma\mathbf{F}_{ei} = \mathbf{F}$, we obtain $\mathbf{F} = \Sigma m_i\ddot{\mathbf{r}}_i$; now writing $\Sigma m_i\mathbf{r}_i = (\Sigma\ m_i)\mathbf{r}_C = M\mathbf{r}_C$, we define by this expression a point C by the position vector $\mathbf{r}_C = (\Sigma m_i\mathbf{r}_i)/M$, where M is the total mass of the body; the point C is the *centre of mass of the body*.

From the expression for $\mathbf{r}_C$, we find

$$\Sigma m_i\ddot{\mathbf{r}}_i = M\ddot{\mathbf{r}}_C$$

Therefore

$$\mathbf{F} = M\ddot{\mathbf{r}}_C \tag{6.1}$$

This is Newton's second law for a rigid body of total mass M acted upon by a resultant force $\mathbf{F}$; $\ddot{\mathbf{r}}_C$ is the *absolute acceleration* of the centre of mass, that is the acceleration of C in an inertial system.

Equation (6.1) is in vector form; analytically it gives the three scalar equations for the motion of the centre of mass:

$$F_x = M\ddot{x}_C \quad F_y = M\ddot{y}_C \quad F_z = M\ddot{z}_C$$

The motion of the centre of mass is then determined by accumulating the total mass in the centre, and moving all forces to act at the centre of mass, keeping the directions in which they act unchanged. The centre then moves as a particle of total mass equal to the mass of the body.

6.3 THE EQUATION OF MOTION FOR A RIGID BODY ROTATING ABOUT A FIXED AXIS

Fig. 6.2 shows a rigid, homogeneous body rotating about a fixed axis which is located in two bearings. The axis of rotation is taken as the z-axis, and the body rotates with instantaneous angular velocity ω and angular acceleration $\dot{\omega}$ due to forces not shown. The position of the body may be given by the angle θ as shown, measured from a fixed line to an axis fixed in the body, both axes perpendicular to the z-axis. Each point in

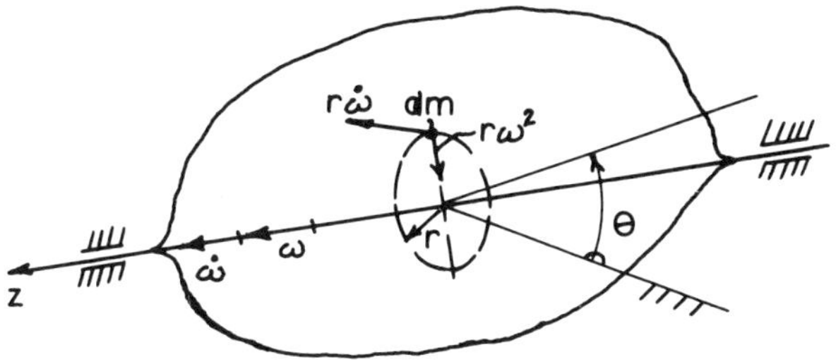

Fig. 6.2.

the body is in circular motion in a plane perpendicular to the z-axis and with the centre of the circle on the z-axis. A representative particle of mass dm is shown at a distance r from the z-axis. The acceleration components of the particle are the normal acceleration $r\omega^2$ towards the centre of the circle, and the tangential acceleration $r\dot{\omega}$ perpendicular to the radius as shown.

The components of the resultant force of all external and internal forces on the particle are now the normal force $(\mathrm{d}m)r\omega^2$ and the tangential force $(\mathrm{d}m)r\dot{\omega}$.

Taking moments of the forces on the particle about the z-axis we obtain $m_z = (\mathrm{d}m)r\dot{\omega}\, r$; the normal force does not contribute to this moment since it intersects the z-axis. Summing up the moment equations for all the particles of the body, the moments of the internal forces cancel, since they occur in equal and opposite pairs, and we are then left with the *moment M_z of all the external forces only* about the z-axis. The result is

$$M_z = \int r^2\dot{\omega}\mathrm{d}m = \dot{\omega}\int r^2\,\mathrm{d}m$$

where the summation is extended over the total volume of the body.

The summation $\int r^2\mathrm{d}m$ is seen to be the moment of inertia of the body about the z-axis; calling this I we find

$$M_z = I\dot{\omega} = I\ddot{\theta} \tag{6.2}$$

Equation (6.2) is the equation of rotational motion of a rigid body about a fixed axis. The equation has the same form as Newton's equation for a particle or a rigid body in the z-direction: $F_z = M\ddot{x}_z$. The equation is called *Euler's equation* after the Swiss mathematician who was the first to develop this equation in 1736. The axis of rotation need not be a principal axis, and the centre of mass of the body need not be on the axis of rotation.

6.4 APPLICATIONS OF EULER'S EQUATION

Rigid bodies that are essentially in one plane may be supported on one bearing only, and we need consider only one bearing reaction on the body. If the body has an appreciable dimension along the z-axis it must be supported in two bearings; in that case we consider the plane through the centre of mass of the body and perpendicular to the axis of rotation as the plane of motion, and if the body is homogeneous and has the plane of motion as a plane of symmetry, we may again combine the bearing reactions to one reaction in the plane of motion. It is important to realise that equation (6.2) is valid for *one* rigid body only, and that M_z is the total external torque about the z-axis of all acting forces, including the gravity force. If the centre of mass is on the z-axis, the gravity torque vanishes.

In the same way as for linear motion of a particle, some simple cases of rotational motion may be determined from the equation of motion $M_z = I_z\dot{\omega}_z$; these are cases where the angle θ or the moment M_z are given functions of time, or where M_z is constant or proportional to θ.

Example 6.1
A rotor of moment of inertia I is rotated by a moment or torque M_z so that the angular position is given by the function $\theta = At^2 + Bt$, where A and B are constants. Determine the necessary moment M_z. The centre of mass is on the z-axis.

Solution
From $\theta = At^2 + Bt$, we find by differentiation that $\dot{\theta} = \omega_z = 2At + B$, and $\ddot{\theta} = \dot{\omega}_z = 2A$. Equation (6.2) then gives the result $M_z = 2AI_z$.

If the acting moment is given as a function of time $M_z = f(t)$, we may again determine the motion $\theta = F(t)$ from equation (6.2) as long as it is possible to integrate the function $f(t)$; otherwise numerical or graphical methods must be applied.

Example 6.2
A rotor of inertia I_z is rotating under the action of a torque $M_z = a \sin kt$, where a and k are constants. Determine the resultant motion if $\omega = \omega_0$ and $\theta = 0$ when $t = 0$. The centre of mass is on the z-axis.

Solution
The equation of motion, from equation (6.2), is $M_z = I_z\dot{\omega} = a \sin kt$, so that the angular acceleration is $\dot{\omega} = (a/I_z) \sin kt$. Integrating gives $\omega = -(a/kI_z) \cos kt + C_1$; when $t = 0$, $\omega = \omega_0$ and $C_1 = \omega_0 + a/kI_z$. A second integration gives

$$\theta = \left(\frac{a}{kI_z} + \omega_0\right)t - \frac{a}{k^2I_z}\sin kt + C_2$$

When $t = 0$, $\theta = 0$ and we find $C_2 = 0$.

In some important cases of rotation, the *applied moment* M_z is *constant*, in which case the equation of motion from (6.2) is

$$\dot{\omega} = M_z/I_z = \alpha \text{ (constant)}$$

assuming that the centre of mass is on the z-axis.

Comparing to the equation in section 2.2.2, $\ddot{x} = F/m = \alpha = \text{constant}$, we see that the two equations are mathematically equivalent, so that the solution may be taken directly from the solution in section 2.2.2:

$$\theta = \tfrac{1}{2}\alpha t^2 + \omega_o t + \theta_0$$

$$\dot{\theta} = \omega = \alpha t + \omega_0$$

$$\ddot{\theta} = \dot{\omega} = M_z/I_z = \alpha$$

In these equations, the angular velocity is ω_0 and the angular displacement θ_0, when $t = 0$.

Example 6.3

A flywheel of moment of inertia $I_z = 4.7\ \text{kg m}^2$ with centre of mass on the axis of rotation is initially rotating at 200 rev/min, when the driving torque is removed. The wheel comes to rest under the action of a constant friction torque of 2 N m. Determine the time taken for the wheel to stop and the number of revolutions performed.

Solution

The constant angular deceleration is

$$\alpha = \frac{-M_z}{I_z} = -\frac{2}{4.7} = -0.43\ rad/s^2$$

The initial angular velocity is $\omega_0 = 200 \times \pi/30 = 20.9$ rad/s.

The equations above now give the result

$$\omega = -0.43t + 20.9 \text{ rad/s} \quad \text{and} \quad \theta = -0.21t^2 + 20.9t = \\ = t(20.9 - 0.215t)$$

where θ_0 has been taken equal to zero when $t = 0$. The wheel stops when $\omega = 0$, so that the time elapsed is $t = 20.9/0.43 = 48.6$ s. The angle of rotation is

$$\theta = 48.6\ (20.9 - 0.215 \times 48.6) = 508 \text{ rad}$$

The number of revolutions is $508/2\pi = 81$.

Since ω is a *linear* function of time, we may also find the angle of revolution:

$$\theta = \omega_{av} \times t = \tfrac{1}{2}(20.6 + 0) \times 48.6 = 508 \text{ rad}$$

We sometimes meet cases of rotational motion, where the applied moment is a *resisting torque proportional to the angle of rotation* $M_z = -k\theta$, where k is a constant. The equation of motion is then, from (6.2),

$$M_z = I_z\ddot{\theta} = -k\theta \quad \text{or} \quad \ddot{\theta} + (K/I_z)\theta = 0$$

Substituting $\omega_0^2 = k/I_z$, we have $\ddot{\theta} + \omega_0^2\theta = 0$. This equation is of the same form as equation (2.6), and the general solution is

$$\theta = A \cos \omega_0 t + B \sin \omega_0 t$$

From the result of the solution of (2.6), we may give the general solution in the form

$$\theta = A_0 \cos (\omega_0 t - \phi)$$

where

$$A_0 = \sqrt{\left(\theta_0^2 + \frac{\dot{\theta}_0^2}{\omega_0^2}\right)} \quad \text{and} \quad \tan \phi = \frac{\dot{\theta}_0}{\theta_0\omega_0}$$

θ_0 and $\dot{\theta}_0$ being the angular displacement and velocity when $t = 0$.

If $\dot{\theta}_0 = 0$, we find $A_0 = \theta_0$ and $\phi = 0$ so that $\theta = \theta_0 \cos \omega_0 t$. This is *a vibratory rotational motion* with *frequency* $f = \omega_0/2\pi = (1/2\pi)\sqrt{(k/I_z)}$ cycles/s, and maximum amplitude θ_0.

Example 6.4

Fig. 6.3 shows a so-called *torsional pendulum* consisting of a uniform circular bar or wire, which is fixed at the upper end and attached to a mass of moment of inertia I at the lower end. The mass performs rotational motion about the vertical central axis.

Determine the equation of motion of the inertia and its solution, if the bar is of length l, diameter d and the modulus of shear of the bar material is G.

Solution

The torque is proportional to the angle of twist θ and may be expressed as $M_0 = -k\theta$, where k is the torque necessary to twist the shaft 1 rad. We call k the *torsional spring constant* of the shaft. It is known from the theory of strength of materials, that $k = GI_p/l$ N m/rad, where I_p is the polar moment of the cross-sectional area. For a circular shaft, $I_p = (\pi/32)d^4$ m^4. If we start the disc vibrating torsionally by

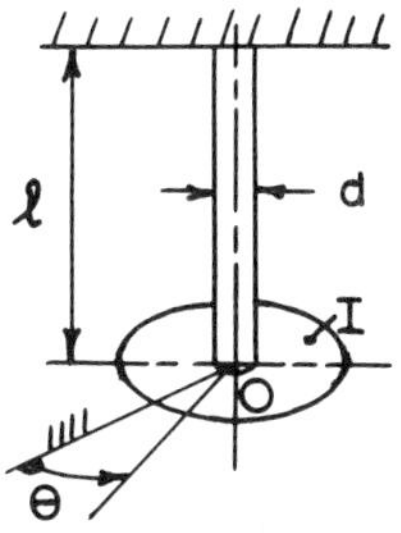

Fig. 6.3.

releasing it at an angle of twist θ_0, the motion is given by $\theta = \theta_0 \cos \omega_0 t$; this is simple harmonic motion with frequency

$$f = \frac{\omega_0}{2\pi} = \frac{1}{2\pi}\sqrt{\frac{k}{I}} = \frac{1}{2\pi}\sqrt{\frac{GI_p}{lI}} \text{ cycles/s}$$

If G, I_p and l are known, we may determine I from this formula by measuring the frequency with a stop-watch. If the system constants are not known, we may still determine I by first replacing the body with a body with a known moment of inertia I_1, usually a cylinder, and measure the frequency f_1 for this system $f_1 = (1/2\pi)\sqrt{(k/I_1)}$. Repeating the process with both bodies we measure a frequency $f_2 = (1/2\pi)\sqrt{[k/(I_1 + I)]}$; eliminating k between these two expressions and solving for I gives

$$I = I_1\left[\left(\frac{f_1}{f_2}\right)^2 - 1\right]$$

The torsional pendulum in Example 6.4 is useful for the determination of moments of inertia where the calculation creates difficulties, for instance in the case of a rotor of an electric motor. Another useful method is the so-called *trifilar pendulum method* shown in the next example.

Example 6.5
Fig. 6.4 shows a trifilar suspension of a symmetrical body. The body of moment of inertia I and mass M is suspended on three evenly spaced vertical wires at a distance r_p = pitch circle radius from C. The body is given an angular displacement θ_0 in the horizontal plane. Determine the equation of motion if the length l of the wires is large, so that the angle ϕ may be assumed to be small and the vertical motion of the centre of mass may be neglected. Determine the frequency of vibration.

Solution
The restoring torque $M_C = -3S \cos(90 - \phi)r_p = -3\, Sr_p \sin \phi$. For a large l and small ϕ, we may take $\sin\phi \simeq r_p\theta/l$, and $\cos\theta \simeq 1$. Since we assume no vertical motion

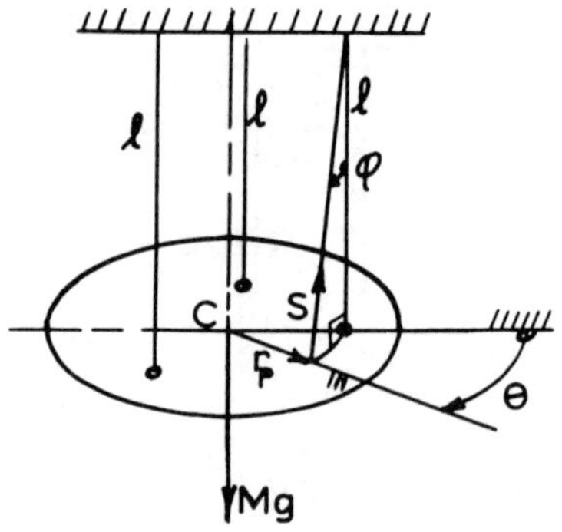

Fig. 6.4.

of the centre of mass, the vertical forces are blanced so that $3S \cos \phi = Mg$, or $S \simeq Mg/3$. The torque is now $M_C = -Mg(r_p^2/l)\theta$, and the equation of motion is $I\ddot{\theta} + Mg(r_p^2/l)\theta = 0$, or $\ddot{\theta} + (Mgr_p^2/lI)\theta = 0$. This is S.H.M. with frequency

$$f = \frac{1}{2\pi}\sqrt{\frac{Mg\,r_p^2}{lI}}\ \text{cps}$$

By measuring the frequency, the moment of inertia I may be determined from this formula. This method is very useful for bodies such as ships' propellers, where an analytical determination of I is impractical.

6.5 KINETIC ENERGY IN ROTATION ABOUT A FIXED AXIS. LANGRANGE'S EQUATIONS

6.5.1 Kinetic energy

For a rigid body rotating about a fixed axis (Fig. 6.2), each mass particle dm, at a distance r from the axis of rotation, has a velocity $v = r\omega$; the kinetic energy of the particle is then $\frac{1}{2}\,\mathrm{d}m\,v^2 = \frac{1}{2}\,\mathrm{d}m\,r^2\omega^2$. Summing up for all particles of the body, we find the total kinetic energy:

$$T = \int \tfrac{1}{2} r^2 \omega^2\,\mathrm{d}m = \tfrac{1}{2}\omega^2 \int r^2\,\mathrm{d}m = \tfrac{1}{2} I\omega^2 \tag{6.3}$$

This expression holds for any rigid body rotating about a fixed axis; the axis need not be principal and the centre of mass need not be on the axis of rotation.

If the centre of mass is a distance x_C from the axis of rotation, we have $v_C = x_C\omega$. If, for an axis through the centre of mass and *parallel* to the axis of rotation, the moment of inertia is I_C, we have from the parallel-axis theorem that $I = I_C + Mx_C^2$, where M is the total mass of the body. Substituting this in (6.3), we have

$$T = \tfrac{1}{2}I_C\omega^2 + \tfrac{1}{2}M\bar{x}_C^2\,\omega^2 = \tfrac{1}{2}I_C\omega^2 + \tfrac{1}{2}Mv_C^2 \tag{6.4}$$

as an alternative form for the kinetic energy.

6.5.2 Work done by a moment or torque

For a couple or moment vector **M**, the two parallel forces forming the couple do no total work during a translation of the couple so that only rotation is important; the work done by a couple during an *infinitesimal* rotation $\mathrm{d}\boldsymbol{\theta}$ is defined, in the same way as the work by a force, by the dot product $\mathbf{M} \cdot \mathrm{d}\boldsymbol{\theta}$. For rotation from an angle θ_1 to an angle θ_2 the work done is

$$\int_{\theta_1}^{\theta_2} \mathbf{M} \cdot \mathrm{d}\boldsymbol{\theta} = \int_1^2 (M_x \mathrm{d}\theta_x + M_y\,\mathrm{d}\theta_y + M_z\,\mathrm{d}\theta_z)$$

The *rate of work done* d(W.D.)/dt *or* the *power* is $\mathbf{M} \cdot \mathrm{d}\boldsymbol{\theta}/\mathrm{d}t = \mathbf{M} \cdot \boldsymbol{\omega}$, for a couple **M**.

6.5.3 Lagrange's equations for rotation about a fixed axis

The number of degrees of freedom n of a body was defined as the number of independent coordinates necessary to specify the position of the body at any time in an inertial reference frame or absolute coordinate system.

For a particle, the degrees of freedom and equations of constraints were discussed in Chapter 3. The maximum value of n was 3.

For a rigid body, $n = 0$ if the body is fixed. For *rectilinear translation or rotation about a fixed axis*, $n = 1$; for a combination of these, $n = 2$, unless there is a fixed relationship between the rotational and longitudinal motion, as in a screw motion, when $n = 1$. For a body in *plane curvilinear translation*, with one point following a curve $y = f(x)$, $n = 1$. If the body is free to translate in a plane, $n = 2$. For *curvilinear translation* with one point following a space curve, $n = 1$. For general curvilinear translation in space, $n = 3$.

A rigid body in *plane motion* has $n = 3$. For *rotation about a fixed point*, $n = 3$; in this case the coordinates of two points other than the fixed point may be given, which determines a triangle fixed in the body and gives six coordinates. However, the length of each side of the triangle is constant, giving three equations of constraint of the form

$$(x_2 - x_1)^2 + (y_2 - y_1)^2 + (z_2 - z_1)^2 = a^2$$

Only three of the six coordinates may be chosen arbitrarily and $n = 3$.

A rigid body *free in space* has $n = 6$. The coordinates may be taken as the three coordinates of a point in the body, together with three angles of rotation of the body about three axes through the point. Any *elastic body* has an infinite number of degrees of freedom.

Practically all dynamics systems in engineering are systems with constraints. Machines may usually be considered as systems of connected rigid bodies with various constraints such as gears, bearings, foundations, cylinders, etc. These constraints limit the number of degrees of freedom of the total system.

Equations of constraint sometimes involve the time t and derivatives of time (moving constraints); in most cases these problems are much more complicated than problems where time is not involved explicitly. Problems with moving constraints are outside the scope of this book.

The body in Fig. 6.2 rotating about a fixed axis has only one degree of freedom. The most convenient generalised coordinate to use is the angle of rotation θ.

The Lagrange equation of motion has the form

$$\frac{\mathrm{d}}{\mathrm{d}t}\frac{\partial T}{\partial \dot{\theta}} - \frac{\partial T}{\partial \theta} = Q_\theta$$

The kinetic energy is, from (6.3),

$$T = \tfrac{1}{2}I\omega^2 = \tfrac{1}{2}I\dot{\theta}^2$$

from which

$$\frac{\mathrm{d}}{\mathrm{d}t}\frac{\partial T}{\partial \dot{\theta}} = I\ddot{\theta} = I\dot{\omega} \quad \text{and} \quad \frac{\partial T}{\partial \theta} = 0$$

The generalised force Q_θ is determined from the expression

$$\delta(\text{W.D.}) = Q_\theta\delta_\theta = M_z\delta_\theta$$

where M_z is the total moment about the z-axis of all external forces on the body, which includes the gravity force and the bearing reactions. The equation of motion is now $M_z = I\dot{\omega}$, which is the same as Euler's equation (6.2).

Example 6.6
Fig 6.5(a) shows three rotors geared together. The centre of mass of each rotor is on its axis of rotation. An external torque M_0 acts on rotor no. 1 as shown. Determine the necessary torque M_0 to give rotor no. 1 an angular acceleration $\dot{\omega}_1$, both by Euler's equation and by Lagrange's equation.

Solution
Euler's equation (6.2) $M_z = I\dot{\omega}$ was developed for one rigid body; to apply it to the system in Fig. 6.5(a), we must first divide the system as shown in Fig. 6.5(b), where the forces acting to drive the rotors have been introduced.

Writing Euler's equation in turn for each of the rotors gives the three equations

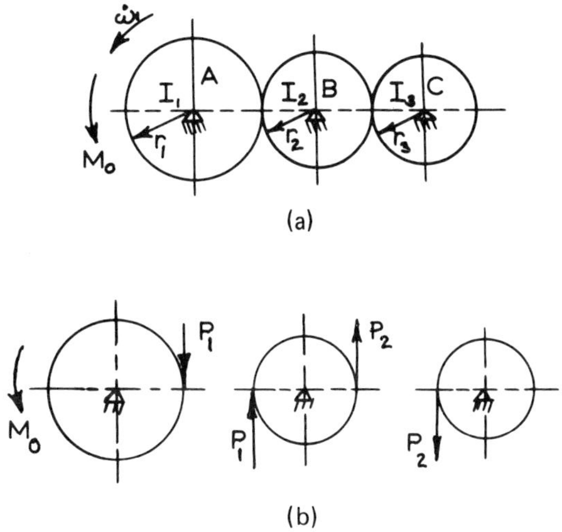

Fig. 6.5.

$$M_0 - P_1 r_1 = I_1 \dot{\omega}_1$$

$$P_1 r_2 - P_2 r_2 = I_2 \dot{\omega}_2$$

$$P_2 r_3 = I_3 \dot{\omega}_3$$

Since there is no slip between the rotors, the peripheral speed is the same for all the rotors, which gives the relationship

$$r_1 \omega_1 = r_2 \omega_2 = r_3 \omega_3$$

or by differentiation

$$r_1 \dot{\omega}_1 = r_2 \dot{\omega}_2 = r_3 \dot{\omega}_3$$

Substituting $\dot{\omega}_2 = (r_1/r_2)\dot{\omega}_1$ and $\dot{\omega}_3 = (r_1/r_3)\dot{\omega}_1$ in the equations, we may eliminate the forces P_1, P_2 and P_3 to finally obtain the result

$$M_0 = \left[I_1 + I_2 \left(\frac{r_1}{r_2} \right)^2 + I_3 \left(\frac{r_1}{r_3} \right)^2 \right] \dot{\omega}_1$$

The systems acts as one rotating inertia

$$I_q = I_1 + I_2\left(\frac{r_1}{r_2}\right)^2 + I_3\left(\frac{r_1}{r_3}\right)^2$$

I_q is called the *equivalent moment of inertia* of the system. The kinetic energy of the system is

$$T = \tfrac{1}{2}[I_1\dot{\theta}_1^2 + I_2\dot{\theta}_2^2 + I_3\dot{\theta}_3^2]$$

Substituting $\dot{\theta}_2 = (r_1/r_2)\dot{\theta}_1$ and $\dot{\theta}_3 = (r_1/r_3\ \dot{\theta}_1$, we obtain

$$T = \tfrac{1}{2}\left[I_1 + I_2\left(\frac{r_1}{r_2}\right)^2 + I_3\left(\frac{r_1}{r_2}\right)^2\right]\dot{\theta}_1^2 = \tfrac{1}{2}I_q\dot{\theta}_1^2$$

and θ_1 is taken as the generalised coordinate of the system which has only one degree of freedom.

$$\frac{\mathrm{d}}{\mathrm{d}t} = \frac{\partial T}{\partial\dot{\theta}_1} = I_q\ddot{\theta}_1 \quad \text{and} \quad \frac{\partial T}{\partial\theta_1} = 0$$

and giving a small increment $\delta\theta_1$ to θ_1, we find

$$\delta(\text{W.D.}) = M_0\ \delta\theta_1 = Q_{\theta_1}\ \delta\theta_1$$

so that Lagrange's equation is $M_0 = I_q\,\dot{\omega}_1$ as above.

Example 6.7

Fig. 6.6 shows a horizontal circular disc which is rotating about its vertical geometri-

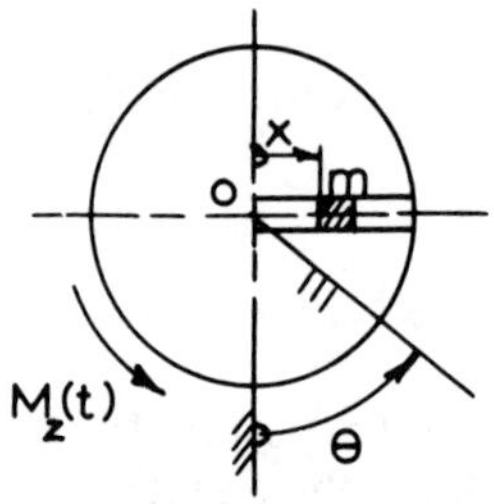

Fig. 6.6.

cal z-axis owing to the action of an external torque $M_z(t)$. The moment of inertia of the disc about the z-axis is I. A radial groove has been cut in the surface of the disc as shown, and a particle of mass m slides without friction in the groove. Determine the equations of motion of the system.

Solution

The system has two degrees of freedom. As generalised coordinates we take the distance x of the particle from the centre of the disc, and the angle θ of rotation of the disc. It is noticeable that x is a relative coordinate and θ an absolute coordinate. Lagrange's equations are

$$\frac{\mathrm{d}}{\mathrm{d}t}\frac{\partial T}{\partial \dot{x}} - \frac{\partial T}{\partial x} = Q_x \quad \text{and} \quad \frac{\mathrm{d}}{\mathrm{d}t}\frac{\partial T}{\partial \dot{\theta}} - \frac{\partial T}{\partial \theta} = Q_\theta$$

We have

$$T = \tfrac{1}{2}mV_{\mathrm{m}}^2 + \tfrac{1}{2}I\omega^2$$

with

$$\mathbf{V}_{\mathrm{m}} = \mathbf{V}_{\mathrm{b}} + \mathbf{V}_{\mathrm{r}} = x\dot{\theta} \uparrow + \dot{x} \rightarrow$$

so that

$$V_{\mathrm{m}}^2 = x^2\dot{\theta}^2 + \dot{x}^2$$

and

$$T = \tfrac{1}{2}mx^2\dot{\theta}^2 + \tfrac{1}{2}m\dot{x}^2 + \tfrac{1}{2}I\dot{\theta}^2$$

The necessary differentiations are

$$\frac{\mathrm{d}}{\mathrm{d}t}\frac{\partial T}{\partial \dot{x}} = m\ddot{x} \quad \text{and} \quad \frac{\partial T}{\partial x} = mx\dot{\theta}^2$$

By fixing θ and giving an increment δx to x, we find

$$\delta(\mathrm{W.D.}) \equiv Q_x\delta x = 0$$

so that $Q_x = 0$.

$$\frac{\partial T}{\partial \dot{\theta}} = mx^2\dot{\theta} + I\dot{\theta} \quad \text{and} \quad \frac{\mathrm{d}}{\mathrm{d}t}\frac{\partial T}{\partial \dot{\theta}} = mx^2\ddot{\theta} + 2mx\dot{x}\dot{\theta} + I\ddot{\theta} \qquad \frac{\partial T}{\partial \theta} = 0$$

By fixing x and giving an increment $\delta\theta$ to θ, we obtain

$$\delta(\mathrm{W.D.}) = Q_\theta\delta\theta = M_z(t)\delta_\theta$$

so $Q_\theta = M_z(t)$. Lagrange's equations are now

$$m\ddot{x} - mx\dot{\theta}^2 = 0 \quad \text{and} \quad mx^2\ddot{\theta} + 2mx\dot{x}\dot{\theta} + I\ddot{\theta} = M_z(t)$$

or by substituting $\dot{\theta} = \omega$ and $\ddot{\theta} = \dot{\omega}$:

$$\ddot{x} - x\omega^2 = 0$$

$$mx^2\dot{\omega} + 2mx\dot{x}\omega + I\dot{\omega} = M_z(t)$$

Writing the second equation in the form

$$M_z(t) = mx(x\dot{\omega} + 2\omega\dot{x}) + I\dot{\omega}$$

the acceleration terms in the brackets may be recognised as the tangential and Coriolis acceleration of the particle; these are perpendicular to the groove, and multiplying by m gives the forces on the particle. The equal and opposite forces act on the disc, and multiplying by the moment arm x as shown gives a retarding moment on the disc which has to be overcome by $M_z(t)$. The term $I\dot{\omega}$ is the usual moment necessary to give the angular acceleration $\dot{\omega}$ to the disc.

If the particle is fixed on the groove a distance $x = b$ from the centre O, we have $\dot{x} = \ddot{x} = 0$. The system has only one degree of freedom with generalised coordinate θ and only the second equation developed in the θ coordinate holds. This equation now gives

$$M_z(t) = (mb^2 + I)\dot{\omega}$$

or the usual Euler equation for a body of inertia $(mb^2 + I)$ about the axis of rotation.

For the special case where $\omega =$ constant, we have $\dot{\omega} = 0$, and the *equation of relative motion* of the particle is $\ddot{x} - x\omega^2 = 0$. The general solution is $x = C_1 \operatorname{sh}\omega t + C_2 \operatorname{ch}\omega t$, where C_1 and C_2 are arbitrary constants. For starting conditions

$$x = x_0 \text{ and } \dot{x} = 0 \quad at\ t = 0$$

the result is $x = x_0 \operatorname{ch}\omega t$. The relative velocity is $\dot{x} = x_0\omega \operatorname{sh}\omega t$; both x and $\dot{x}$ show a rapid increase with time.

For ω constant, the second equation gives $M_z(t) = mx(2\omega\dot{x})$. The necessary torque is, in this case, due to the Coriolis force $2m\omega\dot{x}$ only.

6.6 WORK-ENERGY EQUATION IN ROTATION ABOUT A FIXED AXIS

The equation of motion $M_z = I_z\, \mathrm{d}\omega/\mathrm{d}t$ may be integrated as follows:

$$M_z \mathrm{d}\theta = I_z \mathrm{d}\omega(\mathrm{d}\theta/\mathrm{d}t) = I_z \mathrm{d}\omega(\omega) = \tfrac{1}{2}I_z \mathrm{d}(\omega^2)$$

so that

$$\text{W.D.}\Big|_{\theta_1}^{\theta_2} = \int_{\theta_1}^{\theta_2} M_z \mathrm{d}\theta = \tfrac{1}{2}I_z(\omega_2^2 - \omega_1^2) = T_2 - T_1 \tag{6.5}$$

Where we assume that I is constant.

This equation is called *the work-energy equation*; it states that the work done by the total moment about the axis of rotation from position 1 to position 2 is equal to the *total change* in kinetic energy of the body between the two positions.

If the torque is constant, we find from (6.5) that

$$M_z(\theta_2 - \theta_1) = \tfrac{1}{2}I_z(\omega_2^2 - \omega_1^2) \tag{6.6}$$

the equation is particularly useful in problems where time is not involved in the solution.

Example 6.8

Fig. 6.7 shows a pulley of radius r and moment of inertia I. A mass M is connected to

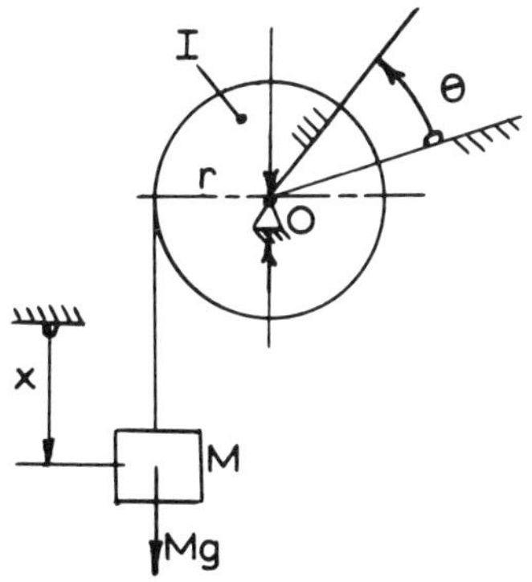

Fig. 6.7.

the pulley as shown. Neglecting friction, determine the velocity of the mass as a function of the distance moved, if the system is released from rest. Determine also the acceleration of the mass M. Solve by work-energy equation and by Lagrange's equation.

Solution

With the mass at a position x as shown, the total kinetic energy is $T = \tfrac{1}{2}MV^2 + \tfrac{1}{2}I\omega^2$; since the kinetic energy at rest is zero, this is also the total change in kinetic energy.

We have now $V^2 = \dot{x}^2$, and $x = r\theta + \chi_0$, or $\dot{x} = r\dot{\theta} = r\omega$, so that $\omega^2 = \dot{x}^2/r^2$, and $T = \frac{1}{2}(M + I/r^2)\dot{x}^2$.

The only force doing work is the gravity force Mg; the work done is Mgx. The work-energy equation now gives the relation

$$Mgx = \tfrac{1}{2}(M + I/r^2)\dot{x}^2.$$

from which

$$\dot{x} = \sqrt{[2Mgx/(M + I/r^2)]}$$

Differentiating the work-energy relationship gives $Mg\dot{x} = [M + (I/r^2)]\dot{x}\ddot{x}$, so that the acceleration is $\ddot{x} = Mg/(M + I/r^2)$.

The system has one degree of freedom. Taking x as the generalised coordinate the kinetic energy is

$$T = \tfrac{1}{2}(M + I/r^2)\dot{x}^2$$

from which

$$\frac{\mathrm{d}}{\mathrm{d}t}\frac{\partial T}{\partial \dot{x}} = (M + I/r^2)\ddot{x} \quad \text{and} \quad \frac{\partial \mathrm{T}}{\partial \mathrm{x}} = 0 \quad \text{and} \quad \delta(\text{W.D.}) = Mg\delta x = $$
$$= Q_x \delta x$$

and Lagrange's equation gives $(M + I/r^2)\ddot{x} = Mg$ as before.

Taking the datum position for the potential energy at $x = 0$, we have $V = -Mgx$, and the Lagrangian $L = T - V$.

Using Lagrange's equation

$$\frac{\mathrm{d}}{\mathrm{d}t}\frac{\partial L}{\partial \dot{x}} - \frac{\partial L}{\partial x} = 0$$

now gives the result quoted above.

Integrating $\ddot{x} = Mg/M + I/r^2 = C$ (constant) gives $\dot{x} = Ct + C_0$; when $t = 0$, $\dot{x} = 0$, so that $C_0 = 0$, and $\dot{x} = Ct$.

Integrating again gives $x = \frac{1}{2}Ct^2$, from which $t \doteq \sqrt{(2x/C)}$ and $\dot{x} = Ct = \sqrt{2C}\,x$ as determined before.

Example 6.9

The system in Fig. 6.8 consists of a body of mass $M = 12.7$ kg which is connected to a rotating drum as shown. The drum has a moment of inertia about its axis of rotation of 16.84 kg m^2.

The drum may be slowed down by applying a force P to the brake handle DB, which connects the brake shoe A to the drum. The coefficient of friction between the

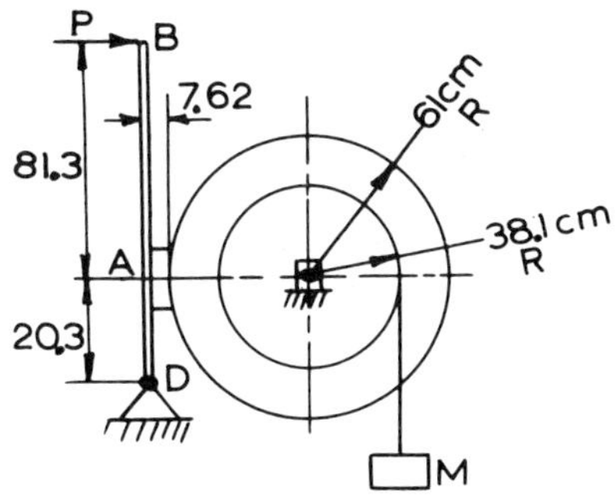

Fig. 6.8.

brake shoe and the drum is 0.40. Friction on the drum shaft may be neglected.

(a) Determine the necessary constant horizontal force P on the brake handle to stop the mass M in 2 s, if the mass is initially moving downwards with a velocity of 4.88 m/s.

(b) Determine the horizontal and vertical reactions on the brake lever at point D during this motion.

Solve by work-energy equation and by Lagrange's equation.

Solution

The initial velocity of the mass M is $V_1 = 4.88$ m/s, from which the initial angular velocity of the drum is

$$\omega_1 = \frac{4.88}{0.381} = 12.8 \text{ rad/s}$$

$$T_1 = \tfrac{1}{2} \cdot 12.7 \cdot 4.88^2 + \tfrac{1}{2} \cdot 16.84 \cdot 12.8^2 = 1531 \, N\,m$$

while $T_2 = 0$. The distance moved by M is

$$\left(\frac{V_1 + V_2}{2}\right)t = \frac{4.88}{2}2 = 4.88\,m = R\theta$$

Therefore $\theta = \dfrac{4.88}{0.381} = 12.8$ rad. Calling the friction force F, the W.D. $= -0.61 \cdot 12.8 \cdot F + 12.7 \cdot 9.81 \cdot 4.88 = -7.81\,F + 608 = T_2 - T_1 = -1531$; $F = \dfrac{2139}{7.81} = 274$ N, and the normal force $N = F/\mu = \dfrac{274}{0.40} = 685$ N.

Taking moments in point D of all the forces acting on the brake lever and brake shoe gives the equation

$$P \cdot 1.016 = N \cdot 0.203 + F \cdot 0.0762$$

from which $\underline{P = 157\,\text{N}}$. Summing all the forces acting on the lever vertically downwards, and horizontally to the right, gives the vertical reaction $Y = F = 274\,\text{N}\downarrow$ and $X = N - P = 528\,\text{N}\rightarrow$. The system has one degree of freedom. Taking the downwards vertical displacement of the mass M as x, we use x as the generalised coordinate.

$$T = \tfrac{1}{2}I\dot{\theta}^2 + \tfrac{1}{2}M\dot{x}^2$$

with $\dot{\theta} = \dot{x}/0.381$, and substituting the values for I and M, we find $T = 64.3546\,\dot{x}^2$

$$\frac{d}{dt}\frac{\partial T}{\partial \dot{x}} = 128.7091\,\ddot{x} \quad \text{and} \quad \frac{\partial T}{\partial x} = 0$$

$$\delta(\text{W.D.}) = Q_x\delta x = Mg\delta x - \frac{FR}{0.381}\delta x$$

$$= (124.587 - 1.6010\,F)\delta x$$

so that

$$Q_x = 124.587 - 1.6010\,F$$

Lagrange's equation now gives

$$\ddot{x} = 0.96797 - 0.01244\,F$$

$$\dot{x} = 0.96797t - 0.01244\,Ft + \dot{x}_0 \quad \text{when } t = 0$$

$$\dot{x} = 4.88\ \text{m/s}$$

and therefore

$$4.88 = \dot{x}_0 \quad \text{and}$$

$$\dot{x} = 0.96797t - 0.01244\,Ft + 4.88$$

When $t = 2\,s$, $\dot{x} = 0$, from which $F = 273.95\,\text{N} \simeq 274\,\text{N}$ as before. The solution may also be determined by writing Newton's second law for the mass M, and Euler's equation for the drum.

6.7 PRINCIPLE OF CONSERVATION OF MECHANICAL ENERGY

The principle of conservation of mechanical energy for a particle was stated in Chapter 2 (2.7) as $V+T=$ constant, or the sum of the potential and kinetic energy is constant for a particle moving under the action of *conservative forces only*.

For a rigid body, the internal forces occur in equal and opposite pairs and produce no work in a summation extended over all particles of the body.

The work-energy equation (6.5) states that the work done (W.D.) $= T_2 - T_1$; dividing the acting forces into conservative forces doing the work $(\text{W.D.})_\text{c}$ and non-conservative forces doing the work $(\text{W.D.})_\text{n}$, we have

$$(\text{W.D.}) = (\text{W.D.})_\text{c} + (\text{W.D.})_\text{n} = T_2 - T_1$$

The potential-energy difference is $V_2 - V_1 = -(\text{W.D.})_\text{c}$, so that

$$(\text{W.D.})_\text{n} - (V_2 - V_1) = T_2 - T_1$$

or

$$(\text{W.D.})_\text{n} = (T_2 - T_1) + (V_2 - V_1)$$

Now *if all the acting forces are conservative*, we have

$$(\text{W.D.})_\text{n} = 0 \quad \text{or} \quad (T_2 - T_1) + (V_2 - V_1) = 0$$

so that

$$T_2 + V_2 = T_1 + V_1$$

or

$$V + T = \text{constant} \tag{6.7}$$

This is *the principle of conservation of mechanical energy* for a rigid body *moving under the action of conservative forces only*. The principle is sometimes useful for establishing the equation of motion of a one-degree-of-freedom system by differentiation with respect to time, since

$$\mathrm{d}V/\mathrm{d}t + \mathrm{d}T/\mathrm{d}t = 0$$

The principle of conservation of mechanical energy holds only for conservative systems, and is not as useful as the more powerful work-energy equation which holds for all systems.

Example 6.10
In Example 6.8, determine the acceleration of the mass M by using the principle of conservation of mechanical energy.

Solution
If we neglect friction on the pulley shaft and air resistance, this is a conservative system.

The reaction from the pulley shaft and the weight force of the pulley do not move and produce no work. The pressures between the string and the pulley and the string tension are internal forces, for which the total work vanishes.

Determining the position of the system by the coordinates x and θ, and taking the datum position at $x=0$, we have $V=-Mgx$. The kinetic energy $T=\frac{1}{2}(M+I/r^2)\dot{x}^2$, as determined in Example 6.8. Equation (6.7) now states that $\frac{1}{2}(M+I/r^2)\dot{x}^2-Mgx=\text{constant}$, and differentiation gives

$$\left(M+\frac{I}{r^2}\right)\dot{x}\ddot{x}-Mg\dot{x} = 0 \quad \text{or} \quad \ddot{x} = \frac{Mg}{(M+I/r^2)}$$

It is always advisable to check a result found in terms of system constants by both a dimensional check and 'logical extremes'.

A dimensional investigation shows that the right-hand side of the result has the dimension of an acceleration as it should have.

There are four logical extremes: if $M \gg I$, $\ddot{x}\rightarrow g$, the pulley has no effect and M is in a free fall. If $M\rightarrow 0$, we find $\ddot{x}\rightarrow 0$. *If* $I \gg M$, $\ddot{x}\rightarrow 0$, the mass is unable to accelerate the pulley. Finally, if $I\rightarrow 0$, $\ddot{x}\rightarrow g$, the mass M is again in a free fall with acceleration g.

6.8 D'ALEMBERT'S PRINCIPLE IN ROTATION. INERTIA TORQUE. DYNAMIC EQUILIBRIUM

Euler's equation $M_0=I\dot{\omega}$ may be stated in the form

$$M_0+(-I\dot{\omega}) = 0 \tag{6.8}$$

This expresses *D'Alembert's principle* for rotation about a fixed axis. The term $(-I\dot{\omega})$ has the dimension of a moment or torque and is called *the inertia torque*. If the inertia torque is introduced in a given situation, we say that D'Alembert's principle has been applied to create *dynamic equilibrium*. The moment equation is then established as in *a statics case*, and may be taken about *any axis* perpendicular to the plane of motion. This may often be an advantage if a point can be found in which unknown forces act, since these forces are eliminated in the moment equation if the point is used as a moment centre.

Example 6.11
Fig. 6.9 shows a lift cage of mass M_2 which is raised by a rope passing over a revolving drum, which carries a counterbalance of mass M_1.

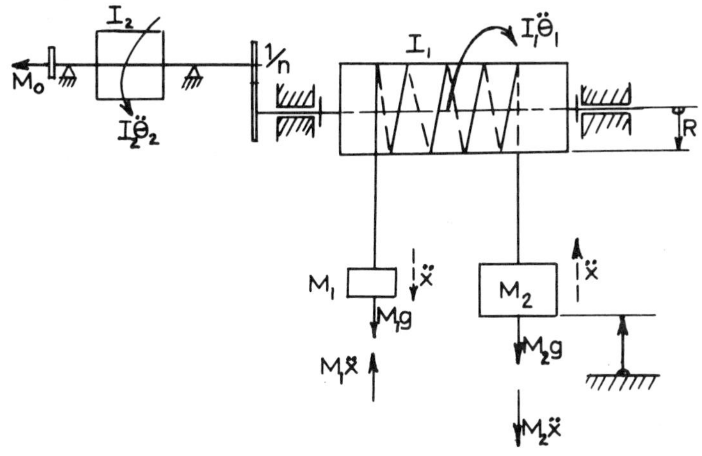

Fig. 6.9.

The radius of the drum is R and the moment of inertia about the axis of rotation is I_1; the drum is geared at a ratio $1/n$ to a second rotating inertia I_2 which is driven by an external torque M_0 as shown.

(a) Determine the equation of motion of the total system in terms of the given system constants.
(b) Determine the necessary starting torque if M_2 has to start upwards with an acceleration of $0.610\ \mathrm{m/s^2}$, and $R = 0.457\ \mathrm{m}$, $n = 5$, $I_1 = 75.5\ \mathrm{kg\ m^2}$, $I_2 = 8.4\ \mathrm{kg\ m^2}$, $M_1 = 3200\ \mathrm{kg}$ and $M_2 = 6100\ \mathrm{kg}$.
(c) Determine the speed of revolution of the driving torque and the necessary power if M_2 has to be lifted at a constant speed of 73.2 m/min.

Solution

(a) The system is established in dynamic equilibrium by introducing all the external acting forces and torques; these are the bearing reactions, the gravity forces $M_1 g$ and $M_2 g$ and the torque M_0. The inertia forces $-M_1\ddot{x}$ and $-M_2\ddot{x}$ and the inertia torques $-I_1\ddot{\theta}_1$ and $-I_2\ddot{\theta}_2$ are introduced as shown. Statics may now be applied.

Writing a torque balance for the right-hand side of the system, we find the torque M_G on the gear wheel to be

$$M_G = (M_2 g + M_2\ddot{x})R + (M_1\ddot{x} - M_1 g)R + I_1\ddot{\theta}_1$$

The torque on the left-hand side of the system is now M_G/n, and a torque balance for the left-hand side gives $M_0 = I_2\ddot{\theta}_2 + M_G/n$. Substituting $\theta_1 = \ddot{x}/R$ and $\ddot{\theta}_2 = n\ddot{x}/R$, we find

$$M_0 = \frac{R}{n}[C\ddot{x} + (M_2 - M_1)g]$$

where

$$C = M_1 + M_2 + (I_1 + n^2 I_2)/R^2$$

(b) With the given figures we find $C = 10667$ kg; and $M_0 = 974.97\ \ddot{x} + 2600.24$. If $\ddot{x} = 0.610$ m/s^2, $M_0 = 3194.97$ N m starting torque.

(c) If $\dot{x} = 73.2$ m/min constant, $\ddot{x} = 0$ and $M_0 = 2600.24$ N m constant. $\dot{\theta}_2 = (n/R)\dot{x} = 800.88$ rad/min. $N = \dot{\theta}_2/2\pi = 127.46$ rpm constant. The power is $M_0\dot{\theta}_2 = 2600.24 + \dfrac{800.88}{60} = 34708$ W $= 34.71$ kW. The power may also be determined as $(M_2 - M_1)g\dot{x} = (6100 - 3200) \cdot 9.81 \cdot \dfrac{73.2}{60} = 34708$ W.

The kinetic energy is $T = \frac{1}{2}M_1\dot{x}^2 + \frac{1}{2}M_2\dot{x}^2 + \frac{1}{2}I_1\dot{\theta}_1^2 + \frac{1}{2}I_2\dot{\theta}_2^2$. We have $R\theta_1 = x$; $\dot{\theta}_1 = \dot{x}/R$; $r\theta_1 = r_2\theta_2$ therefor $\theta_2 = (r_1/r_2)\theta_1 = n\theta_1$ and $\dot{\theta}_2 = n\dot{\theta}_1 = n\dot{x}/R$.

Substituting $\dot{\theta}_1$ and $\dot{\theta}_2$ in T gives $T = \frac{1}{2}C\dot{x}^2$ with the above expression for C.

$$\delta(\text{W.D.}) = Q_x\delta x = M_0\delta\theta_2 + M_1 g\delta x - M_2 g\delta x = (M_0\frac{n}{r} + M_1 g - M_2 g)\delta x$$

so that

$$Q_x = M_0\frac{n}{R} + M_1 g - M_2 g$$

Lagrange's equation now gives the previous result for the equation of motion.

6.9 PRINCIPLE OF VIRTUAL WORK IN ROTATION

For a system *in dynamic equilibrium*, we may state that the virtual work vanishes for any virtual displacement of the system. For rotation this may be stated in *D'Alembert's equation*:

$$(M_0 - I\dot{\omega})\delta\theta = 0 \tag{6.9}$$

D'Alembert's principle in combination with the principle of virtual work is convenient in many problems; for more complicated problems this method is, however, largely superseded by the method of Lagrange.

Example 6.12

Determine the equation of motion in Example 6.11, by using the princple of virtual work.

Solution
The system is in dynamic equilibrium as shown in Fig. 6.9. Giving a virtual displacement δx to the system, the principle of virtual work gives the following equation:

$$\delta(\text{W.D.}) = -(M_2 g + M_2\ddot{x})\delta x + (M_1 g - M_1\ddot{x})\delta x - I_1\ddot{\theta}_1\delta\theta_1 - I_2\ddot{\theta}_2\delta\theta_2 + M_0\delta\theta_2 = 0$$

Introducing $\delta\theta_1 = \delta x/R$ and $\delta\theta_2 = (n/R)\delta x$ together with $\ddot{\theta}_1 = \ddot{x}/R$ and $\ddot{\theta}_2 = (n/R)x$, we find, by cancelling δx and solving for M_0, the same result for M_0 as previously found in Example 6.11.

The main advantage of the principle of virtual work is that the work of internal forces, fixed reactions and normal forces does not enter the expression for virtual work. If friction forces are present, they must be included with the acting forces.

6.10 IMPULSE–MOMENTUM EQUATION

The Euler equation (6.2) of rotational motion $M_0 = I\dot{\omega} = I\, d\omega/dt$ may be integrated directly to give the equation

$$\int_1^2 M_0\, dt = \int_1^2 I\, d\omega = I(\omega_2 - \omega_1) \qquad (6.10)$$

assuming that I is constant. The integral $\int_1^2 M_0\, dt$ is called the *angular impulse* of the moment M_0 and $I\omega$ is called *the angular momentum* about the axis of rotation; the equation then states that the total angular impulse is equal to the total *change* in angular momentum in the same time, for rotation about a fixed axis. This equation is called *the impulse–momentum equation*. If the moment M_0 is *constant*, the equation takes the form

$$M_0(t_2 - t_1) = I(\omega_2 - \omega_1) \qquad (6.11)$$

Example 6.13
Fig. 6.10 shows a shaft with an inertia $I_1 = 16.84$ kg m^2; the right-hand end of the

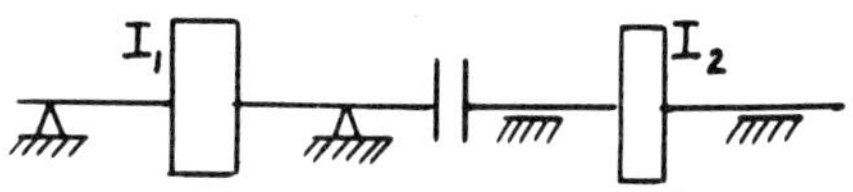

Fig. 6.10.

shaft is connected through a friction clutch to a shaft which carries an inertia $I_2 = 10.5$ kg m^2. The first shaft is originally rotating at 150 rev/min or $\omega_1 =$ 15.7080 rad/s, and the clutch is then engaged to start the other shaft rotating, this being initially at rest.

The driving shaft runs in roller bearings with negligible friction, while the driven shaft runs in journal bearings for which the total friction torque may be assumed *constant* at 13.56 N m.

The clutch is designed to transmit a maximum torque of 67.8 N m before it starts to slip.

Determine the angular velocity of the system just when slipping has ceased, the time of slipping and the number of revolutions of each shaft during slipping. Determine also the energy lost in the clutch during slipping, that is the kinetic energy converted to heat in the clutch.

Solution

The total torque acting on the inertia I_1 during slipping time t is the constant friction torque $M_f = 67.8$ N m; while the total torque on the inertia I_2 is M_f and the resisting friction torque from the journal bearings is 13.56 N m.

Taking $t_1 = 0$ and $t_2 = t$, and taking the combined speed to be ω_2 when slipping ceases, we find the impulse–momentum equations from (6.11):

$$-67.8t = 16.84\,(\omega_2 - 15.7080)$$

$$(67.8 - 13.56)t = 10.5\omega_2$$

The solution is $t = 1.70895$ and $\omega_2 = 8.8277$ rad/s $= 84.2983$ rev/min. Since M_f is constant, we have $M_f = I\,\mathrm{d}\omega/\mathrm{d}t$, or $\mathrm{d}\omega = (M_f/I)\mathrm{d}t$, so that $\omega = (M_f/\mathrm{I})\mathrm{t} + \omega_1$, showing a *linear relationship with time t*.

The total rotation is given by

$$\theta = \int_0^1 \omega\,\mathrm{d}t = \frac{\omega_1 + \omega_2}{2}t$$

For I_1 we find $\theta_1 = (15.708 + 8.8277) \cdot 1.7089/2 = 20.9645$ rad $= 3.3366$ rev.

For I_2 the result is $\theta_2 = 8.8277 \cdot \dfrac{1.7089}{2} = 7.5428$ rad $= 1.2005$ rev.

The total kinetic energy of the system just before the clutch is engaged is $T_1 = \frac{1}{2}I_1\omega_1^2 = \frac{1}{2} \cdot 16.84 \cdot 15.7080^2 = 2077.56$ Nm.

The total kinetic energy just after slipping ceases is $\frac{1}{2}(I_1 + I_2)\omega_2^2 = \frac{1}{2} \cdot 27.34 \cdot 8.8277^2 = 1065.28$ Nm. The decrease is 1012.28 N m, but part of the loss is due to the friction in the journal bearings, and this part amounts to $13.56 \cdot 7.5428 = 102.28$ N m. The total loss in the clutch is thus 910 Nm.

The result may alternatively be determined as the difference in the work done by the friction torque on the two parts of the system, this is $M_f(\theta_1 - \theta_2) = 67.8\,(20.9645 - 7.5428) = 910$ N m.

The problem may also be solved by using the equations of motion; these are (6.2) $M_z = I_z\dot{\omega}$, or

$$M_{z_1} = -67.8 = 16.84\,\dot{\omega}_1$$

$$M_{z_2} = (67.8 - 13.56) = 10.5\,\dot{\omega}_2$$

Integrating the equations leads to

$$\omega_1 = -4.0261t + 15.7080 \quad \text{and} \quad \omega_2 = 5.1657t$$

Equating the angular velocities gives the results determined before. For solutions by Lagrange's equations, we take the angular rotations θ_1 and θ_2 as generalised coordinates.

We have now $T = \frac{1}{2}I_1\dot{\theta}_1^2 + \frac{1}{2}I_2\dot{\theta}_2^2$

Fixing θ_2 we obtain $\delta(\text{W.D.}) = Q_{\theta_1}\delta\theta_1 = -67.8\delta\theta_1$, and fixing θ_1 the result is $\delta(\text{W.D.}) = Q_{\theta_2}\delta\theta_2 = (67.8 - 13.56)\delta\theta_2$.

Lagrange's equations now give the two equations of motion above.

The equation of motion may be derived in a different way by applying the *moment of momentum*.

Consider a particle in Fig. 5.22 of mass m and position vector $\mathbf{r}$. The resultant force is $\mathbf{F}_e + \mathbf{F}_i$, where $\mathbf{F}_e$ is the resultant external force, and $\mathbf{F}_i$ is the resultant internal force. Newton's law gives the equation of motion of the particle $\mathbf{F}_e + \mathbf{F}_i = m\ddot{\mathbf{r}}$. The moment of the resultant force in point O is $\mathbf{r} \times (\mathbf{F}_e + \mathbf{F}_i) = \mathbf{r} \times m\ddot{\mathbf{r}}$. Summing up for all the particles of the body gives

$$\Sigma\mathbf{r} \times \mathbf{F}_e + \Sigma\mathbf{r} \times \mathbf{F}_i = \Sigma\mathbf{r} \times m\ddot{\mathbf{r}}$$

The internal forces occur in equal and opposite pairs and therefore cancel in the moment summation, so that $\Sigma\mathbf{r} \times \mathrm{F}_i = \mathbf{0}$.

Denoting the moment of all *external* forces in point O by $\mathbf{M} = \Sigma\mathbf{r} \times \mathbf{F}_e$, we have

$$\mathbf{M} = \Sigma\mathbf{r} \times \mathbf{m}\ddot{\mathbf{r}} = \frac{\mathrm{d}}{\mathrm{d}t}(\Sigma\mathbf{r} \times m\dot{\mathbf{r}}) - \Sigma\dot{\mathbf{r}} \times m\dot{\mathbf{r}} = \frac{\mathrm{d}}{\mathrm{d}t}(\Sigma\mathbf{r} \times m\dot{\mathbf{r}})$$

since $\Sigma\dot{\mathbf{r}} \times \dot{\mathbf{r}} = \mathbf{0}$,

The vector $m\dot{\mathbf{r}}$ is the *momentum* vector of the particle, and the vector $\mathbf{r} \times m\dot{\mathbf{r}}$ is the *moment of momentum* vector of the particle in point O.

The moment of momentum vector $\mathbf{H}$ of the body in point O was defined as the sum of the moment of momentum vectors of all the particles in point O, so that $\mathbf{H} = \Sigma\mathbf{r} \times m\dot{\mathbf{r}}$.

For a *rotating* body, $\mathbf{H}$ is also called the *angular momentum*. Introducing the expression for $\mathbf{H}$ in the moment equation gives

$$\mathbf{M} = \frac{\mathrm{d}}{\mathrm{d}t}\mathbf{H} = \dot{\mathbf{H}} \qquad (6.12)$$

Consider now an element dm of a rotor with a distance r from the fixed axis of rotation. This element is in circular motion and its velocity is $r\omega$. The moment of momentum of the element *about the* z-axis is now $(\mathrm{d}m)r^2\omega$. Summing up for all elements of the rotor, the total moment of momentum about the z-axis is $\int \omega r^2 \mathrm{d}m$, and from (6.12):

$$M_z = \frac{\mathrm{d}}{\mathrm{d}t}[\int \omega r^2 \, \mathrm{d}m] = \frac{\mathrm{d}}{\mathrm{d}t}[\omega \int r^2 \, \mathrm{d}m]$$

Introducing the *variable* moment of inertia $I_z = \int r^2 \, \mathrm{d}m$, we obtain

$$M_z = \frac{\mathrm{d}}{\mathrm{d}t}(I_z\omega) = I_z\frac{\mathrm{d}\omega}{\mathrm{d}t} + \omega\frac{\mathrm{d}I_z}{\mathrm{d}t} \qquad (6.13)$$

If I_z is *constant*, equation (6.13) gives the usual Euler equation (6.2), that is $M_z = I_z\dot{\omega}$.

The moment of inertia I_z of a rotor may change in magnitude owing to cooling or heating of the rotor or because of a change in the configuration, in which case equation (6.13) must be used. If the external torque $M_z = 0$ (6.13) states that

$$\frac{\mathrm{d}}{\mathrm{d}t}(I_z\omega) = 0 \quad \text{or} \quad I_z\omega = \textit{constant}$$

To change the angular velocity of a rotor when no external torque is acting, the moment of inertia I_z must be changed; this is used by skaters and divers to increase or decrease their angular velocity by changing their body configuration, which changes their moment of inertia.

If we have *several bodies rotating about the same axis* (Fig. 6.11) where two bodies

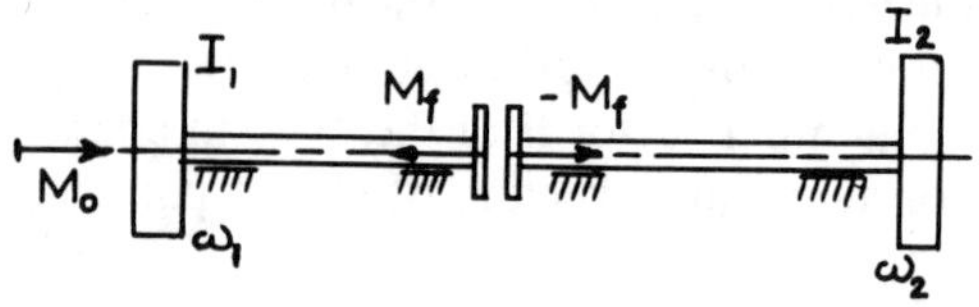

Fig. 6.11.

of inertia I_1 and I_2 are connected by a friction clutch, we have the equations of motion

$$M_0 - M_f = \frac{d(I_1\omega_1)}{dt} \quad \text{and} \quad M_f = \frac{d(I_2\omega_2)}{dt}$$

where M_f is the friction torque. Summing these equations gives

$$M_0 = \frac{d}{dt}(I_1\omega_1 + I_2\omega_2) \tag{6.14}$$

The same type of equation may be established for any number of bodies. It is important to notice that all the bodies must be rotating about the *same axis*. This equation is a statement of the *principle of angular momentum*.

If the *moment* of the external forces with respect to the axis of rotation *vanishes*, so that $M_0 = 0$, we find from (6.14) that the *total angular momentum* $(I_1\omega_1 + I_2\omega_2 + \ldots)$ of the system *remains constant*. This is called the *principle of conservation of angular momentum*.

Example 6.14
Determine the angular velocity ω_2 of the system in Example 6.13, if the friction torque from the journal bearings is neglected.

Solution
With no external torque acting, the total angular momentum remains constant. The initial angular momentum is $I_1\omega_1$ and the final value is $(I_1 + I_2)\omega_2$; equating these gives $\omega_2 = I_1\omega_1/(I_1 + I_2) = 16.84 \cdot 15.7080/27.34 = 9.6753$ rad/s. This is omewhat larger than found in Example 6.13, because the bearing friction has been neglected in this example.

6.11 THE COMPOUND PENDULUM

A rigid body, free to rotate about a fixed axis due to the action of gravity, is called a compound pendulum (Fig. 6.12).

The equation of motion is (6.2); $M_0 = I_0\dot{\omega}$. The moment about the axis of rotation is the moment of the gravity force Mg, and is in the opposite direction to the angular displacement θ, so that $M_0 = -Mgc \sin\theta$, and the equation of motion is

$$-Mgc \sin\theta = I_0\dot{\omega} = I_0\ddot{\theta}$$

For small values of θ, that is $\theta < 20°$, we may substitute θ for $\sin\theta$ with sufficient accuracy, and the equation takes the form $\ddot{\theta} + (Mgc/I_0)\theta = 0$. Introducing $I_0 = Mr_0^2$, where r_0 is the radius of gyration about the z-axis, we obtain $\ddot{\theta} + (gc/r_0^2)\theta = 0$. which indicates a simple harmonic motion with frequency $f = (1/2\pi)\sqrt{(gc/r_0^2)}$. For a simple pendulum of length l, the frequency was found in Example 2.8 to be $f = (1/2\pi)\sqrt{(g/l)}$. Suppose now that we want to determine the length of a simple pendulum so that it has

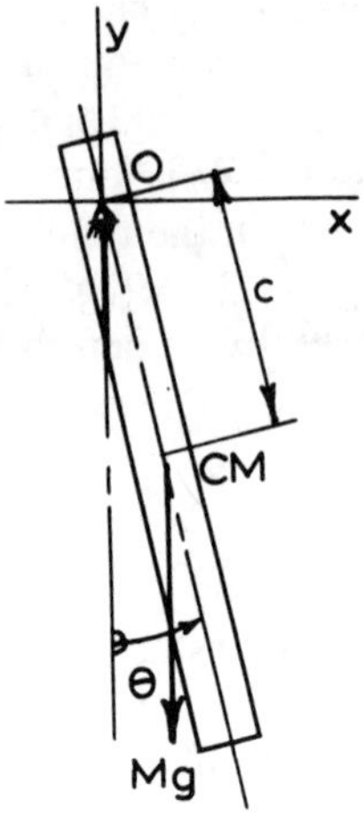

Fig. 6.12.

the same frequency as the compound pendulum; equating the frequencies gives $c/r_0^2 = 1/l$, or $l = r_0^2/c$. This length is called the *equivalent length* of the compound pendulum, and determines the distance from the point of suspension to a point P in which the total mass must be assumed concentrated to keep the frequency unchanged.

For a slender uniform bar of length L, suspended at one end point, we find that $l = r_0^2/c = (L^2/3)/(L/2) = \frac{2}{3}L$. Introducing $I_0 = I_C + Mc^2$, or $Mr_0^2 = Mr_C^2 + Mc^2$, we have $r_0^2 = r_C^2 + c^2$, so that $l = r_C^2/c + c > c$. The point P is evidently further away from the centre of suspension than is the centre of mass. The point P is called the *centre of percussion*. The theory of the compound pendulum may be used to determine the moment of inertia I_C of a rigid body, as shown in the next example.

Example 6.15

Determine the moment of inertia I_C of the connecting rod shown in Fig. 6.13(a)

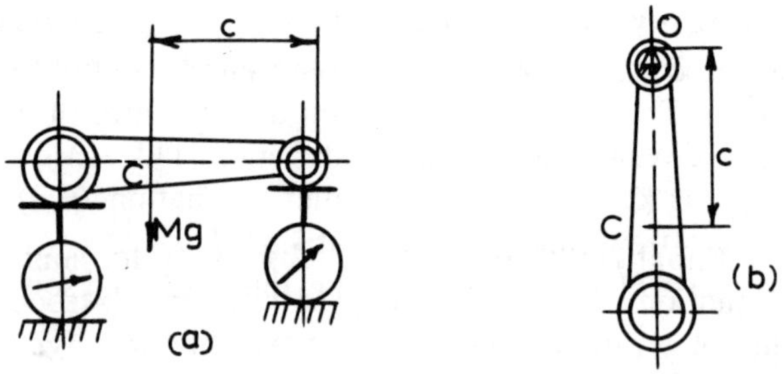

Fig. 6.13.

Solution
The rod is arranged on two scales, with horizontal axis of symmetry, as shown in Fig. 6.13(a). In this way, the total mass M of the rod and the position of the centre of mass C are determined. The distance c as shown is found from statics. Suspending the rod on a knife edge as shown in Fig. 6.13(b), the time for a number of swings of the rod may be measured, and we have $f = (1/2\pi)\sqrt{(cg/r_0^2)}$, which determines r_0^2 and $I_0 = Mr_0^2$. The moment of inertia I_C may now be determined from $I_0 = I_C + Mc^2$.

Example 6.16
Fig. 6.14 shows a compound pendulum at rest. A horizontal force F is suddenly

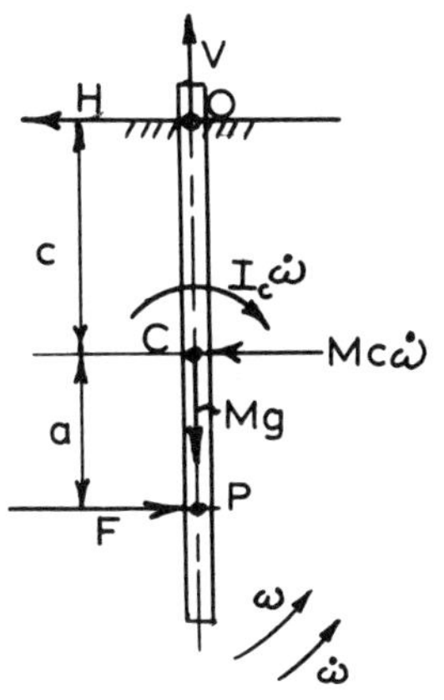

Fig. 6.14.

applied in an impact as shown. Determine the distance a so that the horizontal reaction at O vanishes, and show that P must be the centre of percussion. The vertical plane is a plane of symmetry.

Solution
At the instant considered, the angular velocity $\omega = 0$, while we have a value for the angular acceleration $\dot{\omega}$. The equations of motion of the centre of mass are determined by Newton's law. We find in the vertical direction (1) $V - Mg = Mc\omega^2$, and in the horizontal direction (2) $F - H = Mc\dot{\omega}$. Euler's equation (6.2) leads to (3): $F(c + a) = I_0\dot{\omega}$. Since $\omega \simeq 0$, we find from (1) that $V = Mg$.

Equation (3) gives $\dot{\omega} = F(c + a)/I_0$; substituting this in (2) and solving for H gives the result

$$H = \frac{F}{I_0}[I_0 - Mc(c + a)]$$

Substituting $I_0 = I_C + Mc^2$ leads to

$$H = \frac{F}{I_0}[I_C - Mca]$$

To have $H = 0$, we find $I_C - Mac = 0$, or $a = I_C/Mc = Mr_C^2/Mc = r_C^2/c$. The total distance from point O to the point of impact P is now $c + r_C^2/c$, so that P is the centre of percussion.

In some practical cases of impact-testing machines and hammers, the machinery is designed so that the impact force is close to or through the centre of percussion, in order to minimize the effects of the impact forces on the bearings.

Example 6.17
Fig. 6.15 shows a rigid uniform homogeneous bar of length l and mass M. The bar is

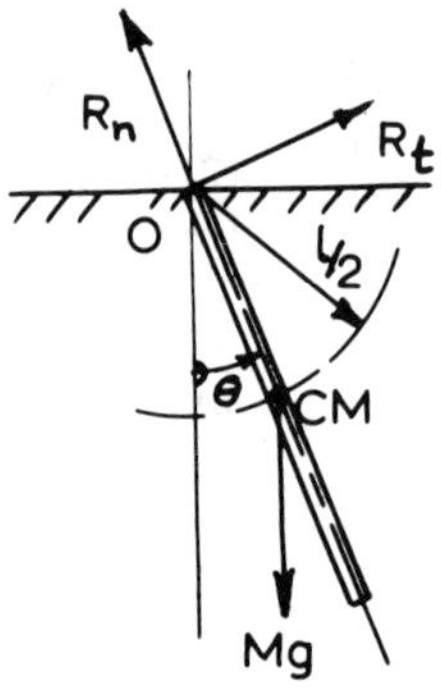

Fig. 6.15.

suspended as a compound pendulum. Determine the components of the reaction on the bar at point O in terms of the angle θ and the mass M. The bar is started from rest when $\theta = \theta_o$.

Solution
The moment of inertia of the bar about the axis of suspension is $I_0 = \frac{1}{3}Ml^2$. The centre of mass is in circular motion with normal acceleration $(l/2)\dot{\theta}^2$ towards O and tangential acceleration $(l/2)\ddot{\theta}$ perpendicular to the bar. Introducing the reaction components at O as R_n along the bar and R_t perpendicular to the bar, Newton's law for the motion of the centre of mass gives the equations

$$(1)\ R_n - Mg\cos\theta = M\frac{l}{2}\dot{\theta}^2$$

$$(2)\ R_t - Mg \sin\theta = M\frac{l}{2}\ddot{\theta}$$

Euler's equation gives the result

$$(3)\ -Mg\frac{l}{2}\sin\theta = \tfrac{1}{3}Ml^2\ddot{\theta}$$

Equation (3) gives $\ddot{\theta} = -(3g/2l)\sin\theta$, and substituting in (2) gives the result $R_t = \tfrac{1}{4}Mg\sin\theta$.

Multiplying (3) by $d\theta$ gives

$$d\theta\frac{d\dot{\theta}}{dt} = -\frac{3g}{2l}\sin\theta\,(d\theta)$$

or

$$d\dot{\theta}^2 = -\frac{3}{l}g\sin\theta\,d\theta \qquad \dot{\theta}^2 = -\frac{3g}{l}\int\sin\theta\,d\theta + C = \frac{3g}{l}\cos\theta + C$$

When $\theta = \theta_0$, $\dot{\theta} = 0$ from which

$$C = -\frac{3g}{l}\cos\theta_0 \quad \text{and} \quad \dot{\theta}^2 = \frac{3g}{l}(\cos\theta - \cos\theta_0)$$

This result may also be determined by the work-energy equation $\text{W.D.}|_1^2 = T_2 - T_1$. Taking position 1 at the start when $\dot{\theta} = 0$, we have $T_1 = 0$; at position 2 at the angle θ we have

$$T_2 = \tfrac{1}{2}I_0\dot{\theta}^2 = \tfrac{1}{6}Ml^2\dot{\theta}^2$$

The only work done is by the gravity force which moves down a distance $(l/2)(\cos\theta - \cos\theta_0)$.

The work-energy equation now gives

$$Mg\frac{l}{2}(\cos\theta - \cos\theta_0) = \tfrac{1}{6}Ml^2\dot{\theta}^2$$

from which $\dot{\theta}^2$ is determined as above. Substituting in (1) leads to

$$R_n = \frac{Mg}{2}(5\cos\theta - 3\cos\theta_0)$$

The vertical and horizontal reactions, if required, are determined by taking the sum of the projections of R_n and R_t on the vertical and horizontal directions.

Example 6.18
Fig. 6.16 shows a slender, rigid rod pinned at O, with a bob of mass m_1 attached at the

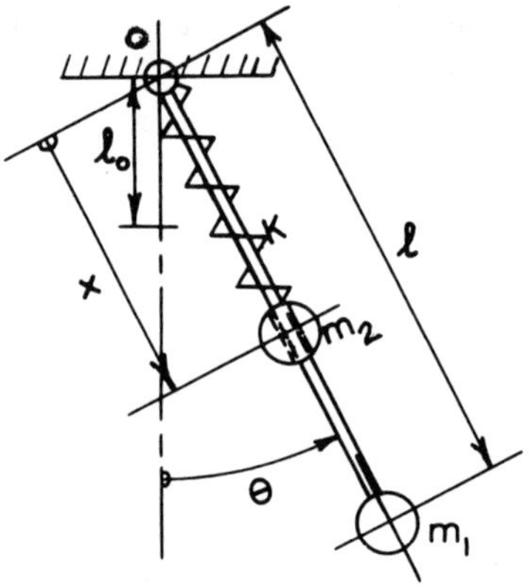

Fig. 6.16.

lower end. The rod may be considered massless, and is free to rotate about O in the vertical plane. A second bob of mass m_2 is free to slide along the smooth rod under the action of gravity and of the spring of constant K. The unstretched length of the spring is l_0.

Determine the equations of motion of the system, and consider the special case where m_2 is fixed to the rod at a distance l_1 from O, and the case where the rod is fixed in a position where $\theta = \beta$.

Solution
Although the equations of motion may be established by applying Newton's law and Euler's equation, it is much simpler to use Lagrange's equations. The generalised coordinates may be taken as x and θ as shown on the figure. The velocity of the bob of mass m_1 is $V_1 = l\dot{\theta}$. The velocity of the other mass is $\mathbf{V}_2 = \mathbf{V}_b + \mathbf{V}_r$; $|\mathbf{V}_b| = x\dot{\theta}$ perpendicular to the bar, and $|\mathbf{V}_r| = \dot{x}$ along the bar so that $V_2^2 = x^2\dot{\theta}^2 + \dot{x}^2$, and

$$T = \tfrac{1}{2}m_1\, l^2\dot{\theta}^2 + \tfrac{1}{2}m_2(\dot{x}^2 + x^2\dot{\theta}^2)$$

Hence

$$\frac{\mathrm{d}}{\mathrm{d}t}\frac{\partial T}{\partial \dot{x}} = m_2\ddot{x} \qquad \frac{\partial T}{\partial x} = m_2\dot{\theta}^2 x$$

$$\frac{\partial T}{\partial \dot{\theta}} = m_1 l^2\dot{\theta} + m_2 x^2\dot{\theta} \qquad \frac{\mathrm{d}}{\mathrm{d}t}\frac{\partial T}{\partial \dot{\theta}} = m_1 l^2\ddot{\theta} + m_2(x^2\ddot{\theta} + 2x\dot{x}\dot{\theta}) \qquad \frac{\partial T}{\partial \theta} = 0$$

To find Q_x, we keep θ constant and find the work done by all external forces on a small increment δx. *The spring force is dealt with by removing the spring and considering the spring force as an external force.* At position x, the spring force is $K(x-l_0)$ and $\delta(\mathrm{W.D.})_s = -K(x-l_0)\delta x$.

The work done by gravity is $\delta(\mathrm{W.D.})_g = m_2 g \cos\theta\, \delta x$. In total then

$$\delta(\mathrm{W.D.}) = -K(x-l_0)\delta x + m_2 g\cos\theta\,\delta x = Q_x\delta x$$

and $Q_x = m_2 g\cos\theta - K(x-l_0)$.

Keeping x constant and giving θ an increment $\delta\theta$, we find $\delta(\mathrm{W.D.}) = -m_2 gx\sin\theta\,\delta\theta - m_1 gl\sin\theta\,\delta\theta = Q_\theta\,\delta\theta$, or $Q_\theta = -m_2 gx\sin\theta - m_1 gl\sin\theta$. Substituting in Lagrange's equations gives the equations of motion

$$m_2\ddot{x} - m_2\dot{\theta}^2 x + K(x-l_0) - m_2 g\cos\theta = 0 \qquad \text{(a)}$$

$$(m_1 l^2 + m_2 x^2)\ddot{\theta} + 2m_2 x\dot{x}\dot{\theta} + g\sin\theta(m_2 x + m_1 l) = 0 \qquad \text{(b)}$$

If m_2 is fixed to the bar a distance l_1 from O, we have $x = l_1$, $\dot{x} = \ddot{x} = 0$. Equation (b), for the coordinate θ, then gives the equation of motion

$$(m_1 l^2 + m_2 l_1^2)\ddot{\theta} + (m_1 l + m_2 l_1)g\sin\theta = 0$$

This is the usual equation for rotation about a fixed axis of a rigid body with moment of inertia $m_1 l^2 + m_2 l_1^2$, and restoring torque $(m_1 l + m_2 l_1)g\sin\theta$.

If $\theta = \beta$, $\dot{\theta} = \ddot{\theta} = 0$, equation (a), for the coordinate x, gives the equation of motion

$$m_2\ddot{x} + K(x-l_0) - m_2 g\cos\beta = 0$$

or the usual type of equation for linear motion along a fixed x-axis. Since the system is conservative, we may also use the potential energy V.

Taking the potential energy $V=0$ when the bar is vertical and the spring unstretched, we find

$$V = m_1 gl(1-\cos\theta) + \tfrac{1}{2}K(x-l_0)^2 - m_2 g(x\cos\theta - l_0)$$

$$L = T - V = \tfrac{1}{2}m_1 l^2\dot{\theta}^2 + \tfrac{1}{2}m_2(\dot{x}^2 + x^2\dot{\theta}^2)$$

$$-m_1gl(1-\cos\theta)-\tfrac{1}{2}K(x-l_0)^2+m_2g(x\cos\theta-l_0)$$

We find from this that

$$\frac{\partial L}{\partial \dot{x}} = m_2\dot{x} \qquad \frac{\mathrm{d}}{\mathrm{d}t}\frac{\partial L}{\partial \dot{x}} = m_2\ddot{x} \qquad \frac{\partial L}{\partial \dot{\theta}} = m_1l^2\dot{\theta}+m_2x^2\dot{\theta}$$

$$\frac{\mathrm{d}}{\mathrm{d}t}\frac{\partial L}{\partial \dot{\theta}} = m_1l^2\ddot{\theta}+m_2x^2\ddot{\theta}+2m_2x\dot{x}\dot{\theta}$$

$$\frac{\partial L}{\partial x} = m_2x\dot{\theta}^2-K(x-l_0)+m_2g\cos\theta$$

$$\frac{\partial L}{\partial \theta} = -m_1gl\sin\theta-m_2gx\sin\theta$$

Substituting in Lagrange's equations gives directly the two equations of motion that were found above.

6.12 GRAVITY TORQUE. STATIC BALANCE OF A ROTOR

A rigid rotor may be placed in just two bearings. Such a rotor is shown in Fig. 6.17(a) in bearings A and B. An enlarged end view of the rotor is shown in Fig. 6.17(b). The

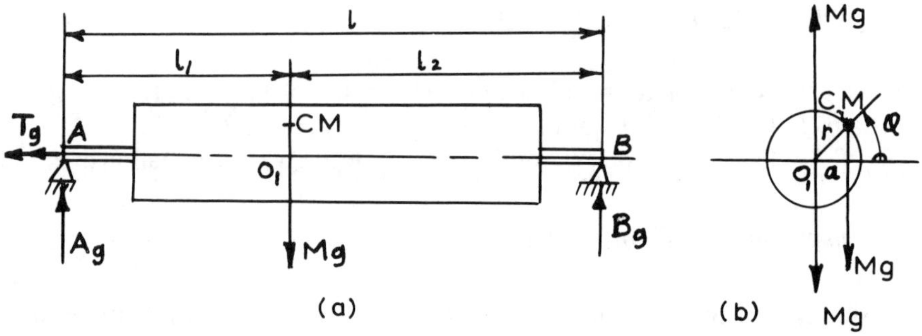

Fig. 6.17.

centre of mass is at a distance r from the axis of rotation, and with total mass of the rotor equal to M, the gravity force Mg produces a moment $T_g = Mga = Mgr\cos\phi$ as shown about the axis of rotation; this moment is clockwise. The gravity force will rotate the rotor until the centre of mass is in the lowest position vertically below O_1. To keep the rotor in the position shown we may apply a pure torque from outside about the axis of rotation equal and opposite to the gravity torque; in the given position this torque would have to be of magnitude $Mgr\cos\phi$ and applied in the anti-clockwise direction. Such a torque does not give any bearing reactions. We may now apply two equal and opposite forces of magnitude Mg at the centre of rotation O_1 as shown in Fig. 6.17(b); taking the gravity force at the centre of mass together with the vertical force upwards at O_1 gives a couple equal to the gravity torque T_g balanced by the externally applied torque; the only force is now the central downward force Mg at O_1, and the reactions at the bearings A and B, to give a force balance for the rotor, may now be determined from simple statics moment equations; the result is $A_g = Mg(l_2/l)$ and $B_g = Mg(l_1/l)$, with $A_g + B_g = Mg$.

A_g and B_g are called the *static bearing reactions*; they are *always vertical* and of *constant* magnitude and independent of the angular position of the rotor. If the centre of mass is *on the axis of rotation* we have $r = 0$ and $T_g = 0$ for all angular positions of the rotor; in this case the rotor will stay in any angular position in which it may be placed, and we say that the rotor is in *static balance*.

6.13 EQUATIONS OF MOTION OF THE CENTRE OF MASS OF A RIGID ROTOR

To rotate the rotor at an angular acceleration $\dot{\omega}$ about the centre line, we find from Euler's equation (6.2) that the necessary external pure torque that must be applied is $T_q = T_g + I_z\dot{\omega}$, where I_z is the moment of inertia of the rotor about the central axis which we take as the z-axis.

To determine the equations of motion of the centre of mass we have Newton's law, and we move all acting forces to the centre of mass; the gravity force is now balanced by the static reactions and the pure torque about the axis may be considered as a couple, that is two equal and opposite forces; all these forces are in balance at the centre of mass, but since the centre of mass is in circular motion there must now be additional bearing reactions to produce the necessary acceleration of the centre of mass.

These additional reactions depend on the angular velocity and acceleration and are called *dynamic reactions*.

To determine the dynamic reactions, we may now *omit* the gravity force, except for its inclusion in the external driving torque. Introducing an xy coordinate system *fixed in the rotor* at bearing B, the situation is as shown in Fig. 6.18, where the dynamic reactions have been introduced by their components.

To determine the acceleration $\mathbf{A}_C$ of the centre of mass, the situation is as shown in Fig. 6.19.

We have $\mathbf{A}_C = \boldsymbol{\omega} \times (\boldsymbol{\omega} \times \mathbf{r}) + \dot{\boldsymbol{\omega}} \times \mathbf{r}$, with $\boldsymbol{\omega} = \omega\mathbf{k}$, $\dot{\boldsymbol{\omega}} = \dot{\omega}\mathbf{k}$ and $\mathbf{r} = x_C\mathbf{i} + y_C\mathbf{j}$.

Using the formula

$$\mathbf{a} \times (\mathbf{b} \times \mathbf{c}) = (\mathbf{a} \cdot \mathbf{c})\mathbf{b} - (\mathbf{a} \cdot \mathbf{b})\mathbf{c}$$

we find for the first term

$$\boldsymbol{\omega} \times (\boldsymbol{\omega} \times \mathbf{r}) = (\boldsymbol{\omega} \cdot \mathbf{r})\boldsymbol{\omega} - (\boldsymbol{\omega} \cdot \boldsymbol{\omega})\mathbf{r} = 0 \cdot \boldsymbol{\omega} - \omega^2(\mathbf{r}) = -x_C\omega^2\mathbf{i} - y_C\omega^2\mathbf{j}$$

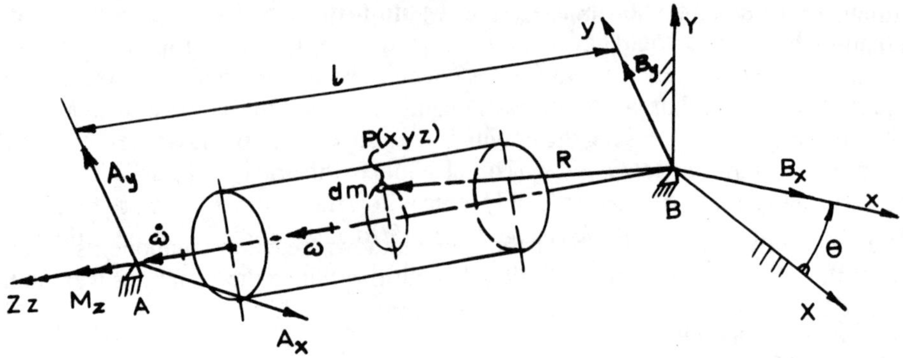

Fig. 6.18.

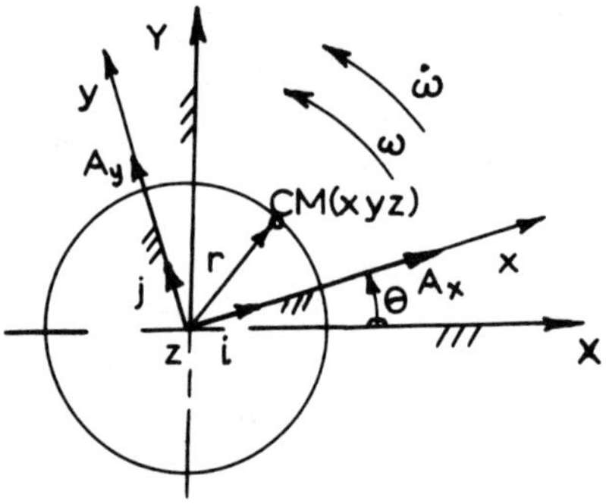

Fig. 6.19.

For the second term in the acceleration we have

$$\dot{\boldsymbol{\omega}} \times \mathbf{r} = \dot{\omega}\mathbf{k} \times (x_C\mathbf{i} + y_C\mathbf{j}) \;=\; x_C\dot{\omega}\mathbf{k} \times \mathbf{i} + y_C\dot{\omega}\mathbf{k} \times \mathbf{j} \;=\; -y_C\dot{\omega}\mathbf{i} + x_C\dot{\omega}\mathbf{j}$$

Accumulating the **i** and **j** terms, the components of $\mathbf{A}_C$ are

$$\left.\begin{aligned} A_{Cx} &= -(x_C\omega^2 + y_C\dot{\omega}) \\ A_{Cy} &= (x_C\dot{\omega} - y_C\omega^2) \\ A_{C_z} &= 0 \end{aligned}\right\} \qquad (6.15)$$

Summing the dynamic reactions in the x and y directions, Newton's law now gives the equations

$$\left.\begin{aligned} A_x + B_x &= -M(x_C\omega^2 + y_C\dot{\omega}) \\ A_y + B_y &= M(x_C\dot{\omega}^2 - y_C\omega^2) \end{aligned}\right\} \qquad (6.16)$$

We have only two equations for the four unknown reaction components, and we must therefore establish two additional equations. Since the equations were determined by *summation* of the *forces*, we determine two additional equations by taking *moments* of the forces about the x- and y-axis respectively. For the moment equations we need to develop a set of equations called *Euler's equations*.

6.14 EULER'S EQUATIONS FOR A RIGID ROTOR

Fig. 6.18 shows an element dm of a rotor at a point P with coordinates (x,y,z), and position vector $\mathbf{R}$ from point B, where $\mathbf{R} = x\mathbf{i} + y\mathbf{j} + z\mathbf{k}$.

The acceleration components of point P are from (6.15),

$$A_x = -(x\omega^2 + y\dot{\omega}) \quad A_y = (x\dot{\omega} - y\omega^2) \quad \text{and} \quad A_z = 0$$

Taking $\mathbf{F}$ as the resultant of all internal and external forces on the element dm, we have from Newton's second law

$$\mathbf{F} = (\mathrm{d}m)\mathbf{A} = A_x(\mathrm{d}m)\mathbf{i} + A_y(\mathrm{d}m)\mathbf{j}$$

Taking the moment $\mathbf{m}_B$ of $\mathbf{F}$ in point B, we obtain

$$\mathbf{m}_B = \mathbf{R} \times \mathbf{F} = \begin{vmatrix} \mathbf{i} & \mathbf{j} & \mathbf{k} \\ x & y & z \\ A_x\mathrm{d}m & A_y\mathrm{d}m & 0 \end{vmatrix}$$

$$= -z\,A_y\mathrm{d}m\mathbf{i} + zA_x\mathrm{d}m\mathbf{j} + (xA_y - yA_x)\mathrm{d}m\,\mathbf{k}$$

The components of $\mathbf{m}_B$ are the moments of $\mathbf{F}$ about the x-, y- and z-axis respectively.

Summing up for all elements of the rotor, the internal forces occur in equal and opposite pairs, and therefore cancel in the summation. The summation then gives the moments of all *external* forces about each axis.

Taking the moments as M_x, M_y and M_z respectively, and extending the summation over all elements, there results

$$M_x = \int(-zA_y)\mathrm{d}m = \int(zy\omega^2 - zx\dot{\omega})\mathrm{d}m = \omega^2\int yz\mathrm{d}m - \dot{\omega}\int xz\ \mathrm{d}m$$

$$M_y = \int zA_x\mathrm{d}m = -\int(xz\omega^2 + yz\dot{\omega})\mathrm{d}m = -\omega^2\int xz\ \mathrm{d}m - \dot{\omega}\int yz\ \mathrm{d}m$$

$$M_z = \int(xA_y - yA_x)\mathrm{d}m = \int(x^2\dot{\omega} - xy\omega^2 + xy\omega^2 + y^2\dot{\omega})\mathrm{d}m$$

$$= \dot{\omega}\int(x^2 + y^2)\mathrm{d}m$$

Introducing the products of inertia $I_{xz} = \int xz\mathrm{d}m$, $I_{yz} = \int yz\mathrm{d}m$ and the moment of inertia $I_z = \int(x^2 + y^2)\mathrm{d}m$, we obtain the three equations

$$\left.\begin{aligned} M_x &= I_{yz}\omega^2 - I_{xz}\dot{\omega} \\ M_y &= -I_{xz}\omega^2 - I_{yz}\dot{\omega} \\ M_z &= I_z\dot{\omega} \end{aligned}\right\} \qquad (6.17)$$

Since the xyz system is *fixed in the rotor*, the products and moments of inertia about those axes are all *constant*. The equations (6.17) are called *Euler's equations* after the Swiss mathematician who developed them in 1736. The xyz system fixed in the rotor is often called an Euler system. If the axis of rotation is a principal axis at point B, we have $I_{xz} = I_{yz} = 0$, and the equations take the simple form

$$M_x = 0 \quad M_y = 0 \quad \text{and} \quad M_z = I_z\dot{\omega}$$

The third equation may be recognised as being the Euler equation (6.2); this is the *equation of motion* of the rotor. The first two equations are necessary to determine the forces that keep the rotor axis stationary.

The Euler equations (6.17) may also be developed by using the moment of momentum vector **H**. For the rotor in Fig. 6.18, we have $\omega_x = \omega_y = 0$, and $\omega_z = \omega$; substituting this in the formula (5.27), we find that the components of **H** are

$$H_x = -I_{xz}\omega \quad H_y = -I_{yz}\omega \quad \text{and} \quad \mathrm{H_z} = \mathrm{I_z}\omega$$

We have now

$$\mathbf{H} = H_x\mathbf{i} + H_y\mathbf{j} + H_z\mathbf{k}$$

and differentiating this in absolute terms we find

$$\dot{\mathbf{H}} = \dot{H}_x\mathbf{i} + H_x\dot{\mathbf{i}} + \dot{H}_y\mathbf{j} + H_y\dot{\mathbf{j}} + \dot{H}_z\mathbf{k} + H_z\dot{\mathbf{k}}$$

From the components of **H** we find

$$\dot{H}_x = -I_{xz}\dot{\omega} \quad \dot{H}_y = -I_{yz}\dot{\omega} \quad \text{and} \quad \dot{H}_z = I_z\dot{\omega}$$

We also have

$$\dot{\mathbf{i}} = \boldsymbol{\omega}\times\mathbf{i} = \omega\mathbf{k}\times\mathbf{i} = \omega\mathbf{j}$$

$$\dot{\mathbf{j}} = \boldsymbol{\omega}\times\mathbf{j} = \omega\mathbf{k}\times\mathbf{j} = -\omega\mathbf{i}$$

and

$$\dot{\mathbf{k}} = \omega\times\mathbf{k} = \omega\mathbf{k}\times\mathbf{k} = \mathbf{0}$$

Substituting these expression we obtain

$$\dot{\mathbf{H}} = (\dot{H}_x - H_y\omega)\mathbf{i} + (\dot{H}_y + H_x\omega)\mathbf{j} + \dot{H}_z\mathbf{k}$$

$$= (-\mathbf{I}_{xz}\dot{\omega} + I_{yz}\omega^2)\mathbf{i} + (-I_{yz}\dot{\omega} - I_{xz}\omega^2)\mathbf{j} + I_z\dot{\omega}\mathbf{k}$$

From equation (6.12) we have $\mathbf{M} = \dot{\mathbf{H}}$, so that the scalar components of the total external moment in point B are

$$\left.\begin{aligned} M_x &= I_{yz}\omega^2 - I_{xz}\dot{\omega} \\ M_y &= -I_{xz}\omega^2 - I_{yz}\dot{\omega} \\ M_z &= I_z\dot{\omega} \end{aligned}\right\}$$

These are the same as the Euler equations (6.17).

6.15 DYNAMIC REACTIONS. DYNAMIC BALANCE

For the rotor in Fig. 6.18, we may now determine the moment M_x of the forces about the x-axis, and the moment M_y about the y-axis. In determining these moments it is convenient to remember that a force has *no moment* about an axis if it is *on* the axis, *parallel* to the axis or *intercepts* the axis.

We find now $M_x = -A_y l$, and $M_y = +A_x l$, where

$$M_x = I_{yz}\omega^2 - I_{xz}\dot{\omega} \quad \text{and} \quad M_y = -I_{xz}\omega^2 - I_{yz}\dot{\omega}$$

from the Euler equations (6.17). We now have the following four equations for the acting dynamic reactions from (6.16) and (6.17):

$$A_x + B_x = -M(x_C\omega^2 + y_C\dot{\omega}$$

$$A_y + B_y = M(x_C\dot{\omega} - y_C\omega^2) \tag{6.18}$$

$$-A_y l = I_{yz}\omega^2 - I_{xz}\dot{\omega}$$

$$A_x l = -I_{xz}\omega^2 - I_{yz}\dot{\omega}$$

If the centre of mass is on the axis of rotation, we have *static balance*; with $x_C = y_C = 0$, we find from (6.18) that $A_x = -B_x$ and $A_y = -B_y$, so that the dynamic reactions form a couple that rotates with the rotor. To eliminate all the dynamic reactions, so that

$$A_x = B_x = A_y = B_y = 0$$

we must have both

$$x_C = y_C = 0 \quad \text{and} \quad I_{xz} = I_{yz} = 0$$

For complete *static* and *dynamic* balance, the axis of rotation must be a *principal axis through the centre of mass*.

The rotating Euler axes fixed in the rotor have the great advantage that all moments and products of inertia are *constant*. A disadvantage is that the dynamic reactions are determined along these axes, while we are usually more interested in the vertical and horizontal components of the bearing reactions.

For the bearing at A, the situation is as shown in Fig. 6.19, where *fixed* horizontal and vertical axes X and Y have been introduced. For the rotor in an angular position θ as shown, we determine the horizontal A_X and vertical A_Y components of the *total* bearing reactions by projecting A_x and A_y on these axes; this gives the result

$$\left.\begin{aligned} A_X &= A_x \cos\theta - A_y \sin\theta \\ A_Y &= A_x \sin\theta + A_y \cos\theta \end{aligned}\right\} \tag{6.19}$$

The *static reaction* A_g must be added to the *vertical* component. If the direction and magnitude of the total reaction in position θ are required, the reactions A_x, A_y and A_g must be added *vectorially*, or the resultant of A_X and A_Y must be determined, including A_g.

For bearing B we use the same procedures as for bearing A. The formula gives the *bearing reactions*, that is the forces as they act *on the rotor*; *if we require the bearing pressures*, all the forces must be reversed in direction, but with the same magnitudes.

Each rotor has, with a given angular velocity, a *unique set* of dynamic reactions rotating with the rotor. These reactions are independent of the position of the *xyz* system, and we therefore fix the system in the rotor in the *most convenient position* where the products of inertia I_{xz} and I_{yz} are known or can easily be determined in the simplest way possible.

Example 6.19
Fig. 6.20 shows a horizontal rotor in two bearings A and B. The rotor consists of a

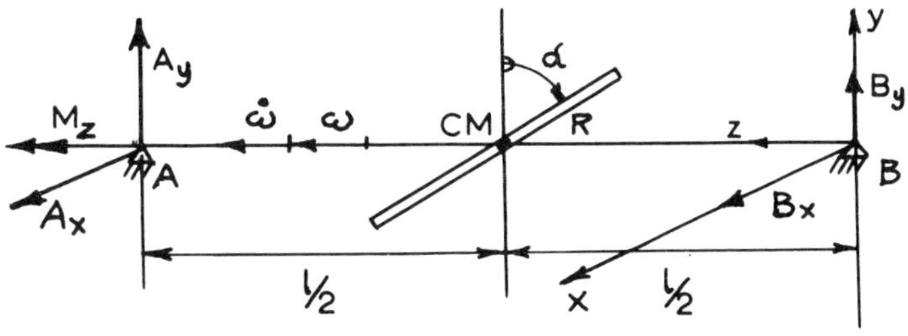

Fig. 6.20.

light, rigid shaft to which a circular disc of mass M and radius R is rigidly attached at the midpoint of the shaft. Instead of being placed square on the rotor, the disc has been rotated through an angle α as shown. The centre of mass of the disc is on the rotor centreline.

A coordinate system *xyz* is fixed in the rotor. The rotor is rotated by an external torque M_z, and at the moment considered the angular velocity is ω and the angular acceleration $\dot{\omega}$, and the x-axis is horizontal.

Determine the *total* vertical and horizontal bearing reactions at this instant.

Solution
The dynamic bearing reactions are determined by equations (6.18). Since the centre of mass is on the axis of rotation we have $x_C = y_C = 0$.

From symmetry we have $I_{xz} = 0$, while I_{yz} and I_z may be taken from Example 5.3, with the result that

$$I_{yz} = \frac{-MR^2}{8}\sin 2\alpha \quad \text{and} \quad I_z = \frac{MR^2}{4}(1+\cos^2\alpha)$$

Substituting in (6.18) results in

$$A_x + B_x = 0 \quad A_y + B_y = 0 \quad A_y l = \frac{MR^2\omega^2}{8}\sin 2\alpha$$

and

$$A_x l = \frac{MR^2\dot{\omega}}{8} \sin 2\alpha$$

The solution is

$$A_x = -B_x = \frac{MR^2\dot{\omega}}{8l}\sin 2\alpha \quad \text{and} \quad A_y = -B_y = \frac{MR^2\omega^2}{8l} \sin \alpha$$

A_x and B_x are the *total horizontal* reactions. To obtain the total *vertical* reactions we must add $A_g = \frac{Mg}{2} = B_g$ to the vertical *dynamic* reactions; the result is

$$A_{yt} = \frac{MR^2\omega^2}{8l} \sin 2\alpha + \frac{Mg}{2} \quad \text{and} \quad B_{yt} = \frac{-MR^2\omega^2}{8l} \sin 2\alpha + \frac{Mg}{2}$$

The instantaneous torque $M_z = I_z\dot{\omega} = \frac{MR^2\dot{\omega}}{4}(1 + \cos^2\alpha)$.

There are two special cases of interest; if $\alpha = 0$, $\sin 2\alpha = 0$ and the only reactions are the *static reactions* $A_y = B_y = Mg/2$. The disc is square on the shaft and the z-axis is a principal axis, so the system is statically and dynamically balanced.

If $\alpha = 90°$, we have again $\sin 2\alpha = 0$ and only the static reactions. The disc is now attached along a diameter and the z-axis is again a principal axis, with the system in static and dynamic balance.

6.16 GENERAL CASE OF DYNAMIC UNBALANCE OF A ROTOR

Fig. 6.21(a) shows a rigid body rotating about a fixed axis. A series of forces, perpendicular to the axis of rotation, is indicated; these forces are due to inhomogeneity of the material, machining tolerances or special parts attached to the rotor. Each mass concentration is in a circular motion with a constant angular velocity. A constant force directed towards the centre of rotation must thus be acting, and the mass elements therefore introduce forces as shown acting *on* the rotor.

We may reduce the force field to a resultant force **R** at an arbitrary point E along the axis of rotation, by moving all forces to this point and adding them vectorially; in doing this we introduce for each force a moment acting in a plane which passes through the centreline AB and through the force. The moment vectors may be

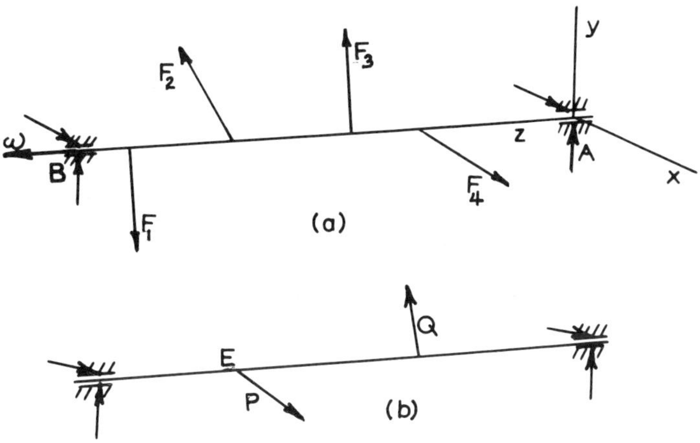

Fig. 6.21.

combined to a resultant moment vector at E; this vector is again perpendicular to AB. A couple may be substituted for the moment vector, and one of the couple vectors combined with **R** to give a force **P** at point E, as shown in Fig. 6.21(b), the other couple vector being **Q**. The force system is thus reduced to two forces **P** and **Q** perpendicular to AB.

Dynamic balance of the rotor may now be obtained by placing two *correction weights* in two arbitrary planes, the correction planes, perpendicular to the axis of rotation, in such a way that the centre of mass of the system falls on the axis of rotation, and the weights set up *equal and opposite* dynamic reactions to those already present.

The position and magnitude of the unbalance of a rotor are generally determined in a balancing machine.

Example 6.20

Fig. 6.22 shows a rigid rotor for which the *dynamic unbalance* has been found to be equivalent to two masses M_1 and M_2 with locations as shown. The rotor is in static balance.

The system is to be dynamically balanced by placing two correction weights of mass M_3 and M_4 in the two correction planes A and B as shown at radial distances r_3 and r_4.

Solution

Placing a coordinate system in the rotor with the z-axis as the axis of rotation, and origin at O, the following system constants are available: M_1, $M_2\, r_1$, $r_2\, \theta_1\, \theta_2$, z_1, z_2. The correction weights are to be placed at known radial distances r_3 and r_4, and at longitudinal positions given by z_3 and, since O is in the correction plane, $z_4 = 0$. The problem now consists in finding the four constants, M_3, M_4, θ_3 and θ_4

We will assume that the correction weights are so small that contributions to I_{xz} and I_{yz} may be calculated as for concentrated masses.

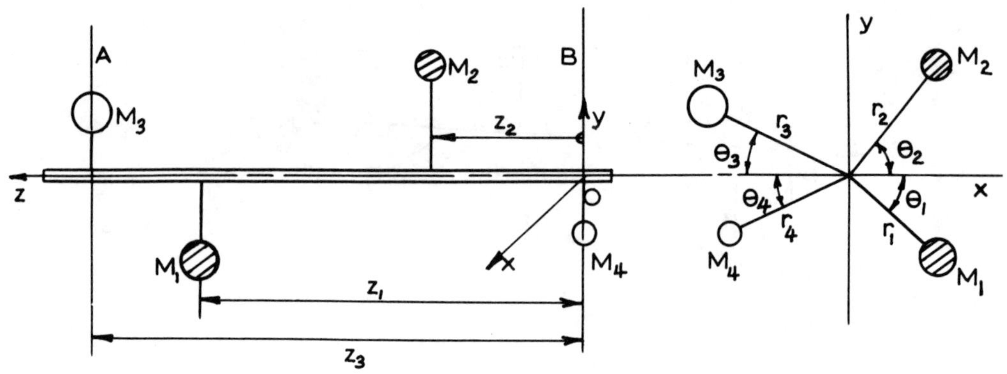

Fig. 6.22.

For dynamical balance of the rotor, the axis of rotation must be a principal axis through the centre of mass, which gives the conditions

$$I_{xz} = 0 \quad \text{or} \quad \sum_1^4 M_n x_n z_n = 0 \tag{a}$$

which gives the equation

$$M_1(r_1 \cos \theta_1)z_1 + M_2(r_2 \cos \theta_2)z_2 - M_3(r_3 \cos \theta_3)z_3 = 0 \tag{a'}$$

$$I_{yz} = 0 \quad \text{or} \quad \sum_1^4 M_n y_n z_n = 0 \tag{b}$$

which gives the result

$$-M_1(r_1 \sin \theta_1)z_1 + M_2(r_2 \sin \theta_2)z_2 + M_3(r_3 \sin \theta_3)z_3 = 0 \tag{b'}$$

These two equations determine M_3 and θ_3.

The condition that the centre of mass must be on the z-axis means that $x_C = 0$ and $y_C = 0$, or $x_C M = \int x \, dm = 0$ and $y_C M = \int y \, dm = 0$.

Since the centre of mass of the system without $M_1, \ldots, M_4$ is already on the z-axis, we need only state that the centre of mass of $M_1, \ldots, M_4$ must be on the z-axis; this gives the equations

$$\sum_1^4 M_n x_n = 0 \tag{c}$$

or

$$M_1 r_1 \cos \theta_1 + M_2 r_2 \cos \theta_2 - M_3 r_3 \cos \theta_3 - M_4 r_4 \cos \theta_4 = 0 \qquad \text{(c}')$$

and

$$\sum_1^4 M_n y_n = 0 \qquad \text{(d)}$$

or

$$-M_1 r_1 \sin \theta_1 + M_2 r_2 \sin \theta_1 + M_3 r_3 \sin \theta_3 - M_4 r_4 \sin \theta_4 = 0 \qquad \text{(d}')$$

These two equations determine M_4 and θ_4.

The theory of balancing works well for short, rigid rotors. For long, flexible rotors, it cannot be applied, since such rotors can generally only be balanced at one particular speed and temperature condition. Balancing is achieved in this case by 'trial and error' methods under field conditions.

PROBLEMS

6.1 Derive the equation of motion for the system in Fig. 6.23. Find the frequency for small vibrations.

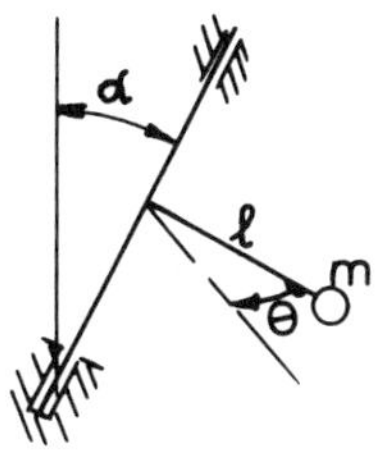

Fig. 6.23.

6.2 The straight pipe in Fig. 6.24 discharges 265 l of water per minute from each end, while rotating with constant angular velocity of 600 rev/min. Determine the magnitude of the driving torque to maintain this uniform rotation.

The weight of water may be taken as 9.78 N/l, and 1 l $= 10^{-3}$ m^3. Wind resistance and friction may be neglected.

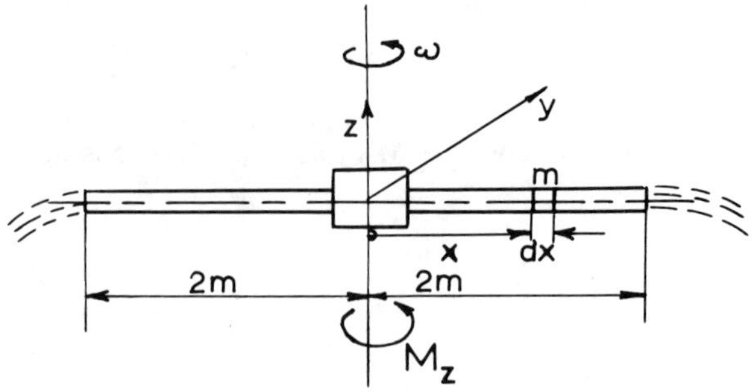

Fig. 6.24.

6.3 A rotor with moment of inertia I_1 (Fig. 6.25) is driven at a constant angular velocity ω_1. It is brought into contact with a second rotor I_2 which is initially at rest. The contact pressure is a constant normal force P between the rotors and the coefficient of friction between the rotors is μ. Find the time of slipping.

If the outside torque is removed from I_1 at the moment of contact, what is the new time of slipping and the final angular velocity of I_1? Neglect bearing friction.

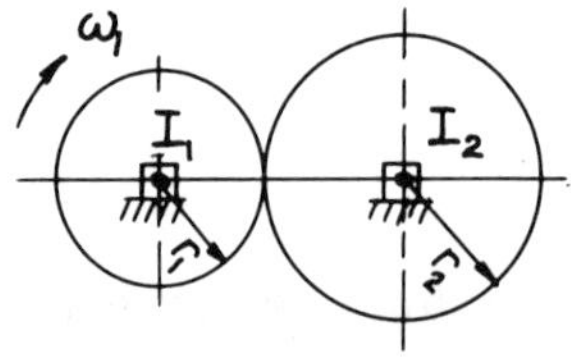

Fig. 6.25.

6.4 A slender rigid uniform bar of length l and mass M is rigidly attached at its midpoint to a horizontal shaft in bearings A and B as shown in Fig. 6.26. Shaft AB may be assumed massless and is secured in the longitudinal direction at bearing A only.

A torque **T** is applied to shaft AB as shown, and at the instant considered the angular velocity of the shaft is $\boldsymbol{\omega}$.

(a) Determine the components of the reactions on the shaft at A and B in the inertial systems shown, expressed as functions of the given system constants and $\boldsymbol{\omega}$ and $\dot{\boldsymbol{\omega}}$.
(b) Find an expression for the torque T as a function of the system constants and $\dot{\omega}$.

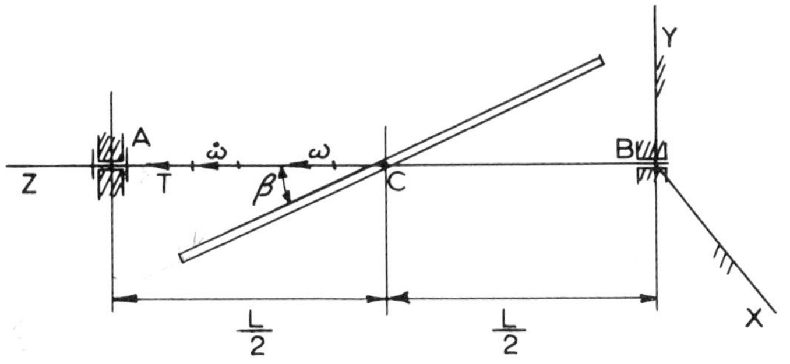

Fig. 6.26.

6.5 In Fig. 6.27 a thin rectangular plate weighing 111 N rotates about axis AB. The plate is restrained in the Z-direction at bearing A only. If a torque of 40.7 N m in the direction of rotation is applied to the shaft on which the plate rotates, what is the angular acceleration at the instant of application of the torque? Find the reactions *on the bearings* at this instant. The plate is in the XZ-plane when the torque is applied and has an angular velocity of 20 rad/s.

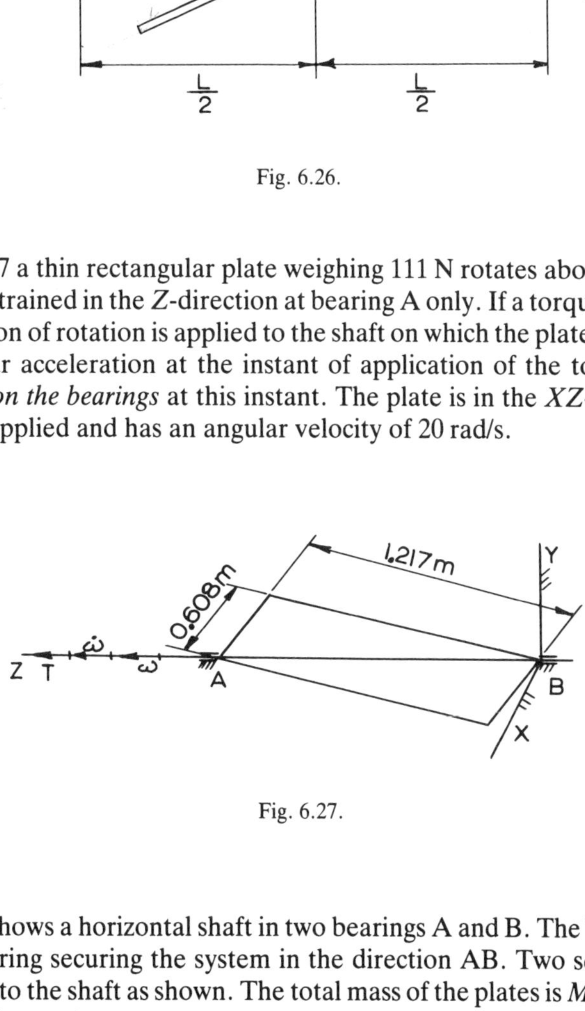

Fig. 6.27.

6.6 Fig. 6.28 shows a horizontal shaft in two bearings A and B. The bearing at A is a thrust bearing securing the system in the direction AB. Two square plates are welded onto the shaft as shown. The total mass of the plates is M, while the mass of the shaft may be neglected.

An external torque **T** rotates the system, and at the instant shown, the plates are in a vertical plane and rotating with angular velocity $\boldsymbol{\omega}$ as shown. Wind resistance and friction may be neglected.

(a) Write Newton's second law for the centre of mass C.
(b) Determine the products of inertia I_{xz} and I_{yz} and the moment of inertia I_z of

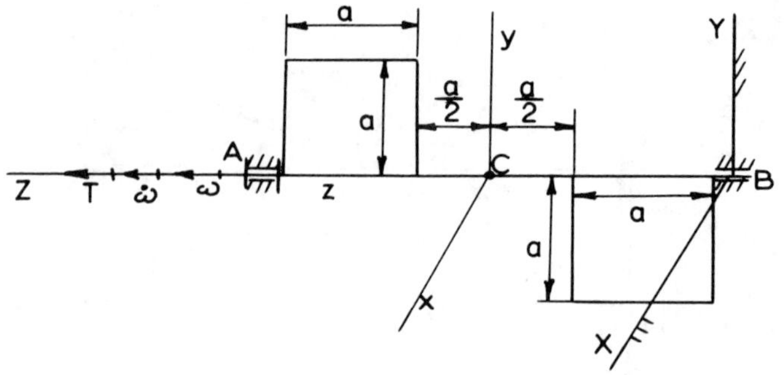

Fig. 6.28.

the system with respect to the Euler system (x, y, z) shown at the centre of mass.

(c) Determine the angular acceleration $\dot{\omega}$ and the components of the bearing reactions in terms of the given system constants T, ω, a and M.

6.7 Fig. 6.29 shows a small rigid rotor in two bearings A and B. The rotor has been examined in a balancing machine, and the unbalance has been found to be equivalent to two masses M_1 and M_2. The mass $M_1 = 113$ g is at a radial distance $r_1 = 12.7$ cm from the rotor centreline, and is located, as shown, in the plane of the figure. $M_2 = 28.4$ g at $r_2 = 15.2$ cm, and M_2 is behind the plane of the paper in Fig. 6.29.

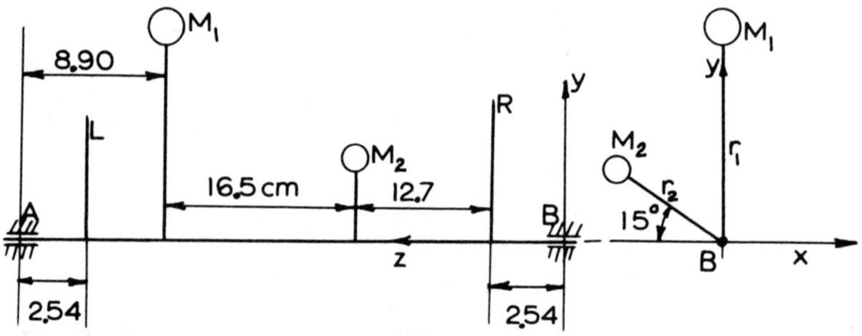

Fig. 6.29.

The rotor is to be dynamically balanced by placing two balance masses M_L and M_R in the balance planes L and R shown in Fig. 6.29; the radial distance is to be $r_L = r_R = 8$ cm. Determine the mass of M_L and M_R and the angular position of each.

7

Plane motion of a rigid body

Plane motion of a rigid body is defined as motion in which a certain plane of the body always remains in a fixed plane. Any point of the body then moves in a plane parallel to the fixed plane. The case of rotation about a fixed axis discussed in Chapter 6 is clearly a special case of plane motion.

Plane motion is perhaps the most important type of motion encountered in engineering dynamics.

71. EQUATIONS OF PLANE MOTION

Fig. 7.1 shows a body in plane motion in an inertial system (X,Y,Z). Because of the

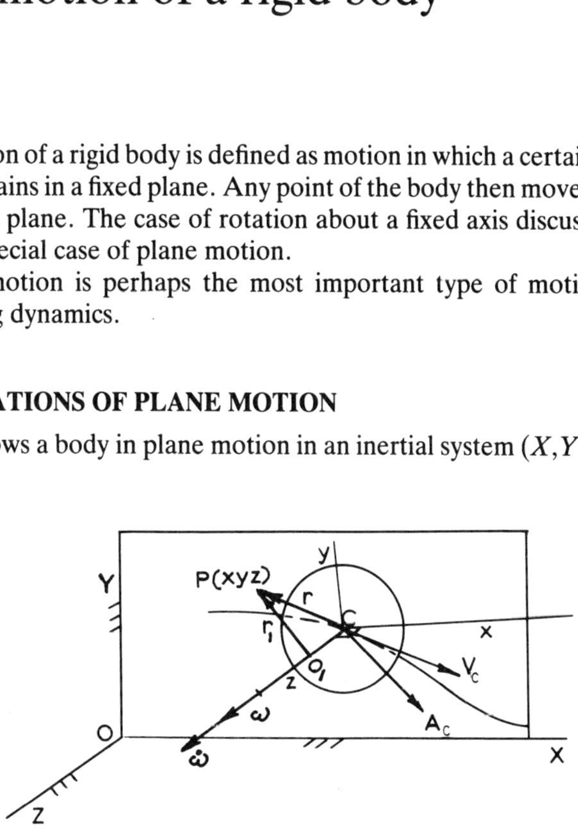

Fig. 7.1.

importance of the centre of mass C of the body, the plane of motion, the XY-plane, has been taken through the centre of mass C.

A body in plane motion has three degrees of freedom: the position of the body may be given, for instance, by the coordinates (X_C,Y_C) of the centre of mass and an angle θ between the X-axis and a line fixed in the body in the XY-plane.

The centre of mass C moves along a curve as shown in the XY-plane, so it is a point in plane curvilinear motion. The angular velocity vector $\boldsymbol{\omega}$ and the angular acceleration vector $\dot{\boldsymbol{\omega}}$ of the body have been located for convenience at the centre of mass, and are both *always perpendicular* to the plane of motion.

The scalar magnitudes of the acceleration components of the centre of mass in the inertial system are $\ddot{X}_C$ and $\ddot{Y}_C$, while $\ddot{Z}_C = 0$. If M is the total mass of the body, Newton's second law gives the force equations of motion

$$F_x = M\ddot{X}_C \quad F_y = M\ddot{Y}_C \quad F_z = 0 \tag{7.1}$$

The first two equations of (7.1) describe the motion of the centre of mass in the XY-plane.

Fixing a coordinate system (x,y,z) in the body as shown with origin at the centre of mass and the z-axis *perpendicular to the plane of motion*, we have

$$\omega_x = \omega_y = 0 \quad \text{and} \quad \boldsymbol{\omega} = \omega_z \mathbf{k}$$

In the same way

$$\dot{\omega}_x = \dot{\omega}_y = 0 \quad \text{and} \quad \dot{\boldsymbol{\omega}} = \dot{\omega}_z \mathbf{k}$$

If the centre of mass has the acceleration $\mathbf{A}_C$ as shown, we have $\mathbf{A}_C = A_{Cx}\mathbf{i} + A_{Cy}\mathbf{j}$, and Newton's second law for the centre of mass may be given the form

$$F_x = MA_{Cx} \quad F_y = MA_{Cy} \quad F_z = 0 \tag{7.2}$$

Considering a point $P(x,y,z)$ in the body as shown in Fig. 7.1, we project this on the z-axis in point O_1.

The vector from O_1 to P is now $\mathbf{r}_1 = x\mathbf{i} + y\mathbf{j}$, and the point O_1 moves exactly as the centre of mass, so that

$$\mathbf{A}_{O_1} = \mathbf{A}_C = A_{Cx}\mathbf{i} + A_{Cy}\mathbf{j}$$

The point P is in plane motion, and using O_1 as a pole, the acceleration $\mathbf{A}$ of point P is $\mathbf{A} = \mathbf{A}_C + \boldsymbol{\omega} \times (\boldsymbol{\omega} \times \mathbf{r}_1) + \dot{\boldsymbol{\omega}} \times \mathbf{r}_1$.

Using the formula $\mathbf{a} \times (\mathbf{b} \times \mathbf{c}) = (\mathbf{a}.\mathbf{c})\mathbf{b} - (\mathbf{a}.\mathbf{b})\mathbf{c}$ on the second term we find $\boldsymbol{\omega} \times (\boldsymbol{\omega} \times \mathbf{r}_1) = (\boldsymbol{\omega}.\mathbf{r}_1)\boldsymbol{\omega} - (\boldsymbol{\omega}.\boldsymbol{\omega})\mathbf{r}_1 = (o)\boldsymbol{\omega} - \omega_z^2(x\mathbf{i} + y\mathbf{j}) = -x\omega_z^2\mathbf{i} - y\omega_z^2\mathbf{j}$.

For the third term we have $\dot{\boldsymbol{\omega}} \times \mathbf{r}_1 = \dot{\omega}_z\mathbf{k} \times (x\mathbf{i} + y\mathbf{j})$

$$= x\dot{\omega}_z\mathbf{k} \times \mathbf{i} + y\dot{\omega}_z\mathbf{k} \times \mathbf{j} = -y\dot{\omega}_z\mathbf{i} + x\dot{\omega}_z\mathbf{j}$$

Accumulating the $\mathbf{i}$ and $\mathbf{j}$ terms, we find the acceleration components of point P:

$$\left.\begin{aligned} A_x &= A_{Cx} - x\omega_z^2 - y\dot{\omega}_z \\ A_y &= A_{Cy} + x\dot{\omega}_z - y\omega_z^2 \\ A_z &= 0 \end{aligned}\right\} \tag{7.3}$$

The resultant of all external and internal forces on an element dm at point P is now $\mathbf{F} = (\mathrm{d}m)\mathbf{A} = (\mathrm{d}m)A_x\mathbf{i} + (\mathrm{d}m)A_y\mathbf{j}$.

The moment of the force $\mathbf{F}$ about the z-axis is

$$m_z = x(\mathrm{d}m)A_y - y(\mathrm{d}m)A_x$$

Summing up for all elements of the body, the internal forces occurring in equal and opposite pairs cancel in the summation and we are left with the moment M_z about the z-axis of all the external forces acting on the body. The result is

$$\begin{aligned} M_z = \int(xA_y - yA_x)\mathrm{d}m = \int(xA_{Cy} + x^2\,\dot{\omega}_z - xy\omega_z^2 - yA_{Cx} \\ + xy\,\omega_z^2 + y^2\,\dot{\omega}_z)\mathrm{d}m = A_{Cy}\int x\mathrm{d}m + \dot{\omega}_z\int(x^2 + y^2)\mathrm{d}m \\ - A_{Cx}\int y\mathrm{d}m \end{aligned}$$

Since the xy are through the centre of mass we have $\int x\mathrm{d}m = \int y\mathrm{d}m = 0$.

The integral $\int(x^2 + y^2)\mathrm{d}m$ is by definition the moment of inertia I_z about the z-axis.

The result is

$$M_z = I_z\dot{\omega}_z \tag{7.4}$$

Equation (7.4) is the same as Euler's equation (6.2) determined for rotation about a fixed axis.

The fact that the equation of rotational motion may be taken exactly as if the body were rotating about a fixed axis through the centre of mass is sometimes called the *principle of independence of translation and rotation in plane motion*. This principle holds only if the axis of rotation is taken through the centre of mass.

7.2 KINETIC ENERGY IN PLANE MOTION. LAGRANGE'S EQUATIONS

The body is shown in Fig. 7.1. Projecting point P on the z-axis in point O_1, we may use O_1 as a pole for the motion of point P. The velocity of point P is now $\mathbf{V} = \mathbf{V}_{\mathrm{O}_1} + \boldsymbol{\omega} \times \mathbf{r}_1$.

Point O_1 has the same velocity as the centre of mass so that

$$\mathbf{V}_{O_1} = \mathbf{V}_C = V_{Cx}\mathbf{i} + V_{Cy}\mathbf{j}$$

$$\boldsymbol{\omega} \times \mathbf{r}_1 = \omega_z\mathbf{k} \times (x\mathbf{i} + y\mathbf{j}) = x\omega_z\mathbf{k} \times \mathbf{i} + y\omega_z\mathbf{k} \times \mathbf{j} = -y\omega_z\mathbf{i} + x\omega_z\mathbf{j}$$

We have now

$$\mathbf{V} = (V_{Cx} - y\omega_z)\mathbf{i} + (V_{Cy} + x\omega_z)\mathbf{j}$$

and

$$\begin{aligned} V^2 &= (V_{Cx} - y\omega_z)^2 + (V_{Cy} + x\omega_z)^2 \\ &= V_C^2 + (x^2 + y^2)\omega_z^2 + 2(xV_{Cy} - yV_{Cx})\omega_z \end{aligned}$$

The total kinetic energy is thus

$$T = \tfrac{1}{2} V_C^2 \int \mathrm{d}m + \tfrac{1}{2} \omega_z^2 \int (x^2 + y^2)\, \mathrm{d}m + V_{Cy}\omega_z \int x\, \mathrm{d}m - V_{Cx}\omega_z \int y\, \mathrm{d}m$$

Since $\int x\, \mathrm{d}m = \int y\, \mathrm{d}m = 0$, we have finally

$$\mathrm{T} = \tfrac{1}{2}MV_C^2 + \tfrac{1}{2}I_z\omega_z^2 \qquad (7.5)$$

The first term in (7.5) represents the kinetic energy of translation of the total mass with the instantaneous velocity V_C of the centre of mass, and the second term represents the kinetic energy of rotation about the centre of mass. To use (7.5) we *must* use the centre of mass of the body, otherwise the expression takes a more complicated form.

The formula (7.5) corresponds to the formula for rotation about a fixed axis.

Example 7.1

Fig. 7.2 shows a circular disc of radius r and mass m which rolls without slipping on a cylindrical surface of radius R. The disc is pin-connected to a rigid, slender uniform bar OA of length l and mass M. The instantaneous angular velocity of OA is ω. Determine the total kinetic energy of the system in terms of the system constants, l, ω, M and m. Calculate the numerical value of the kinetic energy if $l = 0.40$ m, $M = m = 6.5$ kg and $\omega = 10$ rad/s.

Solution

The velocity of the centre A of the disc is $V_A = l\omega = r\omega_1$, so $\omega_1 = (l/r)\omega$. The kinetic energy of the bar is $T_B = \tfrac{1}{2}I_0\omega^2 = \tfrac{1}{2}(\tfrac{1}{3}Ml^2)\omega^2 = \tfrac{1}{6}Ml^2\omega^2$. For the disc we obtain from (7.5)

$$T_D = \tfrac{1}{2}mV_A^2 + \tfrac{1}{2}I_A\omega_1^2 = \tfrac{1}{2}ml^2\omega^2 + \tfrac{1}{2}(\tfrac{1}{2}mr^2)\,\frac{l^2}{r^2}\,\omega^2 = \tfrac{3}{4}ml^2\omega^2$$

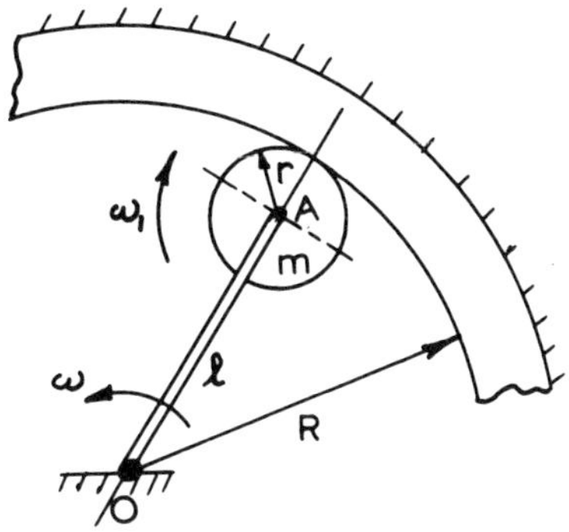

Fig. 7.2.

so that

$$T = \tfrac{1}{6}Ml^2\omega^2 + \tfrac{3}{4}ml^2\omega^2 = \frac{l^2\omega^2}{12}\,(2M + 9m)$$

The numerical value is

$$T = \frac{0.40^2 \times 10^2}{12} \times 11 \times 6.5 = 95.2\ J$$

The body in Fig. 7.1 in plane motion has three degrees of freedom. For generalised coordinates of the body we may choose the coordinates (X, Y) of the centre of mass, and the angle of rotation θ, which may be measured in the plane of motion from *any* fixed direction in the plane of motion to *any* line *fixed* on the body in the plane of motion and therefore rotating with the body. We have thus $\dot{\theta} = \omega_z$ and $\ddot{\theta} = \dot{\omega}_z$.

Using the formula (7.5) for the kinetic energy we have

$$T = \tfrac{1}{2}M\,(\dot{X}^2 + \dot{Y}^2) + \tfrac{1}{2}\,I_z\dot{\theta}^2$$

Lagrange's equations of motion are

$$\frac{\mathrm{d}}{\mathrm{d}t}\frac{\partial T}{\partial \dot{q}_i} - \frac{\partial T}{\partial q_i} = Q_i$$

with $q_1 = X$, $q_2 = Y$ and $q_3 = \theta$.

We find

$$\frac{\mathrm{d}}{\mathrm{d}t}\frac{\partial T}{\partial \dot{X}} = M\ddot{X} \quad \frac{\partial T}{\partial X} = 0$$

and similar expressions for Y and θ.

The acting forces may be reduced to a resultant force **F** at the centre of mass and a torque M_z. Using components F_X and F_Y of the force we find

$$\delta(\text{W.D.})_{\delta X} = Q_X\ \delta X = F_X\delta X \quad \delta(\text{W.D.})_{\delta Y} = Q_Y\delta Y = \\ = F_Y\delta Y \text{ and } \delta\ (\text{W.D.})_{\delta\theta} = Q_\theta\delta\theta = M_z\delta\theta$$

Substituting in Lagrange's equations gives the equations of motion

$$F_X = M\ddot{X} \quad F_Y = M\ddot{Y} \text{ and } M_z = I_z\ddot{\theta} = I_z\dot{\omega}_z$$

These are the equations of motion (7.1) and (7.4) determined before.

7.3 TRANSLATION OF A RIGID BODY

Translation of a rigid body is defined as a motion in which all lines in the body remain parallel to their original directions. This means that the body has no rotation, so the angular velocity and acceleration are zero during the motion.

Translation is the simplest form of rigid-body motion. Although it need not be plane motion, we shall consider only translation in a plane. The equations of motion in translation are the force equations (7.1) and (7.2), and the moment equation (7.4); since we have $\dot{\omega} = 0$, the moment equation takes the simple form

$$M_z = 0 \tag{7.6}$$

An example will illustrate the general procedure for the solution of problems of this nature.

Example 7.2

Fig. 7.3 shows a sliding door of mass M_1 which is carried on two small rollers at A and B. The rollers run on a horizontal track as shown. The door is opened by releasing a mass M_2 which is connected to the door by a string over a pulley of radius r and inertia I.

Determine the acceleration of the door, the reactions at A and B and the string tensions. The inertia of the rollers and friction on the rollers and on the pulley shaft are to be neglected. The string does not slip on the pulley.

Solution

All the acting forces on the door, on the falling mass and on the pulley have been shown on the figure.

Applying the force equations (7.1) to the door gives the equations

$$F_x = S_1 = M_1\ddot{x} \tag{a}$$
$$F_y = A + B - M_1 g = M_1\ddot{y} = 0 \tag{b}$$

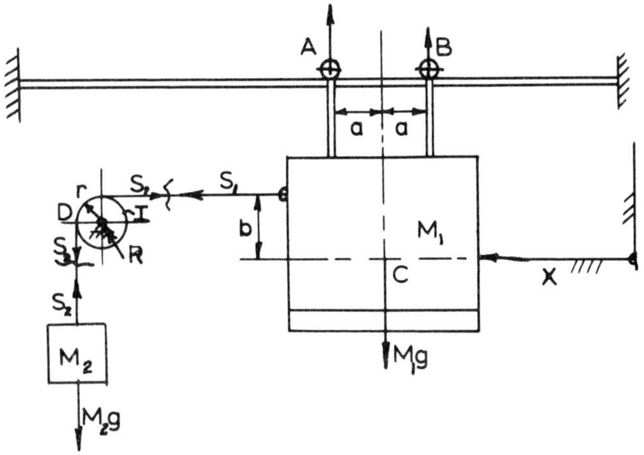

Fig. 7.3.

The Euler equation from (7.4) for the door is

$$M_z = Aa - Ba - S_1 b = 0 \tag{c}$$

and for the pulley

$$(S_2 - S_1)r = I\ddot{\theta} = \frac{I\ddot{x}}{r} \tag{d}$$

The force equation for the falling body is

$$M_2 g - S_2 = M_2 \ddot{x} \tag{e}$$

Substituting S_1 from (a) and $S_2 = M_2 g - M_2 \ddot{x}$ from (c) in equation (d) and solving for $\ddot{x}$ gives the acceleration of the door and the falling body

$$\ddot{x} = \frac{M_2 g}{M_1 + M_2 + I/r^2}$$

Substituting the expression for $\ddot{x}$ back in (a) and (e) gives the string tensions

$$S_1 = \frac{M_1 M_2 g}{M_1 + M_2 + I/r^2} \quad \text{and} \quad \frac{M_2(M_1 + I/r^2)g}{M_1 + M_2 + I/R^2} = S_2$$

It is noticeable that $S_2 > S_1$, which is necessary to give the torque on the pulley. The equations (b) and (c) are

$$A + B = M_1 g \quad \text{and} \quad A - B = S_1\, b/a$$

with the solution $A = M_1 g/2 + S_1 b/2a$ and $B = M_1 g/2 - S_1 b/2a$, where S_1 is given above. The reaction A is always positive, that is directed as shown, while for the reaction B we have $B \gtreqless 0$, depending on the values of the system constants.

The system has one degree of freedom. Taking x as the generalised coordinate we have

$$\begin{aligned} T &= \tfrac{1}{2}M_1\dot{x}^2 + \tfrac{1}{2}M_2\dot{x}^2 + \tfrac{1}{2}I\,\dot{x}^2/r^2 \\ &= \tfrac{1}{2}[M_1 + M_2 + I/r^2]\dot{x}^2 = \tfrac{1}{2}C\dot{x}^2 \end{aligned}$$

where $C = [M_1 + M_2 + I/r^2]$. Giving an increment δx to x, we find $\delta(\text{W.D.}) = Q_x\,\delta x = M_2 g\,\delta x$, and Lagrange's equation

$$\frac{\mathrm{d}}{\mathrm{d}t}\frac{\partial T}{\partial \dot{x}} - \frac{\partial T}{\partial x} = Q_x$$

now gives the result $A\ddot{x} = M_2 g$ or $\ddot{x} = M_2 g/A$ as above.

The string tensions may now be determined by writing Lagrange's equations for two of the subsystems consisting of the door, the falling mass and the pulley, these equations, however, are here simply Newton's law and Euler's equation as given above.

7.4 GENERAL PLANE MOTION

As a case of general plane motion consider the following. Fig. 7.4 shows a

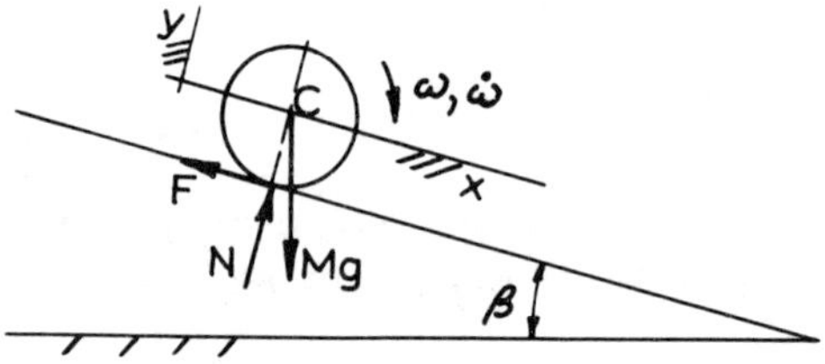

Fig. 7.4.

homogeneous circular disc of radius R, which is rolling down an inclined plane. The mass of the body is M, and the moment of inertia about the geometric axis is $I_C = M(R^2/2)$.

To describe the motion of the body, we introduce a fixed coordinate system (x,y) in the most suitable position, with the x-axis through the centre of mass and parallel to the inclined plane; the motion of the centre C is then rectilinear along the x-axis.

The forces acting are the gravity force Mg and the reaction from the plane, and this reaction is taken for convenience in two components, a normal force N and a friction force F as shown. As long as there is no relative motion between the body and the plane, the friction force is of unknown magnitude and $F<\mu N$, where μ is the coefficient of friction between the body and the plane. If the body is sliding on the plane, the friction force is at a *maximum* $F=\mu N$.

The force equations (7.1) are

$$F_y = N - Mg \cos \beta = M\ddot{y} = 0 \quad \text{and} \quad \text{therefore } N = Mg \cos \beta$$

$$F_x = Mg \sin \beta - F = M\ddot{x} \tag{a}$$

The Euler equation (7.4) is

$$M_C = FR = I_C\dot{\omega} \tag{b}$$

We have two equations (a) and (b) with three unknown quantities $\ddot{x}$, F and $\dot{\omega}$. To obtain a third equation we must consider the constraints on the system, so we divide the investigation in two parts, depending on whether the body is in pure rolling or sliding.

Rolling without sliding

In this case the relationship $x = R\theta + x_0$ is valid, where x is the distance moved of the centre C, and θ is the angle of rotation; x_0 is a constant which is the value of x when $\theta = 0$; differentiating this equation twice gives an equation

$$\ddot{x} = R\ddot{\theta} = R\dot{\omega} \tag{c}$$

The equations (a), (b) and (c) may now be solved with the result that

$$\ddot{x} = \frac{2g \sin \beta}{3} \quad \dot{\omega} = \frac{\ddot{x}}{R} \quad F = \frac{Mg \sin \beta}{3}$$

The acceleration and the friction force are constant.

Rolling with sliding

In this case the equations of motion (a) and (b) are still valid, but (c) *no longer holds*. The friction force is *now a maximum*, and a new equation (c′) is then

$$F = \mu N = \mu Mg \cos \beta \tag{c′}$$

The solution in this case is

$$\ddot{x} = g(\sin \beta - \mu \cos \beta) \quad \dot{\omega} = \frac{2g\mu}{R} \cos \beta \quad F = \mu Mg \cos \beta$$

In any given problem it is necessary to investigate whether there is pure rolling or sliding. *Pure rolling* requires

$$F = \frac{Mg \sin \beta}{3} < F_{max} = \mu Mg \cos \beta \quad \text{or} \quad \tan \beta < 3\mu$$

otherwise we have rolling with sliding. For instance for a solid cylinder, if $\mu = 1/3$ the cylinder is in pure rolling if $\tan \beta < 1$, or $\beta < 45°$. The cylinder will roll and slide for this value of μ if $\beta > 45°$.

If the disc is in *pure rolling*, it has *one degree of freedom*, and we may take the coordinate x of the centre of mass as the generalised coordinate. $T = \frac{1}{2} MV_C^2 + \frac{1}{2} I_C^2$ from (7.5) and substituting $V_C = \dot{x}$, $I_C = MR^2/2$ and $\omega = \dot{\theta} = \dot{x}/R$, we find $T = \frac{3}{4} M\dot{x}^2$. During the motion of the disc the normal force N does no work, and the friction force acting through the instananeous centre does no work since there is no sliding at the point of contact; only the component $Mg \sin \beta$ of the gravity force down the plane does work and we find

$$\delta(\text{W.D.}) = Q_x\, \delta x = Mg \sin \beta (\delta x)$$

Substituting in Lagrange's equation gives $\frac{3}{2} M\ddot{x} = Mg \sin \beta$, or $\ddot{x} = \frac{2}{3} g \sin \beta$ as above.

For the case of *rolling and sliding*, we have *two degrees* of freedom and we may take the coordinate x of the centre of mass and the angle of rotation θ as generalised coordinates. The kinetic energy $T = \frac{1}{2} M\dot{x}^2 + (MR^2/2)\dot{\theta}^2/2$; keeping θ constant we find

$$\delta(\text{W.D.}) = -F\delta x + (Mg \sin \beta)\, \delta x = Q_x\, \delta x$$

$$Q_x = -F + Mg \sin \beta = -\mu N + Mg \sin \beta = -\mu Mg \cos \beta + Mg \sin \beta$$
$$= Mg(\sin \beta - \mu \cos \beta)$$

The first Lagrange's equation is now $M\ddot{x} = Mg(\sin \beta - \mu \cos \beta)$. Keeping x constant we find for an incremental rotation $\delta\theta$ that Mg and N do no work, and $\delta(\text{W.D.}) = FR\delta\theta = Q_\theta \delta\theta$, or $Q_\theta = FR$.

The second Lagrange's equation is now $(MR^2/2)\ddot{\theta} = FR$, or $\ddot{\theta} = \dot{\omega} = 2F/MR = 2\mu g \cos \beta/R$ as before.

Example 7.3

Fig. 7.5 shows a uniform slender rigid bar of length l and mass M. The bar moves in the vertical plane with the ends of the bar sliding on frictionless walls as shown. Given that the bar starts from rest in the vertical position after a slight disturbance, determine the reactions A and B as functions of the angle θ. Determine also the angle at which the bar leaves the vertical wall.

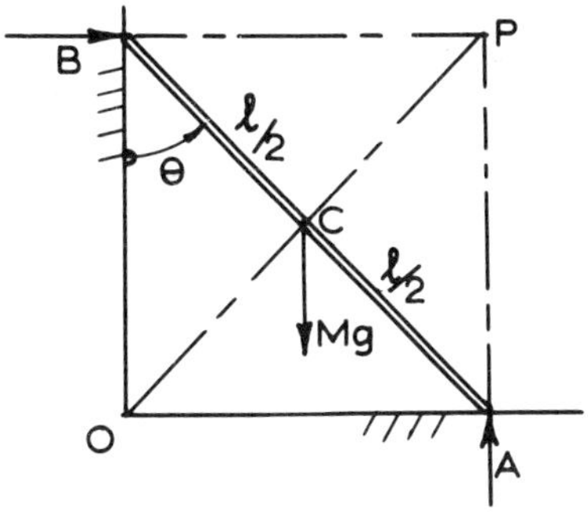

Fig. 7.5.

Solution

The distance OC is equal to $l/2$ for any position of the bar in contact with both walls. The centre of mass C of the bar then moves on a circle with centre O and radius $l/2$. The acceleration of C is thus $\frac{1}{2}l\dot{\theta}^2$ towards O, and $\frac{1}{2}l\ddot{\theta}$ perpendicular to OC. The equations of motion of C are thus in the normal and tangential direction respectively:

$$Mg\cos\theta - B\sin\theta - A\cos\theta = \tfrac{1}{2}Ml\dot{\theta}^2 \qquad \text{(a)}$$

$$Mg\sin\theta + B\cos\theta - A\sin\theta = \tfrac{1}{2}Ml\ddot{\theta} \qquad \text{(b)}$$

The Euler equation (7.4) is $M_C = I_C\ddot{\theta}$, or

$$A\frac{l}{2}\sin\theta - B\frac{l}{2}\cos\theta = \frac{Ml^2}{12}\ddot{\theta}$$

from which

$$A\sin\theta - B\cos\theta = \frac{Ml}{6}\ddot{\theta} \qquad \text{(c)}$$

From equation (b) we have

$$A\sin\theta - B\cos\theta = Mg\sin\theta - \frac{Ml}{2}\ddot{\theta}$$

and substituting this in (c) we find

$$\ddot{\theta} = (3g/2l)\sin\theta$$

We may write

$$d\theta \frac{d\dot{\theta}}{dt} = \ddot{\theta}\, d\theta = \tfrac{1}{2}\, d(\dot{\theta})^2 = \frac{3g}{2l} \sin\theta\, d\theta$$

so that

$$d(\dot{\theta}^2) = \frac{3g}{l} \sin\theta\, d\theta \quad \dot{\theta}^2 = -\frac{3g}{l} \cos\theta + C$$

When $\theta = 0$, $\dot{\theta} = 0$; therefore $C = 3g/l$ and $\dot{\theta}^2 = (3g/l)(1 - \cos\theta)$. Substituting the expressions for $\ddot{\theta}$ and $\dot{\theta}^2$ in equations (a) and (b) gives two equations in A and B. The solution is

$$A = \frac{Mg}{4} (3\cos\theta - 1)^2 \quad \text{and} \quad B = 3\frac{Mg}{4} (3\cos\theta - 2) \sin\theta$$

This result is valid only as long as the bar is in contact with both walls. The bar loses contact with the vertical wall when $B = 0$; therefore $3\cos\theta - 2 = 0$, or $\cos\theta = \frac{2}{3}$, $\theta = 48.19°$. The above expressions for A and B are then valid for $0 < \theta < 48.19°$.

Taking θ as generalised coordinate, we have $T = \frac{1}{2} MV_C^2 + \frac{1}{2} I_C\omega^2$. The centre of mass moves on a circle so $V_C = \frac{l}{2}\dot{\theta}$, and we obtain $T = \frac{1}{2} M \frac{l^2}{4} \dot{\theta}^2 + \frac{1}{2}\left(\frac{Ml^2}{12}\right)\dot{\theta}^2 = \frac{1}{6} Ml^2 \dot{\theta}^2$.

$$\delta(\text{W.D.}) = Q_\theta \delta\theta = Mg\, \delta y_C$$

we have $y_C = \frac{l}{2}\cos\theta$ and $|dy_C| = \frac{l}{2}\sin\theta\, d\theta$, so that $Q_\theta\, \delta\theta = Mg\frac{l}{2}\sin\theta\delta\theta$.

Lagrange's equation is now

$$\tfrac{1}{3}Ml^2\, \ddot{\theta} = Mg\frac{l}{2}\sin\theta \quad \text{or} \quad \ddot{\theta} = \frac{3g}{2l}\sin\theta$$

and the rest of the problem may be solved as before.

For more complicated systems, Lagrange's equations are a considerable advantage in establishing the equations of motion, as shown in the next two examples.

Example 7.4
Determine the acceleration of the body of mass m_1 in Fig. 7.6 down the plane. The coefficient of friction between the body and the plane is μ. The pulleys may be treated as solid flat discs of radius r and mass M; there is no slip of the string on the pulleys.

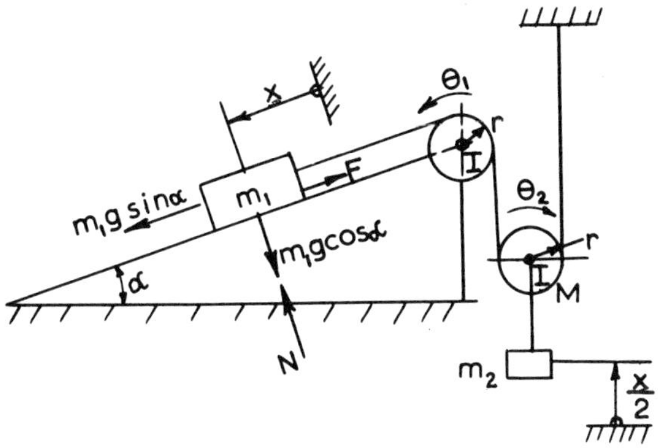

Fig. 7.6.

Solution

Taking the distance x as generalised coordinate, the equation of motion is

$$\frac{\mathrm{d}}{\mathrm{d}t}\frac{\partial T}{\partial \dot{x}} - \frac{\partial T}{\partial x} = Q_x$$

The kinetic energy of the system is

$$T = \tfrac{1}{2}m_1\dot{x}^2 + \tfrac{1}{2}I\dot{\theta}_1^2 + \left(\tfrac{1}{2}I\dot{\theta}_2^2 + \tfrac{1}{2}M\,\frac{\dot{x}^2}{4}\right) + \tfrac{1}{2}\,m_2\,\frac{\dot{x}^2}{4}$$

We have $I = \tfrac{1}{2}Mr^2$, and for no slip on the pulleys, $r\theta_1 = x$, $r\theta_2 = x/2$, from which $\dot{\theta}_1 = \dot{x}/r$ and $\dot{\theta}_2 = \dot{x}/2r$; substituting this, we find

$$T = \tfrac{1}{2}[m_1 + \tfrac{7}{8}M + \tfrac{1}{4}m_2]\,\dot{x}^2 = \tfrac{1}{2}\,A\dot{x}^2$$

$$\frac{\mathrm{d}}{\mathrm{d}t}\frac{\partial T}{\partial \dot{x}} = A\ddot{x} \quad \text{and} \quad \frac{\partial T}{\partial x} = 0$$

Giving an increment δx to x, the work done is

$$Q_x\delta x = m_1g(\sin\alpha)\,\delta x - m_1gu(\cos\alpha)\,\delta x - Mg(\delta x/2) - m_2g(\delta x/2)$$

and

$$Q_x = m_1 g(\sin\alpha - \mu\cos\alpha) - \tfrac{1}{2}(M + m_2)g = B$$

The equation of motion is $A\ddot{x} = B$, and the acceleration of the mass is $\ddot{x} = B/A$, with the above expressions for the constants A and B.

Example 7.5
Fig. 7.7 shows a horizontal uniform bar of mass $2m$ and centre of mass C. The bar

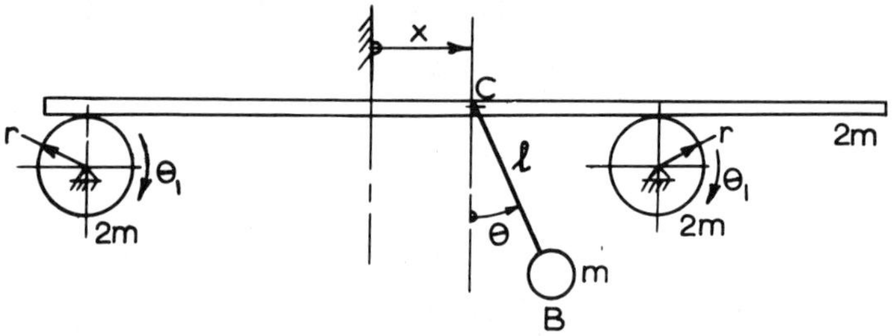

Fig. 7.7.

rolls without sliding on two identical cylindrical rotors, each of radius r and mass $2m$.

A pendulum of length l and with a bob of mass m is attached at C and swings in the vertical plane.

(a) Determine the equations of motion of the system.
(b) Linearise the equations for small displacement and velocities.
(c) Determine the natural frequency of the system for small motions.

Solution

(a) The moment of inertia of a roller is $I = \tfrac{1}{2}(2m)r^2 = mr^2$. Since there is no slip, $x = r\theta_1$, and $\theta_1 = x/r$, $\dot{\theta}_1 = \dot{x}/r$. The velocity of the bob is $\mathbf{V}_B = \mathbf{V}_C + \mathbf{V}_{cir} = \dot{x}\rightarrow + l\dot{\theta}\uparrow$, so that

$$V_B^2 = \dot{x}^2 + l^2\dot{\theta}^2 + 2l\dot{x}\dot{\theta}\cos\theta$$

The kinetic energy is

$$T = \tfrac{1}{2}(2m)\dot{x}^2 + \tfrac{1}{2}m(\dot{x}^2 + l^2\dot{\theta}^2 + 2l\dot{x}\dot{\theta}\cos\theta)$$

$$+ 2(\tfrac{1}{2}mr^2)\dot{\theta}_1^2 = \tfrac{5}{2}m\dot{x}^2 + \tfrac{1}{2}ml^2\dot{\theta}^2 + ml\dot{x}\dot{\theta}\cos\theta$$

The potential energy is $V = mgl(1 - \cos\theta)$, and

$$L = T - V = \tfrac{5}{2}\,m\dot{x}^2 + \tfrac{1}{2}\,ml^2\dot{\theta}^2 + ml\dot{x}\dot{\theta}\cos\theta - mgl\,(1 - \cos\theta)$$

$$\frac{\partial L}{\partial \dot{x}} = 5m\dot{x} + ml\dot{\theta}\cos\theta \quad \frac{\partial L}{\partial x} = 0 \quad \frac{\mathrm{d}}{\mathrm{d}t}\frac{\partial L}{\partial \dot{x}}$$
$$= 5m\ddot{x} + ml\ddot{\theta}\cos\theta - ml\dot{\theta}^2\sin\theta$$

$$\frac{\partial L}{\partial \dot{\theta}} = ml^2\dot{\theta} + ml\dot{x}\cos\theta \quad \frac{\partial L}{\partial \theta} = -ml\dot{x}\dot{\theta}\sin\theta - mgl\sin\theta$$

$$\frac{\mathrm{d}}{\mathrm{d}t}\frac{\partial L}{\partial \dot{\theta}} = ml^2\ddot{\theta} + ml\ddot{x}\cos\theta - ml\dot{x}\dot{\theta}\sin\theta$$

Substituting in Lagrange's equations, we obtain

$$5\ddot{x} + l(\cos\theta)\ddot{\theta} - l(\sin\theta)\dot{\theta}^2 = 0 \tag{a}$$

$$(\cos\theta)\,\ddot{x} + l\ddot{\theta} + g\sin\theta = 0 \tag{b}$$

(b) For small displacements and velocities, we may substitute $\sin\theta \to \theta$, $\cos\theta \to 1$ and $\dot{\theta}^2 \to 0$. Hence

$$5\ddot{x} + l\ddot{\theta} = 0 \tag{a$'$}$$

$$\ddot{x} + l\ddot{\theta} + g\theta = 0 \tag{b$'$}$$

(c) Substituting $\ddot{x} = -(l/5)\ddot{\theta}$ from (a′) in (b′), we obtain $\ddot{\theta} + (5g/4l)\dot{\theta} = 0$; this is simple harmonic motion with frequency $\omega = \tfrac{1}{2}\sqrt{(5g/4l)}$.

For Langrange's equations in the form involving the kinetic energy only, we find

$$\delta(\text{W.D.}) = Q_x\delta x = 0$$

and

$$\delta(\text{W.D.}) = Q_\theta\delta\theta = -mgl\sin\theta\delta\theta$$

or

$$Q_x = 0 \quad \text{and} \quad Q_\theta = -mgl\sin\theta$$

7.5 WORK-ENERGY EQUATION FOR A RIGID BODY IN PLANE MOTION

It is shown in statics that any plane force system may be reduced to a resultant force and a couple in the plane acting at any particular point. The forces and couples acting on a rigid body in plane motion may, therefore, always be reduced to a *resultant force* F and a *resultant moment* M_C acting at the centre of mass C of the body.

Fig. 7.8 shows the body with the components F_x and F_y of the resultant force

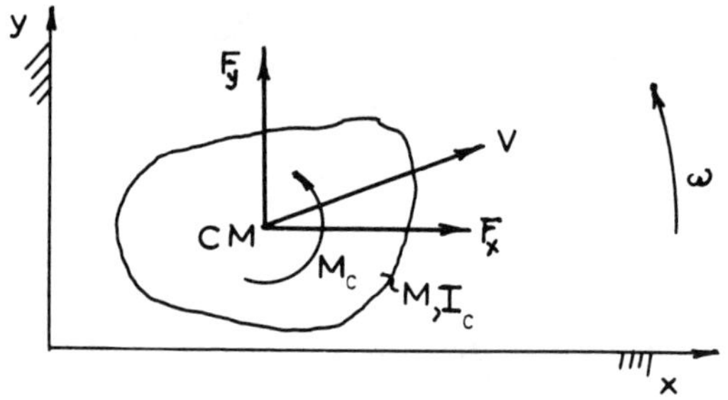

Fig. 7.8.

acting at the centre of mass, and the moment M_C about the centre of mass.

For small displacements dx, dy and dθ of the body we have

$$d(\text{W.D.}) = F_x\,dx + Fy\,dy + M_C\,d\theta$$

Introducing Newton's second law for the centre of mass, we have

$$F_x = M\frac{d\dot{x}}{dt} \quad \text{and} \quad F_y = M\frac{d\dot{y}}{dt}$$

and the work done by the resultant force is now

$$d(\text{W.D.})_1 = M\left(\frac{d\dot{x}}{dt}\,dx + \frac{d\dot{y}}{dt}\,dy\right) = M(\dot{x}\,d\dot{x} + \dot{y}\,d\dot{y}) = \tfrac{1}{2}M\,(d\dot{x}^2 + d\dot{y}^2)$$

Integrating between two positions of the body we find

$$\text{W.D.}_1\,\big|_1^2 = \tfrac{1}{2}M\int_{\dot{x}_2}^{\dot{x}_2} d\dot{x}^2 + \tfrac{1}{2}M\int_{\dot{y}_1}^{\dot{y}_2} d\dot{y}^2 = \tfrac{1}{2}M\,[(\dot{x}_2^2 + \dot{y}_2^2) - (\dot{x}_1^2 + \dot{y}_1^2)]$$
$$= \tfrac{1}{2}M(V_2^2 - V_1^2)$$

Introducing Euler's equation $M_C = I_C\,(d\dot{\theta}/dt)$, the work done by the moment is

$$d(\text{W.D.})_2 = I_C \frac{d\dot{\theta}}{dt}\, d\theta = I_C\, \dot{\theta}\, d\dot{\theta} = \tfrac{1}{2} I_C\, d\dot{\theta}^2$$

Integrating gives

$$\text{W.D.}_2 \Big|_1^2 = \tfrac{1}{2} I_C \int_{\theta_1}^{\theta_2} d\dot{\theta}^2 = \tfrac{1}{2} I_C (\dot{\theta}_2^2 - \dot{\theta}_1^2) = \tfrac{1}{2} I_C (\omega_2^2 - \omega_1^2)$$

The total work done is now

$$\text{W.D.} \Big|_1^2 = \tfrac{1}{2}\,(MV_2^2 + I_C\omega_2^2) - \tfrac{1}{2}(MV_1^2 + I_C\omega_1^2)$$

or

$$\text{W.D.} \Big|_1^2 = T_2 - T_1 \qquad (7.7)$$

This is the work-energy equation for a rigid body in plane motion. The equation states that the *total work* done on the body in the motion from a position 1 to a position 2 is equal to the *total change in the kinetic energy* of the body between the two positions.

Example 7.6
Consider the disc in Fig. 7.4 rolling without sliding. Determine the velocity $\dot{x}$ of the centre of mass as a function of the distance moved x, if the body starts from rest. Determine also the acceleration $\ddot{x}$.

Solution
Taking the position $x = 0$ as the first position, we have $T_1 = 0$, since the body is at rest. For any position x we have from previous results

$$T_2 = \frac{3M}{4}\dot{x}^2$$

The normal force N does no work, and the friction force F is always directed through the instantaneous centre and does no work. The only work done is by the gravity force component along the inclined plane, so that W.D. $= Mg(\sin\beta)x$.

$$\text{W.D.} = mg(\sin\beta)x = T_2 - T_1 = \tfrac{3}{4}M\dot{x}^2$$

therefore

$$\dot{x} = [\tfrac{3}{4}\, x\, g \sin \beta]^{\frac{1}{2}}$$

Differentiating the equation gives $(Mg \sin \beta)\dot{x} = \tfrac{3}{2}M\dot{x}\ddot{x}$, or $\ddot{x} = \tfrac{2}{3}\, g \sin \beta$ as found previously.

Example 7.7
Consider Example 7.3. By using the work-energy equation, determine the angular velocity and acceleration of the bar as a function of the angle θ if the bar starts from rest.

Solution
Since the bar starts from rest, $T_1 = 0$. From previous results

$$T_2 = \tfrac{1}{6}Ml^2\dot{\theta}^2$$

Only the gravity force does work, so that

$$\text{W.D.} = \tfrac{1}{2}Mgl(1 - \cos \theta) \quad \text{and} \quad \tfrac{1}{6}Ml^2\dot{\theta}^2 = \tfrac{1}{2}Mgl(1 - \cos \theta)$$

This gives the result $\dot{\theta}^2 = (3g/l)(1 - \cos \theta)$, and by differentiation $2\dot{\theta}\ddot{\theta} = (3g/l)(\sin \theta)\dot{\theta}$, or $\ddot{\theta} = (3g/2l)\sin \theta$. Both results are in agreement with the results found in Example 7.3.

7.6 PRINCIPLE OF CONSERVATION OF MECHANICAL ENERGY IN PLANE MOTION

The principle of conservation of mechanical energy for a particle was stated in equation (2.9) as $V + T =$ constant; that is the sum of the potential and kinetic energy is constant for a particle moving under the action of conservative forces only.

For a rigid body, the principle was stated in (6.7) with the same form

$$V + T = \text{constant} \tag{7.8}$$

This is the principle of conservation of mechanical energy for a rigid body under the *action of conservative external forces only*. This principle is not as powerful as the work-energy equation, since that equation also includes friction forces and all non-conservative forces.

The principle (7.8) is sometimes useful for establishing the equation of motion for a system with one degree of freedom by differentiation, since $\mathrm{d}V/\mathrm{d}t + \mathrm{d}T/\mathrm{d}t = 0$.

Example 7.8
Fig. 7.9 shows *a rocking pendulum*, which consists of a body with mass M and moment of inertia with respect to an axis perpendicular to the plane of motion

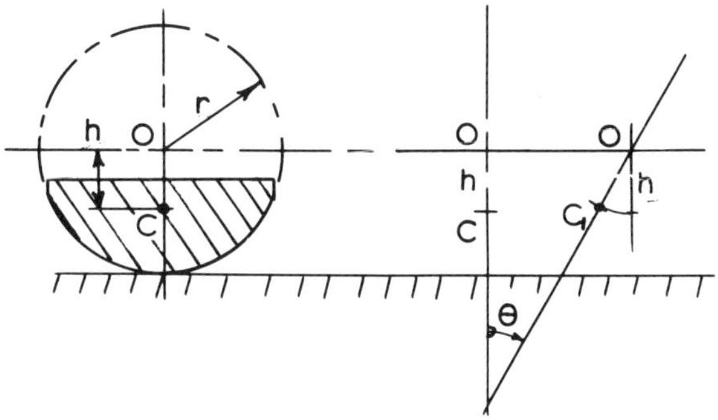

Fig. 7.9.

through the centre of mass $I_C = Mr_C^2$. The body rolls *without slipping* on a horizontal plane. Determine the equation of motion of the body by using the principle of conservation of mechanical energy. Determine also the frequency of small-displacement rocking.

Solution

This is a conservative system, since only the gravity force does work during the motion. For a rotation θ of the body, the centre of mass increases its height above the plane by a distance $h(1 - \cos\theta)$, so that $V = Mgh(1 - \cos\theta)$. The velocity relationship is $\mathbf{V}_C = \mathbf{V}_0 + \mathbf{V}_{cir}$, where $\mathbf{V}_0 = r\dot{\theta}\rightarrow$ and $\mathbf{V}_{cir} = h\dot{\theta}\uparrow$ at $(180° - \theta°)$ to the horizontal, from which

$$V_C^2 = (r^2 + h^2 - 2rh\cos\theta)\dot{\theta}$$

The kinetic energy is $T = \frac{1}{2}MV_C^2 + \frac{1}{2}I_C\omega^2 = \frac{1}{2}M(r^2 + h^2 - 2rh\cos\theta + r_C^2)\dot{\theta}^2$. The principle now gives the equation

$$Mgh(1 - \cos\theta) + \tfrac{1}{2}M(r^2 + h^2 - 2rh\cos\theta + r_C^2)\dot{\theta}^2 = \text{constant}$$

Differentiating and cancelling M and $\dot{\theta}$ gives the equation of motion

$$(r^2 + h^2 - 2rh\cos\theta + r_C^2)\ddot{\theta} + (rh\sin\theta)\dot{\theta}^2 + gh\sin\theta = 0$$

For small displacements we take $\sin\theta \sim \theta$, $\cos\theta \sim 1$ and also $\dot{\theta}$ small, so that the term $(rh\sin\theta)\dot{\theta}^2 \sim rh\theta\dot{\theta}^2$ may be neglected. The equation then takes the form

$$\ddot{\theta} + \left[\frac{gh}{(r-h)^2 + r_C^2}\right]\theta = 0$$

This represents simple harmonic motion with

$$f=\frac{1}{2\pi}\sqrt{\frac{gh}{(r-h)^2+r_C^2}}$$

Using Lagrange's equation

$$\frac{\mathrm{d}}{\mathrm{d}t}\frac{\partial L}{\partial\dot\theta}-\frac{\partial L}{\partial\theta}=0$$

we have $L=T-V$ as given above, from which

$$\frac{\mathrm{d}}{\mathrm{d}t}\frac{\partial L}{\partial\dot\theta}=M(r^2+h^2-2rh\cos\theta+r_C^2)\ddot\theta+(2rh\,M\sin\theta)\dot\theta^2$$

and

$$\frac{\partial L}{\partial\theta}=(Mrh\sin\theta)\dot\theta^2-Mgh\sin\theta$$

from which we find the equation of motion. Using

$$\frac{\mathrm{d}}{\mathrm{d}t}\frac{\partial T}{\partial\dot\theta}-\frac{\partial T}{\partial\theta}=Q_\theta$$

we find that the increase in height above the plane of the centre of mass for an increment $\delta\theta$ is $h\delta\theta\sin\theta$, with $\delta(\mathrm{W.D.})=Q_\theta\delta\theta=-Mgh\,\delta\theta\sin\theta$, from which we find the above equation of motion. To determine this equation by Euler's equation, we have to involve the two Newton equations for the centre of mass, which makes the determination of the equation much more difficult.

Example 7.9

Fig. 7.10 shows *a rolling pendulum*. This consists of a cylinder of radius r and mass M rolling *without slipping* on a cylindrical surface.

Determine the equation of motion of the cylinder and the frequency for small displacements ϕ from the vertical line OA.

Solution

Only gravity does work on the motion. The potential energy is $V=Mg(R-r)(1-\cos\phi)$.

If the angle of rotation of the cylinder is θ, the velocity of the centre C is $V_C=(R-r)\dot\phi=r\dot\theta$, so that $\dot\theta=\omega=[(R-r)/r]\dot\phi$. With $I_C=Mr^2/2$, the kinetic energy is

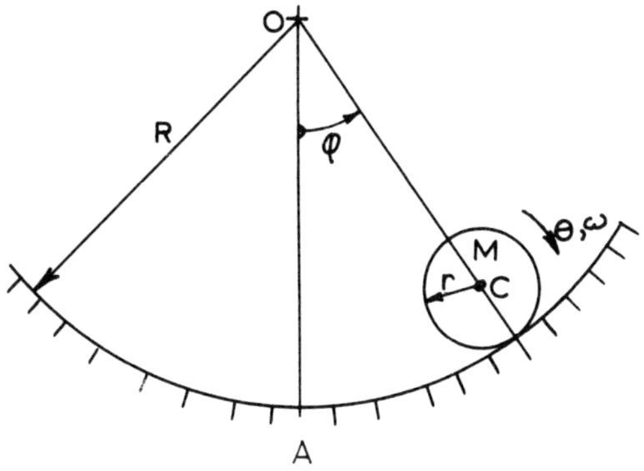

Fig. 7.10.

$$T = \tfrac{1}{2}M(R-r)^2\dot{\phi}^2 + \tfrac{1}{2}\left(\frac{Mr^2}{2}\right)\frac{(R-r)^2}{r^2}\dot{\phi}^2 = \tfrac{3}{4}M(R-r)^2\dot{\phi}^2$$

We have then $Mg(R-r)\,(1-\cos\phi) + \tfrac{3}{4}M(R-r)^2\dot{\phi}^2 = \text{constant}$; differentiating and cancelling the factors M, $(R-r)$ and $\dot{\phi}$ gives the equation of motion

$$\ddot{\phi} + \frac{2g}{3(R-r)}\sin\phi = 0$$

For small values of ϕ the equation is

$$\ddot{\phi} + \frac{2g}{3(R-r)}\phi = 0$$

This represents simple harmonic motion with frequency

$$f = \frac{1}{2\pi}\sqrt{\frac{2g}{3(R-r)}}$$

Using the Lagrange's equation

$$\frac{\mathrm{d}}{\mathrm{d}t}\frac{\partial L}{\partial\dot{\phi}} - \frac{\partial L}{\partial\phi} = 0$$

we have $L = T - V$ from above, with

$$\frac{\mathrm{d}}{\mathrm{d}t}\frac{\partial L}{\partial \dot{\phi}} = \tfrac{3}{2}M(R-r)^2\ddot{\phi} \quad \text{and} \quad \frac{\partial L}{\partial \phi} = -Mg\,(R-r)\sin\phi$$

from which we obtain the above equation of motion.

Using the Lagrange's equation

$$\frac{\mathrm{d}}{\mathrm{d}t}\frac{\partial T}{\partial \dot{\phi}} - \frac{\partial T}{\partial \phi} = Q_\phi$$

we find that the centre of mass moves up a distance $(R-r)\delta\phi\,\sin\phi$ for an increment $\delta\phi$, so that $\delta(\text{W.D.}) = -Mg(R-r)\delta\phi\,\sin\phi = Q_\phi\delta\phi$.

The equation of motion follows directly from this.

The result may also be established by using Euler's equation and the two Newton equations for the centre of mass; this involves a somewhat greater effort than the above.

7.7 D'ALEMBERT'S PRINCIPLE. DYNAMIC EQUILIBRIUM IN PLANE MOTION

The equations of motion for a rigid body in plane motion were $F_x = M\ddot{X}$, $Fy = M\ddot{Y}$ and $M_z = I_z\dot{\omega}$, where X and Y are the coordinates of the centre of mass. We may now apply D'Alembert's principle to these equations; that is we write the equations in terms of the inertia forces, inertia torques and applied external forces and moments as follows:

$$\left.\begin{aligned} &F_x + (-M\ddot{X}) = 0 \\ &F_y + (-M\ddot{Y}) = 0 \\ &M_z + (-I_z\dot{\omega}) = 0 \end{aligned}\right\} \tag{7.9}$$

This is D'Alembert's form of the equations of motion. The forces and moments acting on the body, together with the inertia forces and torques, may now be taken as a system of *plane forces in equilibrium*, so that the body is in *dynamic equilibrium* and equations of statics may be used for the equilibrium conditions. The most important of these is the condition that moments may be taken about any axis perpendicular to the plane of motion, or more briefly about any point in the plane, while in Euler's equation the moment centre was limited to the centre of mass.

Example 7.10

Solve Example 7.9 by D'Alembert's principle.

Solution
Fig. 7.11 shows the rolling cylinder in dynamic equilibrium. The acting forces are

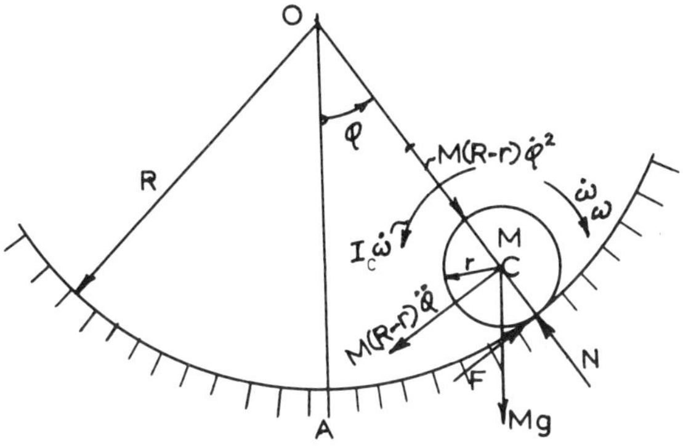

Fig. 7.11.

Mg, N and F; the inertia forces are $M(R-r)\dot{\phi}^2$ and $M(R-r)\ddot{\phi}$ directed as shown opposite to the accelerations; the inertia torque is $I_C\dot{\omega}$ as shown.

Treating the system as a static system in equilibrium, we may take a moment equation about the contact point to eliminate F and N; this leads to

$$I_C\dot{\omega} + M(R-r)\ddot{\phi}r + Mgr\sin\phi = 0$$

Substituting

$$I_C = \frac{Mr^2}{2} \quad \text{and} \quad \dot{\omega} = \ddot{\theta} = \left(\frac{R-r}{r}\right)\ddot{\phi}$$

leads to the previous result in Example 7.9.

7.8 PRINCIPLE OF VIRTUAL WORK IN PLANE MOTION

Virtual displacements $\delta\mathbf{r}$ of a particle were defined in seciton 2.11. Any force system acting on a rigid body in statics may be reduced to a single resultant force $\mathbf{F}$ and a couple vector $\mathbf{M}$ at any chosen point. If the body is given a virtual displacement $\delta\mathbf{r}$ and a virtual rotation $\delta\boldsymbol{\theta}$, the work done by all acting forces is the virtual work $\delta(\text{W.D.}) = \mathbf{F}\cdot\delta\mathbf{r} + \mathbf{M}\cdot\delta\boldsymbol{\theta}$; if the force system is in equilibrium, it will also be in equilibrium during this virtual displacement according to the definition of virtual displacements, and we have consequently $\delta(\text{W.D.}) = 0$ for any virtual displacement. This is the *principle of virtual work in statics*.

In dynamics the acting forces are generally not in equilibrium, but if D'Alembert's principle is applied we have dynamic equilibrium and the principle of virtual work may then be applied as in a static case.

If the force system acting on a body in plane motion is reduced to a resultant force **F** at the centre of mass C and a resultant couple $\mathbf{M}_C$ about C, the principle of virtual work in dynamics for a rigid body in plane motion may be expressed as follows from equation (7.9):

$$
\begin{aligned}
&[F_X+(-M\ddot{X})]\delta x=0\\
&[F_Y+(-M\ddot{Y})]\delta y=0\\
&[M_z+(-I_z\dot{\omega})]\ \delta\theta=0
\end{aligned}
\tag{7.10}
$$

Although D'Alembert's principle combined with the principle of virtual work is a powerful method in many problems, it has generally been superseded by the more useful Lagrange method.

Example 7.11

Fig. 7.12 shows a two-step pulley of total mass M_1, and radius of gyration with respect

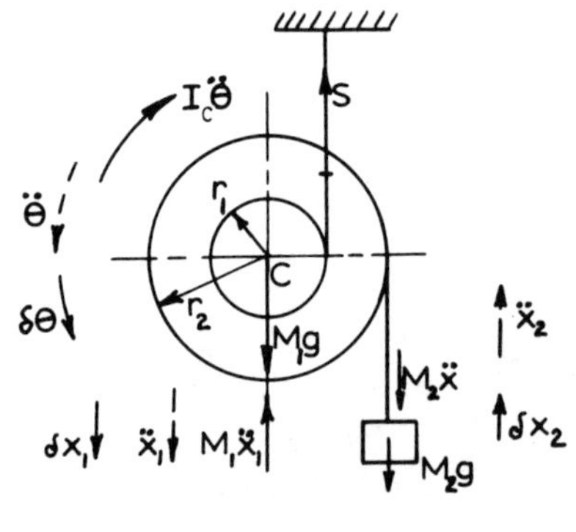

Fig. 7.12.

to a central axis through C equal to r_C. The pulley carries a mass M_2 as shown. A second hub and vertical string on the far side make the vertical plane a plane of symmetry. The mass M_2 is accelerating upwards.

Determine the acceleration of the mass M_2 by applying the principle of virtual work.

Solution

The system is shown in dynamic equilibrium with all inertia forces and external forces and torques.

Giving the centre C a virtual displacement δx_1 as shown, the virtual work is

$$\delta(\text{W.D.}) = (M_1 g - M_1 \ddot{x})\delta x_1 - I_C \ddot{\theta}\delta\theta - (M_2 g + M_2 \ddot{x})\,\delta x_2 = 0$$

Substituting

$$\delta\theta = \frac{\delta x_1}{r_1} \quad \delta x_2 = r_2\,\delta\theta - r_1\delta\theta = (r_2 - r_1)\,\frac{\delta x_1}{r}$$

and cancelling δx_1 we obtain

$$(M_1 g - M_1 \ddot{x}_1) - M_1 r_C^2\,\frac{\ddot{\theta}}{r_1} - (M_2 g + M_2 \ddot{x})\,\frac{r_2 - r_1}{r_1} = 0$$

Substituting

$$\ddot{x}_1 = \frac{r_1}{r_2 - r_1}\,\ddot{x}_2 \quad \ddot{\theta} = \frac{\ddot{\mathrm{x}}_1}{\mathrm{r}_1} = \frac{\ddot{\mathrm{x}}_2}{\mathrm{r}_2 - \mathrm{r}_1}$$

and solving for $\ddot{x}_2$ gives the result

$$\ddot{x}_2 = \frac{[(M_1 + M_2)r_1 - M_2 r_2](r_2 - r_1)g}{M_1(r_1^2 + r_C^2) + M_2(r_2 - r_1)^2}$$

The advantage of the principle of virtual work is, as in statics, that the work of internal forces, fixed reactions and normal forces does not enter the expression for vitual work. If friction forces are present, they must be included with the other acting forces.

Using x_2 as the generalised coordinate, we have

$$T = \tfrac{1}{2}\,M_2 \dot{x}_2^2 + \tfrac{1}{2}\,M_1\,\dot{x}_1^2 + \tfrac{1}{2}\,I_C \dot{\theta}^2$$

Substituting

$$\dot{x}_1 = \frac{r_1}{r_2 - r_1}\,\dot{x}_2 \quad \dot{\theta} = \frac{\dot{x}_2}{r_2 - r_1} \quad \text{and} \quad I_C = M_1 r_C^2$$

leads to

$$T = \tfrac{1}{2}\left[M_2 + M_1\left(\frac{r_1}{r_2 - r_1}\right)^2 + \frac{M_1 r_C^2}{(r_2 - r_1)^2}\right]\dot{x}_2^2 = \tfrac{1}{2}A\,\dot{x}_2^2$$

$$\frac{\mathrm{d}}{\mathrm{d}t}\frac{\partial T}{\partial \dot{x}_2} = A\,\ddot{x}_2 \quad \text{and} \quad \frac{\partial T}{\partial x_2} = 0$$

$$\delta(\text{W.D.}) = Q_{x_2}\,\delta x_2 = -M_2 g\,\delta x_2 + M_1 g\,\delta x_1$$
$$= \left[-M_2 g + M_1 g\,\frac{r_1}{r_2 - r_1}\right]\delta x_2$$

Substituting in Lagrange's equation, we find

$$A\ddot{x}_2 = \left[M_1\,\frac{r_1}{r_2 - r_1} - M_2\right] g$$

or the previous result.

7.9 IMPULSE–MOMENTUM EQUATIONS IN PLANE MOTION

The two force equations and the Euler equation may be integrated directly to give an alternative form of the equations of motion for a body in plane motion.

Taking the coordinates of the centre of mass as (x,y), introducing $\omega_z = \omega$ and assuming that the mass M and moment of inertia I_z are constants, we find

$$\int_1^2 F_x\,\mathrm{d}t = \int_1^2 M\,\mathrm{d}\dot{x} = M(\dot{x}_2 - \dot{x}_1)$$
$$\int_1^2 F_y\,\mathrm{d}t = \int_1^2 M\,\mathrm{d}\dot{y} = M(\dot{y}_2 - \dot{y}_1) \qquad (7.11)$$
$$\int_1^2 M_z\,\mathrm{d}t = \int_1^2 I_z\,\mathrm{d}\omega = I_z(\omega_2 - \omega_1)$$

The equations are *the impulse–momentum equations in plane motion*. The two linear impulse–momentum equations state that the impulse of a force in a certain time and direction is equal to the *total change* in linear momentum of the body in the same time and direction.

The third equation states that the total angular impulse about the z-axis through the centre of mass is equal to the *total change* in angular momentum about that axis in the same time.

Equations (7.11) are simply a different form of the original equations of motion, and therefore contain the same information as then. The new equations are, however, convenient in certain problems.

If the force components and the moment M_z *are constants* during a time t, equations (7.11) take the simpler form

$$\left.\begin{aligned} F_x t &= M(\dot{x}_2 - \dot{x}_1) \\ F_y t &= M(\dot{y}_2 - \dot{y}_1) \\ M_z t &= I_z(\omega_2 - \omega_1) \end{aligned}\right\} \tag{7.12}$$

If there is no impulse in the x direction in a certain time interval, we have $\int_1^2 F_x \, dt = 0$, and $M\dot{x}_2 = M\dot{x}_1$, or the linear momentum in that direction *is conserved in* that time interval. In the same manner, if the angular impulse in a certain time is $\int_1^2 M_z dt = 0$, $I_z\omega = \text{constant}$, so that the angular momentum is conserved during that time; this is called *the principle of conservation of angular momentum.*

Example 7.12
Fig. 7.13 shows a solid, homogeneous right circular cylinder of mass M and radius r.

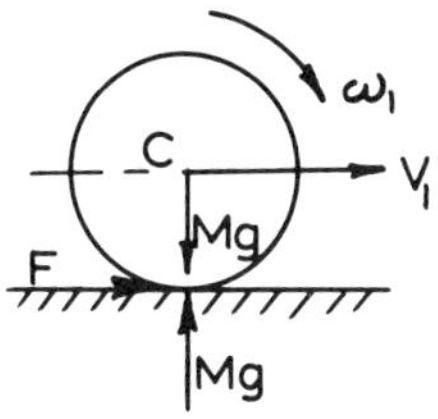

Fig. 7.13.

The cylinder is initially rotating about its axis with an angular velocity ω_1 directed as shown, while the linear velocity of the centre C is $V_1 = 0$. The cylinder is placed on a horizontal plane with coefficient of friction μ between the cylinder and the plane. Determine the angular velocity of the cylinder when pure rolling starts, the time of slipping and the distance moved by the centre C in this time.

Solution
As long as slipping occurs, the friction force $F = \mu Mg$. This force accelerates the centre C and slows down the rotation; pure rolling occurs when the angular velocity is ω_2 and the linear velocity is $V_2 = r\omega_2$. If the slipping time is t, the impulse–momentum equations (7.12) give the following results:

$$Ft = M(V_2 - V_1)$$
$$-(Fr)t = I(\omega_2 - \omega_1) = \tfrac{1}{2}Mr^2(\omega_2 - \omega_1)$$

Substituting $F=\mu Mg$, $V_1=0$ and $V_2=r\omega_2$ leads to

$$\mu Mgt = Mr\omega_2$$
$$\mu Mgrt = \tfrac{1}{2} Mr^2(\omega_1-\omega_2)$$

from which $r\omega_2=\frac{1}{2}r(\omega_1-\omega_2)$, or $\omega_2=\frac{1}{3}\omega_1$. The linear velocity when pure rolling starts is $V_2=r\omega_2=\frac{1}{3}r\omega_1$. The time of slipping is $t=r\omega_2/\mu g=r\omega_1/3\mu g$. Since the friction force is constant, the distance moved by the centre is

$$x=\tfrac{1}{2}(V_1-V_2)t=\tfrac{1}{2}V_2t=\tfrac{1}{6}r\omega_1 t=r^2\omega_1^2/18\ \mu g$$

Taking the displacement of the centre C as coordinate x, and the angle of rotation as θ, the *equations of motion* are

$$F=\mu Mg=M\ddot{x} \quad \text{and} -Fr=-\mu Mgr=\frac{Mr^2}{2}\ddot{\theta}$$

so that $\ddot{x}=\mu g$ and $\ddot{\theta}=-2\mu g/r$.

Integrating and using $V_1=0$ and $\omega=\omega_1$ at $t=0$ leads to $\dot{x}=\mu gt$ and $\dot{\theta}=-2\mu gt/r+\omega_1$.

When $\dot{\theta}=\omega_2$, $\dot{x}=V_2=r\omega_2$ so that $\mu gt=r\omega_2$ and $-2\mu gt/r+\omega_1=\omega_2$, these two equations are the same as the two impulse–momentum equations. Solving by *Lagrange's equations* we have

$$T=\tfrac{1}{2}M\dot{x}^2+\tfrac{1}{2}\frac{Mr^2}{2}\dot{\theta}^2$$

with

$$\frac{\mathrm{d}}{\mathrm{d}t}\frac{\partial T}{\partial \dot{x}}=M\ddot{x} \quad \text{and} \quad \frac{\partial T}{\partial x}=0$$

Fixing θ, we find $\delta(\text{W.D.})=Q_x\,\delta x=F\delta x=\mu Mg\delta x$, so that $M\ddot{x}=\mu Mg$ as before.

$$\frac{\mathrm{d}}{\mathrm{d}t}\frac{\partial T}{\partial \dot{\theta}}=\frac{Mr^2}{2}\ddot{\theta} \quad \text{and} \quad \frac{\partial T}{\partial \theta}=0$$

Fixing x, we find $\delta(\text{W.D.})=Q_\theta\delta\theta=-F_1r\delta\theta=-\mu Mgr\delta\theta$, and $(Mr^2/2)\ddot{\theta}=-\mu Mgr$, as before.

The greatest use of the impulse–momentum equation in plane motion is, perhaps, in *impact problems* in plane motion to be considered next.

7.10 IMPACT IN PLANE MOTION

For impact between rigid bodies in plane motion, we apply the same rules as for impact between particles (section 2.9). These rules are as follows.

1. The impact forces are so great that other forces such as gravity, friction, etc., may be neglected during impact.
2. The time of impact is so short that no appreciable motion can take place. The configuration just *after* impact may therefore be assumed to be the same as just *before* impact.

Since the impact forces and the impact time are generally unknown, equations of motion during impact cannot be established, but an *approximate* solution may be obtained by applying the above rules, together with the concept of *moment of momentum*.

Fig. 7.14 shows a rigid body in *general motion*. The instantaneous velocity of the centre of mass is $\mathbf{V}_C$ and the instantaneous angular velocity vector is $\boldsymbol{\omega}$ as shown.

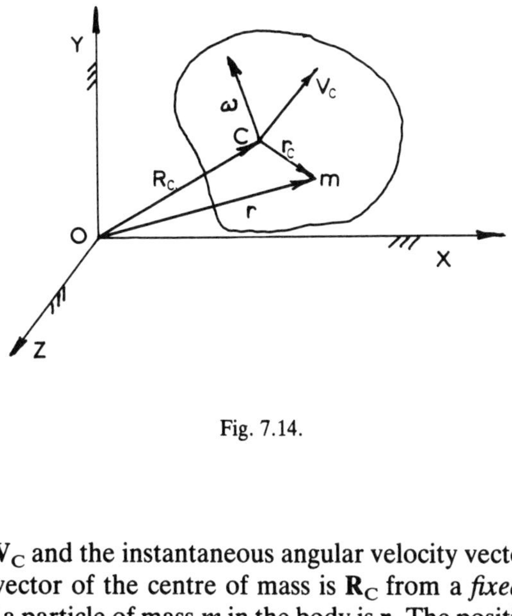

Fig. 7.14.

The position vector of the centre of mass is $\mathbf{R}_C$ from a *fixed point* O, and the position vector of a particle of mass m in the body is $\mathbf{r}$. The position vector from the centre of mass to the particle is $\mathbf{r}_C$. The velocity of the particle is $\dot{\mathbf{r}}$ and its momentum is $m\dot{\mathbf{r}}$. The moment of momentum of the particle in point O is $\mathbf{h}_0 = \mathbf{r} \times m\dot{\mathbf{r}}$, and for the rigid body

$$\mathbf{H}_0 = \Sigma \mathbf{r} \times m\dot{\mathbf{r}} \tag{7.13}$$

if the resultant of external and internal forces on the particle is $\mathbf{F}'$, the moment of $\mathbf{F}'$ in point O is $\mathbf{m}_0 = \mathbf{r} \times \mathbf{F}'$.

Summing up for all particles, the internal forces vanish in the summation, and we obtain the moment of all *external* forces $\mathbf{M}_0$ in point O, with $\mathbf{M}_0 = \Sigma \mathbf{r} \times \mathbf{F}'$. From Newton's law we have $\mathbf{F}' = m\ddot{\mathbf{r}}$ so that

$$\mathbf{M}_0 = \Sigma \mathbf{r} \times m\ddot{\mathbf{r}} = \frac{\mathrm{d}}{\mathrm{d}t}(\Sigma \mathbf{r} \times m\dot{\mathbf{r}}) - \Sigma \dot{\mathbf{r}} \times m\dot{\mathbf{r}} = \frac{\mathrm{d}}{\mathrm{d}t}(\Sigma \mathbf{r} \times m\dot{\mathbf{r}})$$

Comparing this to (7.13) we have

$$\mathbf{M}_0 = \frac{\mathrm{d}}{\mathrm{d}t}\mathbf{H}_0 = \dot{\mathbf{H}}_0 \tag{7.14}$$

valid for *any fixed point* O.

The moment of momentum in the centre of mass C is

$$\mathbf{H}_\mathrm{C} = \Sigma \mathbf{r}_\mathrm{C} \times m\dot{\mathbf{r}} \tag{7.15}$$

From the figure we have $\mathbf{r} = \mathbf{R}_\mathrm{C} + \mathbf{r}_\mathrm{C}$, from which $\dot{\mathbf{r}} = \dot{\mathbf{R}}_\mathrm{C} + \dot{\mathbf{r}}_\mathrm{C}$. Substituting this gives $\mathbf{H}_\mathrm{C} = \Sigma \mathbf{r}_\mathrm{C} \times m(\dot{\mathbf{R}}_\mathrm{C} + \dot{\mathbf{r}}_\mathrm{C}) = \Sigma \mathbf{r}_\mathrm{C} \times m\dot{\mathbf{R}}_\mathrm{C} + \Sigma \mathbf{r}_\mathrm{C} \times m\dot{\mathbf{r}}_\mathrm{C}$. The term $\Sigma \mathbf{r}_\mathrm{C} \times m\dot{\mathbf{R}}_\mathrm{C} = (\Sigma m\mathbf{r}_\mathrm{C}) \times \dot{\mathbf{R}}_\mathrm{C} = \mathbf{0}$, since $\Sigma m\mathbf{r}_\mathrm{C} = \mathbf{0}$, when $\mathbf{r}_\mathrm{C}$ is measured from C.

We have now

$$\mathbf{H}_\mathrm{C} = \Sigma \mathbf{r}_\mathrm{C} \times m\dot{r}_\mathrm{C} \tag{7.16}$$

Comparing to (7.15) shows that $\mathbf{H}_\mathrm{C}$ may be determined by using the *absolute* velocity $\dot{\mathbf{r}}$ of each particle, or by using the *relative* velocity $\dot{\mathbf{r}}_\mathrm{C}$ of each particle, the result is the same.

We may relate $\mathbf{H}_0$ to H_C by substituting $\mathbf{r} = \mathbf{R}_\mathrm{C} + \mathbf{r}_\mathrm{C}$, and $\dot{\mathbf{r}} = \dot{\mathbf{R}}_\mathrm{C} + \dot{r}_\mathrm{C}$ in (7.13), which gives

$$\begin{aligned}\mathbf{H}_0 &= \Sigma(\mathbf{R}_\mathrm{C} + \mathbf{r}_\mathrm{C}) \times m\,(\dot{\mathbf{R}}_\mathrm{C} + \dot{\mathbf{r}}_\mathrm{C}) \\ &= \Sigma \mathbf{R}_\mathrm{C} \times m\dot{\mathbf{R}}_\mathrm{C} + \Sigma \mathbf{R}_\mathrm{C} \times m\dot{\mathbf{r}}_\mathrm{C} + \Sigma \mathbf{r}_\mathrm{C} \times m\dot{\mathbf{R}}_\mathrm{C} + \Sigma \mathbf{r}_\mathrm{C} \times m\dot{\mathbf{r}}_\mathrm{C}\end{aligned}$$

The second term is

$$\Sigma \mathbf{R}_\mathrm{C} \times m\dot{\mathbf{r}}_\mathrm{C} = \mathbf{R}_\mathrm{C} \times (\Sigma m\dot{\mathbf{r}}_\mathrm{C}) = \mathbf{0}$$

The third term is

$$\Sigma \mathbf{r}_\mathrm{C} \times m\dot{\mathbf{R}}_\mathrm{C} = (\Sigma m\mathbf{r}_\mathrm{C}) \times \dot{\mathbf{R}}_\mathrm{C} = \mathbf{0}$$

We have now

$$\mathbf{H}_0 = (\Sigma m)\ \mathbf{R}_C \times \dot{\mathbf{R}}_C + \Sigma \mathbf{r}_C \times m\dot{\mathbf{r}}_C$$

Substituting

$$\Sigma m = M \quad \dot{\mathbf{R}}_C = \mathbf{V}_C \quad \text{and} \quad \Sigma \mathbf{r}_C \times m\dot{\mathbf{r}}_C = \mathbf{H}_C$$

we obtain finally

$$\mathbf{H}_0 = \mathbf{R}_C \times M\mathbf{V}_C + \mathbf{H}_C \tag{7.17}$$

For a rigid body we have from (7.15) by summing up for all elements that

$$\mathbf{H}_C = \int \mathbf{r}_C \times \dot{\mathbf{r}}\ \mathrm{d}m = \int \mathbf{r}_C \times \mathbf{V}\ \mathrm{d}m$$

where $\mathbf{V}$ is the velocity of the element dm.

Using the centre of mass as a pole we have $\mathbf{V} = \mathbf{V}_C + \boldsymbol{\omega} \times \mathbf{r}_C$, so that

$$\begin{aligned}\mathbf{H}_C &= \int \mathbf{r}_C \times (\mathbf{V}_C + \boldsymbol{\omega} \times \mathbf{r}_C)\ \mathrm{d}m \\ &= \int \mathbf{r}_C \times \mathbf{V}_C\ \mathrm{d}m + \int \mathbf{r}_C \times (\boldsymbol{\omega} \times \mathbf{r}_C)\ \mathrm{d}m\end{aligned}$$

The first term is $\int \mathbf{r}_C \times \mathbf{V}_C\ \mathrm{d}m = [\int \mathbf{r}_C\ \mathrm{d}m] \times \mathbf{V}_C = \mathbf{0}$, so that

$$\mathbf{H}_C = \int \mathbf{r}_C \times (\boldsymbol{\omega} \times \mathbf{r}_C)\ \mathrm{d}m \tag{7.18}$$

For a rigid body in *plane motion*, we fix a coordinate system xyz in the body with origin at the centre of mass and the z-axis perpendicular to the plane of motion; we then have

$$\mathbf{r}_C = x\mathbf{i} + y\mathbf{j} + z\mathbf{k}$$

for an element dm, and $\omega = \omega_z\mathbf{k}$. Using the formula

$$\mathbf{a} \times (\mathbf{b} \times \mathbf{c}) = (\mathbf{a}.\mathbf{c})\mathbf{b} - (\mathbf{a}.\mathbf{b})\mathbf{c}$$

we obtain

$$\mathbf{r}_C \times (\boldsymbol{\omega} \times \mathbf{r}_C) = (\mathbf{r}_C \cdot \mathbf{r}_C)\boldsymbol{\omega} - (\mathbf{r}_C \cdot \boldsymbol{\omega})\mathbf{r}_C$$

$$= (x^2 + y^2 + z^2)\omega_z\mathbf{k} - (z\omega_z)(x\mathbf{i} + y\mathbf{j} + z\mathbf{k})$$

$$= -xz\omega_z\mathbf{i} - yz\omega_z\mathbf{j} + (x^2 + y^2)\omega_z\mathbf{k}$$

Multiplying by dm and summing up for all elements, we find the components of $\mathbf{H}_C$ as

$$\left.\begin{aligned} H_{Cx} &= -\omega_z \int xz \, dm = -I_{xz}\omega_z \\ H_{Cy} &= -\omega_z \int yz \, dm = -I_{yz}\omega_z \\ H_{Cz} &= \omega_z \int (x^2+y^2) \, dm = I_z\omega_z \end{aligned}\right\} \tag{7.19}$$

valid for *plane motion* of a rigid body.

If the z-axis is a *principal axis*, we have $I_{xz} = I_{yz} = 0$ and (7.19) simplifies to

$$\left.\begin{aligned} H_{Cx} &= 0 \\ H_{Cy} &= 0 \\ H_{Cz} &= I_z\omega_z \end{aligned}\right\} \tag{7.20}$$

and in this case $\mathbf{H}_C$ is along the z-axis with magnitude $H_{Cz} = I_z\omega_z$.

In the formula (7.17), $\mathbf{H}_0 = \mathbf{R}_C \times M\mathbf{V}_C + \mathbf{H}_C$; in the special case of *plane motion*, the vectors $\mathbf{R}_C$ and $\mathbf{V}_C$ are in the plane of motion, and consequently $\mathbf{R}_C \times M\mathbf{V}_C$ is perpendicular to the plane of motion, and is the moment of the vector $M\mathbf{V}_C$ in point O. The magnitude of $\mathbf{R}_C \times M\mathbf{V}_C$ is now the scalar moment of the vector $M\mathbf{V}_C$ about the Z-axis through point O, with magnitude MV_Ca, where a is the perpendicular distance from point O to the line of action of $\mathbf{V}_C$. The total magnitude of $\mathbf{H}_0$ is now

$$H_0 = MV_Ca + I_z\omega_z \tag{7.21}$$

The formula (7.21) is valid for a rigid body in *plane motion* with the z-axis as a *principal* axis through the centre of mass and *perpendicular* to the plane of motion; point O is a *fixed point* in the plane of motion.

The terms in (7.21) must be taken with *proper signs*, where the positive direction of rotation is anti-clockwise. Sometimes the two terms have opposite signs.

For a fixed point O, we have the equation (7.14): $\mathbf{M}_0 = \dot{\mathbf{H}}_0$. If $\mathbf{M}_0 = \mathbf{0}$ during a certain time, we have $\dot{\mathbf{H}}_0 = \mathbf{0}$, or $\mathbf{H}_0$ is a vector in a fixed direction and with constant magnitude during the same time. In impact problems we consider only the impact force acting during impact; these forces act through point O and by taking moments about the point of impact O, we have then $\mathbf{M}_0 = \mathbf{0}$, so that $\mathbf{H}_0$ is constant during impact, which means that H_0 *is the same* just *before and after impact*.

Example 7.13

Fig. 7.15 shows a right circular disc rolling without slipping on a horizontal plane. The velocity of the centre of mass C is V, and the angular velocity is ω. The disc is

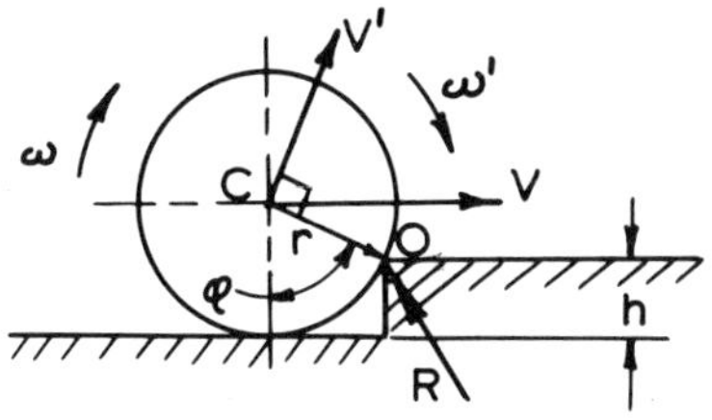

Fig. 7.15.

shown at the moment of impact with a step of height h. Assuming that there is no slip or rebound, determine the velocity of C just after impact. Investigate the conditions under which the cylinder will continue along the higher plane.

Solution

The moment of inertia of the cylinder is $I = Mr^2/2$ and we have $V = r\omega$. Taking moments about point O, and considering only the impact force R, we have $\mathbf{M}_0 = \mathbf{0} = \dot{\mathbf{H}}_0$, so that H_0 is constant during impact. *Just before* impact we find from (7.21) that

$$H_0 = (MV)r \cos \phi + I\omega = MVr(\cos \phi + \tfrac{1}{2})$$

Just after impact, the point O becomes the new instantaneous centre, so that V' is directed as shown. We find then $H_0' = MV'r + I\omega'$, with $V' = r\omega'$. This becomes $H_0' = \tfrac{3}{2}MV'r$, and so taking $H_0' = H_0$, we find $V' = \tfrac{1}{3}V(1 + 2\cos\phi)$, which determines the motion just after impact. If $\phi = 0$, there is no step and we obtain $V' = V$; if $\phi = 90°$, the result is $V' = V/3$.

The work-energy equation may be used to determine whether the cylinder will continue along the upper plane. The kinetic energy after impact is

$$T_1 = \tfrac{1}{2}MV'^2 + \tfrac{1}{2}I\omega'^2$$

while the work done in moving up on the step is W.D. = $-Mgh$, so that

$$T_2 - T_1 = -Mgh \quad \text{or} \quad T_2 = T_1 - Mgh$$

If this expression is positive, the cylinder will roll along the upper plane; the velocity in this motion may be determined from $T_2 = \tfrac{1}{2}MV_2^2 + \tfrac{1}{2}I\omega_2^2$.

The problem may also be solved by using the impulse–momentum equations (7.11). This method has the advantage of giving expressions for the acting impulses. Resolving the impact force R in a normal component R_n along the radial line OC, and a tangential component R_t perpendicular to OC, we find, by writing the impulses

$$\int_0^t R_n \, dt = R_n' \quad \int_0^t R_t \, dt = R_t' \quad \text{and} \quad \int_0^t R_t r \, dt = R_t' r$$

the equations

$$-R'_n = M(0 - V \sin \phi)$$

$$R'_t = M(V' - V \cos \phi)$$

$$-R'_t r = I_C(\omega' - \omega) = \frac{Mr^2}{2}(\omega' - \omega)$$

Dividing the third equation by r and adding to the second leads to

$$(V' - V \cos \phi) + \frac{r}{2}(\omega' - \omega) = 0$$

Substituting $\omega = V/r$, $\omega' = V'/r$ and solving for V' gives the results

$$V' = \frac{V}{3}(1 + 2 \cos \phi) \quad \text{and} \quad \omega' = \frac{V}{3r}(1 + 2 \cos \phi)$$

as before. Substituting back in the equations gives the impulses

$$R'_n = MV \sin \phi \quad R'_t = \frac{MV}{3}(1 - \cos \phi)$$

If the disc hits a vertical wall, $\phi = 90°$ and $V' = V/3$; the impulses are $R'_n = MV$, which changes the horizontal momentum to zero; $R'_t = MV/3$, and the angular impulse $-R'_t r = -MV/3r$; this is the same as the change in angular momentum $Mr^2/2$ $(\omega' - \omega) = MV/3r$.

Example 7.14
A slender rigid uniform bar of length l and mass M is falling without rotation in a vertical plane, as shown in Fig. 7.16. The lower end of the bar comes into contact with a *smooth* horizontal plane, and the velocity of the centre of mass C at this instant is V. Determine the motion of the bar after impact. Assume that there is no rebound at the impact.

Solution
Before, during, and after the impact, all forces acting are vertical; there is therefore no horizontal acceleration of the centre of mass C, and the horizontal component of the velocity of C, which is zero before impact, must remain zero; this means that C moves down in a vertical line, so that the velocity V' after impact is vertical.

During the impact, $M_p = 0 = \dot{H}_p$, or $H_p = \text{constant}$.

Before impact, $H_p = \frac{1}{2} MVl \cos \alpha + I\omega = \frac{1}{2} MVl \cos \alpha$.

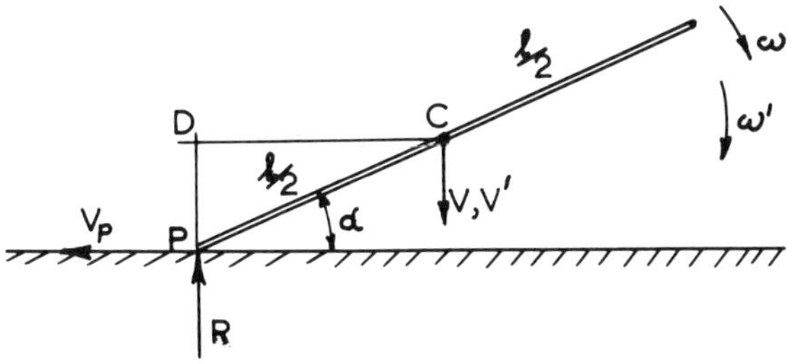

Fig. 7.16.

After impact, $H'_p = \frac{1}{2} MV'l \cos\alpha + I\omega' = \frac{1}{2}MV'l\cos\alpha + \frac{1}{12}Ml^2\omega'$.
Equating the two expressions leads to the result

$$\omega' = (6/l)\,(V - V')\cos\alpha$$

Since this expression contains two unknown quantities ω' and V', the conservation of angular momentum is not sufficient to solve the problem. A second equation may be established kinematically by using the instantaneous centre D after impact; this gives the relationship $V' = (\frac{1}{2}l\cos\alpha)\,\omega'$, and the solution is

$$V' = \frac{3V\cos^2\alpha}{1 + 3\cos^2\alpha} \qquad \omega' = \frac{6V\cos\alpha}{l(1 + 3\cos^2\alpha)}$$

If $\alpha = 90°$, the bar is vertical, and $V' = \omega' = 0$. The solution does not apply for $\alpha = 0$, since then the bar would be in impact along its length; for $\alpha \rightarrow 0$ the result is $V' \rightarrow \frac{3}{4}V$, $\omega' \rightarrow \frac{3}{2}V/l$.

Taking

$$\int_0^t R\,\mathrm{d}t = R' \quad \text{and} \quad \int_0^t R\frac{l}{2}\cos\alpha\,\mathrm{d}t = I' = \frac{l}{2}\cos\alpha\,R'$$

the impulse–momentum equations (7.11) lead to

$$-R' = M(V' - V)$$

and

$$I' = \frac{l}{2}\cos\alpha \quad R' = I_C(\omega' - \omega) = \frac{Ml^2}{12}\,\omega'$$

Dividing the second equation by $l/2\cos\alpha$ and adding the equations, we obtain the equation

$$(V' - V) + \frac{l}{6 \cos \alpha} \omega' = 0$$

or

$$\omega' = \frac{6}{l} (V' - V) \cos \alpha$$

as before. The rest of solution is as before by using the instantaneous centre.

The linear impulse is

$$R' = M(V - V') = \frac{MV}{1 + 3 \cos^2 \alpha}$$

This is the magnitude of the change in linear momentum. The angular impulse is

$$I' = MVl \cos \alpha/2(1 + 3 \cos^2 \alpha)$$

This is the change in angular momentum.

7.11 EULER'S EQUATIONS FOR A RIGID BODY IN PLANE MOTION

In all the cases in plane motion of a rigid body considered so far, the body was homogeneous and symmetrical with respect to the plane of motion through the centre of mass, so that the products of inertia I_{xz} and I_{yz} was zero, and the z-axis of rotation through the centre of mass and perpendicular to the plane of motion was a *principal* axis. The acting forces were all in the plane of motion or symmetrically distributed with respect to the plane of motion. In these cases the motion could be completely determined by the two Newton equations for the motion of the centre of mass and the Euler equation for the rotational motion, and no additional forces were necessary to hold the body in the plane of motion. The majority of cases in engineering dynamics are of this type. Sometimes we meet more complicated cases,where the z-axis is *not* a principal axis, and where the acting forces are not symmetrical with respect to the plane of motion. In these cases, additional forces are necessary to hold the body in the plane of motion, and we need additional equations to determine the forces. One additional equation available is the third Newton equation for the motion of the centre of mass. Since there is no motion perpendicular to the plane of motion we have $F_Z = F_z = 0$, or the forces perpendicular to the plane of motion must be balanced. Two additional equations may be established by taking moments of the forces about the x- and y-axis; these moment equations may be developed as follows.

For the body in Fig. 7.1 the acceleration components of a point in the body were determined by the equations (7.3)

$$A_x = A_{Cx} - x\omega_z^2 - y\dot{\omega}_z$$

$$A_y = A_{Cy} + x\dot{\omega}_z - y\omega_z^2$$

$$A_z = 0$$

The resultant of all external and internal forces on an element dm at point P is $\mathbf{F} = \mathrm{d}m\mathbf{A} = \mathbf{d}m\ A_x\mathbf{i} + \mathrm{d}m\ A_y\mathbf{j}$.

Introducing a position vector $\mathbf{r} = x\mathbf{i} + y\mathbf{i} + z\mathbf{k}$ from the centre of mass C to point P in Fig. 7.1, the moment of $\mathbf{F}$ in the centre of mass is

$$\mathbf{m}_C = \mathbf{r} \times \mathbf{F} = \mathbf{r} \times \mathrm{d}m\mathbf{A} = (\mathrm{d}m)\begin{vmatrix} \mathbf{i} & \mathbf{j} & \mathbf{k} \\ x & y & z \\ A_x & A_y & 0 \end{vmatrix}$$

$$= (dm)\ [-zA_y\mathbf{i} + zA_x\mathbf{j} + (xA_y - yA_x\mathbf{k}]$$

The components of $\mathbf{m}_C$ are the moments of $\mathbf{F}$ about the x-, y- and z-axes respectively. Summing up for all particles, the internal forces cancel in the summation and we are left with the moments M_x, M_y and M_z about the x-, y- and z-axes of all *external* forces. We have now

$$\begin{aligned} M_x &= \int -zA_y\ \mathrm{d}m = \int(-z\ A_{Cy} - zx\ \dot{\omega}_z + zy\ \omega_z^2)\ \mathrm{d}m \\ &= -A_{Cy}\int z\ \mathrm{d}m - \dot{\omega}_z\int xz\ \mathrm{d}m + \omega_z^2\int yz\ \mathrm{d}m \end{aligned}$$

$$\begin{aligned} M_y &= \int zA_x\ \mathrm{d}m = \int(z\ A_{Cx} - xz\ \omega_z^2 - yz\ \dot{\omega}_z)\ \mathrm{d}m \\ &= A_{Cx}\int z\ \mathrm{d}m - \omega_z^2\int xz\ \mathrm{d}m - \dot{\omega}_z\int yz\ \mathrm{d}m \end{aligned}$$

$$\begin{aligned} M_z &= \int(x\ Ay - yA_x)\ \mathrm{d}m = \int(x\ A_{Cy} + x^2\ \dot{\omega}_z - xy\ \omega_z^2 - \\ &\qquad yA_{Cx} + xy\ \omega_z^2 + y^2\ \dot{\omega}_z)\ \mathrm{d}m \\ &= A_{Cy}\int x\ \mathrm{d}m + \dot{\omega}_z\int(x^2 + y^2)\ \mathrm{d}m - A_{Cx}\int y\ \mathrm{d}m \end{aligned}$$

Since the x-, y- and z-axes are located at the centre of mass, we have

$$\int x\ \mathrm{d}m = \int y\ \mathrm{d}m = \int z\ \mathrm{d}m = 0$$

We also have

$$\int xz\ \mathrm{d}m = I_{xz} \quad \int yz\ \mathrm{d}m = I_{yz}$$

and

$$\int(x^2+y^2)\,\mathrm{d}m=I_z$$

Substituting this in the above equations results in the following three equations:

$$\begin{aligned}M_x&=-I_{xz}\,\dot{\omega}_z+I_{yz}\,\omega_z^2\\M_y&=-I_{xz}\,\omega_z^2-I_{yz}\,\dot{\omega}_z\\M_z&=I_z\,\dot{\omega}_z\end{aligned}\qquad(7.22)$$

These equations are called Euler's equations after the Swiss mathematician who was the first to develop these equations in 1736. The third equation may be seen to be the equation of rotational motion; the first two equations are additional equations to determine the necessary forces that must act to keep the body in plane motion if the z-axis is *not* principal and if the acting forces are *not* symmetrical with respect to the plane of motion. If the z-axis is principal we have $I_{xz}=I_{yz}=0$, and the first two equations simplify to $M_x=0$ *and* $M_y=0$; these equations are still necessary for non-symmetrical applied forces.

Euler's equations may also be developed by using the concept of *moment of momentum*.

Fig. 7.14 shows a rigid body in *general* motion. If the resultant of external and internal forces is $\mathbf{F}'$ on the particle with position vector $\mathbf{r}_C$, the moment of $\mathbf{F}'$ in the centre of mass is $\mathbf{m}_C=\mathbf{r}_C\times\mathbf{F}'$. Summing up for all particles the internal forces cancel in the summation and we obtain the moment $\mathbf{M}_C$ of all the *external* forces in the centre of mass, with

$$\mathbf{M}_C=\Sigma\mathbf{r}_C\times\mathbf{F}'$$

From Newton's second law we have $\mathbf{F}'=m\ddot{\mathbf{r}}$, so that

$$\mathbf{M}_C=\Sigma\mathbf{r}_C\times m\ddot{\mathbf{r}}=\frac{\mathrm{d}}{\mathrm{d}t}\,(\Sigma\mathbf{r}_C\times m\dot{\mathbf{r}})-\Sigma\dot{\mathbf{r}}_C\times m\dot{\mathbf{r}}$$

Introducing $\mathbf{r}=\mathbf{R}_C+\mathbf{r}_C$ from Fig. 7.14, we have $\dot{\mathbf{r}}=\dot{\mathbf{R}}_C+\dot{\mathbf{r}}_C$ and the second term in $\mathbf{M}_C$ becomes

$$\begin{aligned}\Sigma\dot{\mathbf{r}}_C\times m\dot{\mathbf{r}}&=\Sigma\dot{\mathbf{r}}_C\times m\,(\dot{\mathbf{R}}_C+\dot{\mathbf{r}}_C)=\Sigma\dot{\mathbf{r}}_C\times m\dot{\mathbf{R}}_C+\Sigma\dot{\mathbf{r}}_C\times m\dot{\mathbf{r}}_C=\\&=(\Sigma m\dot{\mathbf{r}}_C)\times\dot{\mathbf{R}}_C=\mathbf{0}\end{aligned}$$

From (7.15) we have $\mathbf{H}_C=\Sigma\mathbf{r}_C\times m\dot{\mathbf{r}}$; substituting this gives the result

$$\mathbf{M}_C = \frac{d}{dt}\,\mathbf{H}_C = \dot{\mathbf{H}}_C \tag{7.23}$$

(7.23) is valid for a rigid body in *general motion*. For a rigid body in *plane motion*, we have the components of $\mathbf{H}_C$ from (7.19), so that

$$\mathbf{H}_C = (-I_{xz}\,\omega_z)\mathbf{i} + (-I_{yz}\omega_z)\mathbf{j} + (I_z\omega_z)\mathbf{k}$$

Differentiating this we find

$$\dot{\mathbf{H}}_C = -I_{xz}\dot{\omega}_z\mathbf{i} - I_{yz}\dot{\omega}_z\mathbf{j} + I_z\dot{\omega}_z\mathbf{k} - I_{xz}\omega_z\dot{\mathbf{i}} - I_{yz}\omega_z\dot{\mathbf{j}} + I_z\omega_z\dot{\mathbf{k}}$$

We have now

$$\dot{\mathbf{i}} = \boldsymbol{\omega} \times \mathbf{i} = \omega_z\mathbf{k} \times \mathbf{i} = \omega_z\mathbf{j}$$

$$\dot{\mathbf{j}} = \boldsymbol{\omega} \times \mathbf{j} = \omega_z\mathbf{k} \times \mathbf{j} = -\omega_z\mathbf{i}$$

and

$$\dot{\mathbf{k}} = \boldsymbol{\omega} \times \mathbf{k} = \omega_z\mathbf{k} \times \mathbf{k} = \mathbf{0}$$

Substituting this we obtain

$$\dot{\mathbf{H}}_C = (-I_{xz}\dot{\omega}_z + I_{yz}\omega_z^2)\mathbf{i} + (-I_{yz}\dot{\omega}_z - I_{xz}\omega_z^2)\mathbf{j} + (I_z\dot{\omega}_z)\mathbf{k}$$

Since $\mathbf{M}_C = \dot{\mathbf{H}}_C$, the three scalar components of $\mathbf{M}_C$ are

$$\begin{aligned} M_x &= -I_{zx}\dot{\omega}_z + I_{yz}\omega_z^2 \\ M_y &= -I_{xz}\omega_z^2 - I_{yz}\dot{\omega}_z \\ M_z &= I_z\dot{\omega}_z \end{aligned}$$

These are the three Euler equations (7.22).

Problems where the z-axis is not a principal axis and the acting forces are not symmetrical with respect to the plane of motion are not common in engineering.

If they are encountered we have the three Newton equations for the centre of mass, and the three Euler equations for their solution. These types of problem are considered to be outside the scope of this book.

PROBLEMS

7.1 For the system shown in Fig. 7.17, determine the acceleration of the falling mass

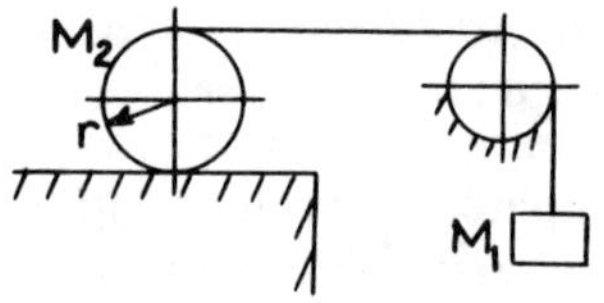

Fig. 7.17.

M_1 and the string tension. The cylinder rolls without slip and the fixed pulley is frictionless.

7.2 Fig. 7.18 shows a homogeneous spool consisting of two cylindrical discs of radius

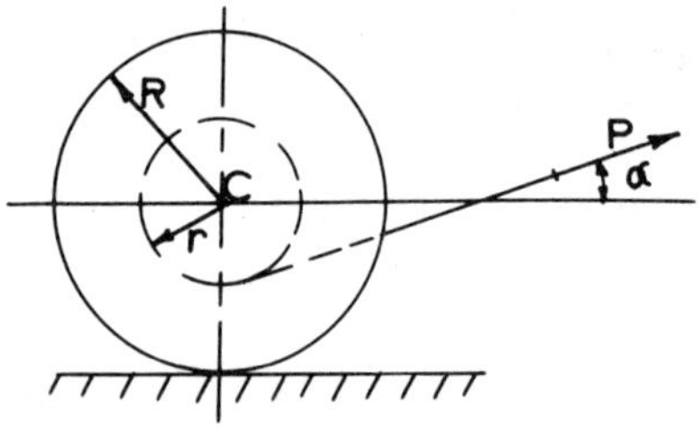

Fig. 7.18.

R rigidly connected by a cylindrical section of radius r. The spool rolls without slipping on a horizontal plane owing to the pull P in a string which is wrapped around the spool.

The force P acts in the vertical plane of symmetry of the spool and is inclined at an angle α to the horizontal direction.

The total mass of the spool is M and the moment of inertia with respect to the geometrical axis is $I = Mr_C^2$, where r_C is the radius of gyration.

(a) Determine the acceleration of the centre of mass of the spool in terms of the given system constants and the force P.
(b) Determine the limits of the angle α for the spool to roll to the right or to the left, if $r = R/2$.
(c) Determine the magnitude of the acceleration of the centre of mass of the spool, if $P = 44.5$ N; $M = 9.1$ kg; $R = 15.2$ cm; $r = R/2$; $r_C = R/3$ and $\alpha = 30°$.
(d) Determine the minimum value of the coefficient of friction between the spool and the plane, for the conditions given under (c).

7.3 A thin steel hoop of mass M and radius r starts from rest at A (Fig. 7.19) and rolls down along a circular cylindrical surface of radius a. Determine the angle θ_0 defining the position B where the hoop starts to slip, if the coefficient of friction $\mu = \frac{1}{3}$.

7.4 Find the period of oscillation of a homogeneous right semi-circular cylinder of radius 30.5 cm for small amplitudes of rolling without slip on a horizontal plane. Repeat for a homogeneous hemisphere of radius 30.5 cm.

7.5 The thin cylindrical shell in Fig. 7.20 approaches the inclined plane with velocity V, rolls up this plane and eventually rolls back to the left. Find the velocity of the shell to the left. Assume no slip or rebound.

7.6 An arrow of length 76.2 cm travelling with a velocity $V = 15.25$ m/s strikes a smooth, hard wall (Fig. 7.21). Assuming that the end A slides downward without friction or rebound, find the angular velocity of the arrow just after impact if $\alpha = 30°$. Determine also the velocity of the centre of mass just after impact.

7.7 A homogeneous right circular cylinder of mass m and radius r is rolling without slipping on a triangular prism also of mass m (Fig. 7.22). The prism of angle α is sliding without friction on a horizontal plane. Find the vertical and horizontal acceleration of the centre of gravity of the cylinder, and the acceleration of the prism, in terms of the given quantities.

7.8 A right circular homogeneous cylinder (Fig. 7.23) of radius r and mass m rolls without slipping in a semi-circular groove of radius R cut in block of mass M which is constrained to move without friction in a vertical guide. The block is supported on a spring with spring constant K. Assume that the cylinder and block are always in contact.

(a) Find the equations of motion of the system without assuming small displacements.
(b) Assuming small displacements and velocities, find the simplified equations of motion.
(c) Find the natural frequency of the rolling cylinder if the block is fixed.
(d) Find the natural frequency of the system if the cylinder is fixed to the block.

7.9 Fig. 7.24 shows a vibrating system consisting of a mass M which is sliding without friction on a horizontal plane. The mass M is connected by a light spring of spring constant K to a vertical wall. A simple pendulum consisting of a light string of length l and a concentrated mass m is attached to M as shown. A force $P(t)$ acts upon the mass M in the positive x direction.

Establish the equations of motion of the system without assuming that θ is small. Simplify the equations, if θ is assumed small, and terms of higher order

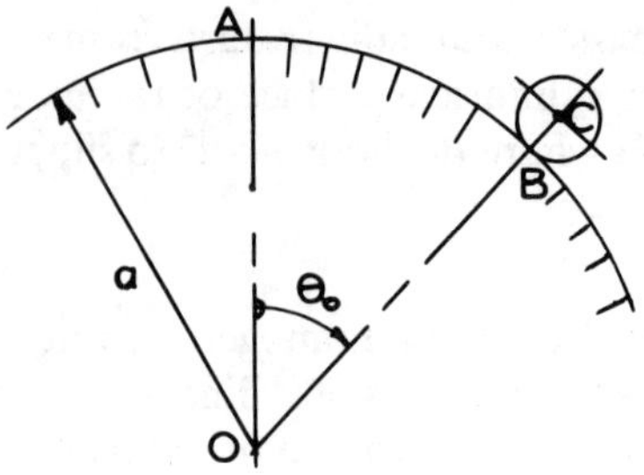

Fig. 7.19.

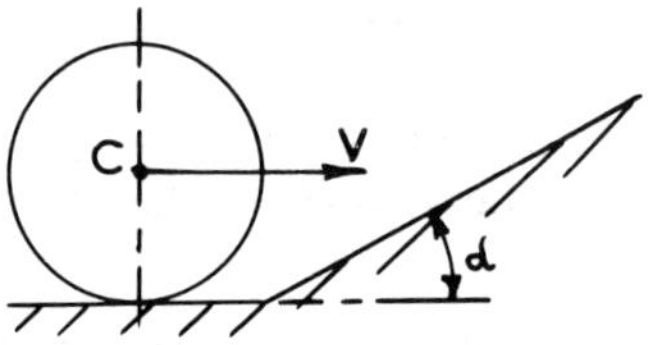

Fig. 7.20.

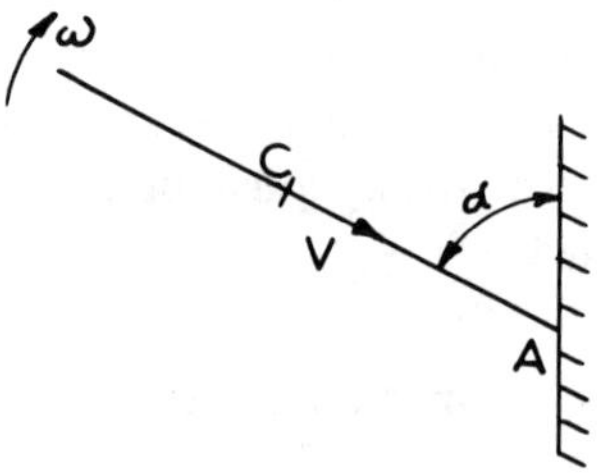

Fig. 7.21.

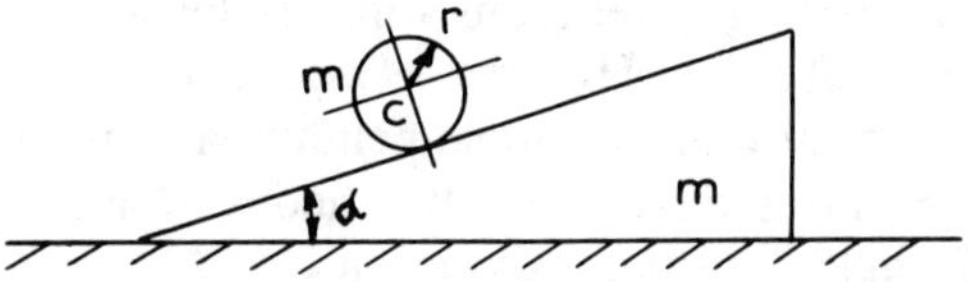

Fig. 7.22.

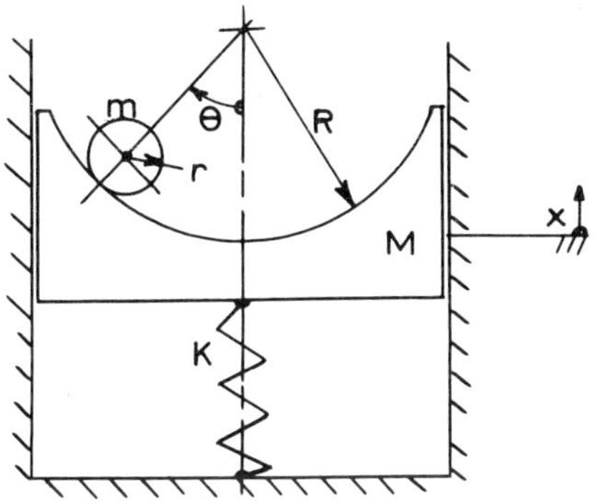

Fig. 7.23.

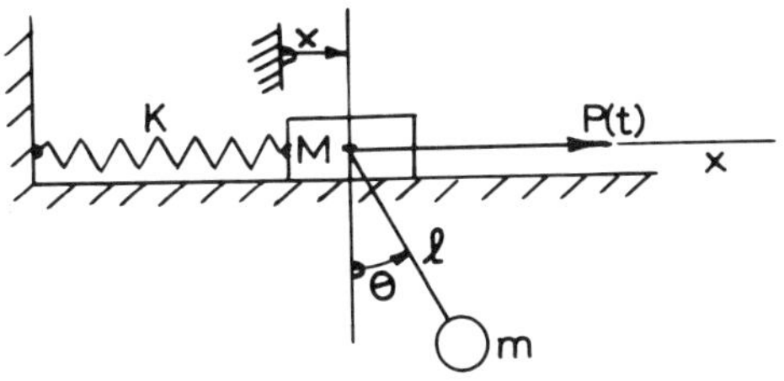

Fig. 7.24.

than the second in θ and $\dot{\theta}$ are neglected.

Establish the equation of rotational motion of the pendulum for small angular displacements for the particular case of $M = 0$ and $P(t) = 0$ and discuss the motion in detail.

8

Plane gyroscopic motion

8.1 INTRODUCTION

A symmetrical gyroscope in plane motion is here defined as a homogeneous body of revolution which is rotating about its central axis and, at the same time, this axis is rotating in a plane. Fig. 8.1 shows the gyroscope and the plane of motion which is a

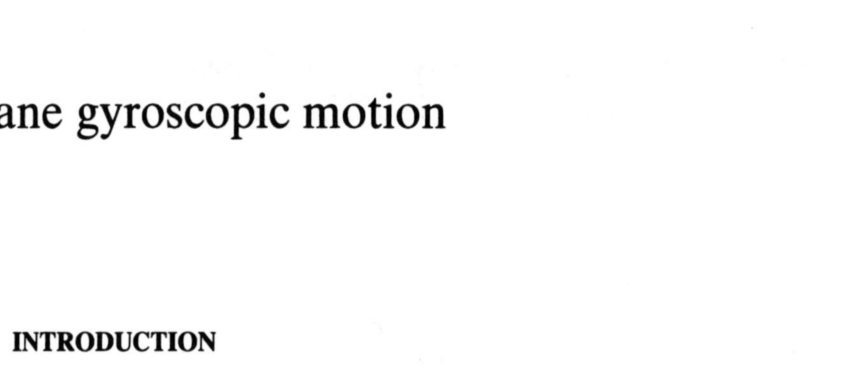

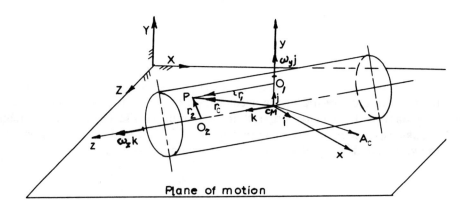

Fig. 8.1.

plane through the spin-axis. A coordinate system xyz is located in the gyroscope and with the z-axis along the spin-axis. The xyz system follows the body in its motion in the plane in such a way that the x- and z-axes stay in the plane of motion and the y-axis is a normal to the plane. The system does not, therefore, partake in the rotational motion of the body about the z-axis.

For a homogeneous body of revolution about the z-axis, the centre of mass is on the z-axis, and from symmetry the products of inertia $I_{xy} = I_{xz} = I_{yz} = 0$, so that the x-, y- and z-*axes are principal axes* at the centre of mass. The moments of inertia I_x, I_y and I_z stay constant, although the body rotates about the z-axis. We also have the further simplification that $I_y = I_x$.

Certain other bodies such as a car wheel and a three-bladed propeller fulfill the same conditions and may be classified as symmetrical gyroscopes, although they are not homogeneous bodies.

8.2 THE ACCELERATION OF A POINT IN THE GYROSCOPE

To determine the acceleration of a point in the gyroscope, consider a point P(x, y, z) in Fig. 8.1, in circular motion in a groove cut in a solid block (not shown). The xyz system is fixed in this block, and together they are taken as a base for the motion. The circular groove is in a plane perpendicular to the z-axis. The acceleration **A** of point P is determined by the well-known formula.

$\mathbf{A} = \mathbf{A}_b + \mathbf{A}_r + \mathbf{A}_{cor}$, where $\mathbf{A}_b$ is the base point acceleration, $\mathbf{A}_r$ is the relative acceleration, and $\mathbf{A}_{cor}$ is the Coriolis acceleration.

Base point acceleration $\mathbf{A}_b$

The base point is the point P fixed in the groove at the instant considered; its motion is in a plane parallel to the plane of motion of the centre of mass. The angular velocity and acceleration of the base are $\boldsymbol{\omega}_y = \omega_y\mathbf{j}$ and $\dot{\boldsymbol{\omega}}_y = \dot{\omega}_y\mathbf{j}$. Projecting P on the y-axis in the point O_1 (Fig. 8.1), we take O_1 as a pole and the vector $\mathbf{O_1P} = \mathbf{r}_1 = x\mathbf{i} + z\mathbf{k}$.

The base point acceleration $\mathbf{A}_b$ is now

$$\mathbf{A}_b = \mathbf{A}_{O_1} + \boldsymbol{\omega}_y \times (\boldsymbol{\omega}_y \times \mathbf{r}_1) + \dot{\boldsymbol{\omega}}_y \times \mathbf{r}_1$$

The first term is $\mathbf{A}_{O_1} = \mathbf{A}_{CM} = A_{C_x}\mathbf{i} + A_{C_z}\mathbf{k}$. Applying the vector formula $\mathbf{a} \times (\mathbf{b} \times \mathbf{c}) = (\mathbf{a}\cdot\mathbf{c})\mathbf{b} - (\mathbf{a}\cdot\mathbf{b})\mathbf{c}$ to the second term in $\mathbf{A}_b$ leads to $\boldsymbol{\omega}_y \times (\boldsymbol{\omega}_y \times \mathbf{r}_1) = (\boldsymbol{\omega}_y\cdot\mathbf{r}_1)\boldsymbol{\omega}_y - (\boldsymbol{\omega}_y\cdot\boldsymbol{\omega}_y)\mathbf{r}_1 = (0)\boldsymbol{\omega}_y - \omega_y^2(x\mathbf{i} + z\mathbf{k}) = -x\omega_y^2\mathbf{i} - z\omega_y^2\mathbf{k}$. The third term in $\mathbf{A}_b$ is $\dot{\boldsymbol{\omega}}_y \times \mathbf{r}_1 = \dot{\omega}_y\mathbf{j} \times (x\mathbf{i} + z\mathbf{k}) = x\dot{\omega}_y\mathbf{j} \times \mathbf{i} + z\dot{\omega}_y\mathbf{j} \times \mathbf{k} = -x\dot{\omega}_y\mathbf{k} + z\dot{\omega}_y\mathbf{j}$.

Accumulating the **i**, **j** and **k** terms gives the components of $\mathbf{A}_b$:

$$A_{b_x} = A_{C_x} - x\omega_y^2 + z\dot{\omega}_y \quad A_{b_y} = 0 \quad \text{and} \quad A_{b_z} = A_{C_z} - z\omega_y^2 - x\dot{\omega}_y$$

Relative acceleration $\mathbf{A}_r$

The relative motion of the point P is determined by *fixing the base*. The relative motion is then circular motion around the fixed z-axis, with relative angular velocity $\boldsymbol{\omega}_z = \omega_z\mathbf{k}$ and relative angular acceleration $\dot{\boldsymbol{\omega}}_z = \dot{\omega}_z\mathbf{k}$. Projecting point P on the z-axis in the point O_2 (Fig. 8.1), we take O_2 a a pole for the motion, and the vector $\mathbf{O_2P} = \mathbf{r}_2 = x\mathbf{i} + y\mathbf{j}$.

The relative acceleration $\mathbf{A}_r$ is now

$$\mathbf{A}_r = \mathbf{A}_{O_2} + \boldsymbol{\omega}_z \times (\boldsymbol{\omega}_z \times \mathbf{r}_2) + \dot{\boldsymbol{\omega}}_z \times \mathbf{r}_2$$

The first term is $\mathbf{A}_{O_2} = \mathbf{O}$. The second term is $\boldsymbol{\omega}_z \times (\boldsymbol{\omega}_z \times \mathbf{r}_2) = (\boldsymbol{\omega}_z \cdot \mathbf{r}_2)\boldsymbol{\omega}_z - (\boldsymbol{\omega}_z \cdot \boldsymbol{\omega}_z)\mathbf{r}_2 = (0)\boldsymbol{\omega}_z - \omega_z^2(x\mathbf{i} + y\mathbf{j}) = -x\omega_z^2\mathbf{i} - y\omega_z^2\mathbf{j}$. The third term is $\dot{\boldsymbol{\omega}}_z \times \mathbf{r}_2 = \dot{\omega}_z\mathbf{k} \times (x\mathbf{i} + y\mathbf{j}) = x\dot{\omega}_z\mathbf{k} \times \mathbf{i} + y\dot{\omega}_z\mathbf{k} \times \mathbf{j} = x\dot{\omega}_z\mathbf{j} - y\dot{\omega}_z\mathbf{i}$. Accumulating terms, the components of $\mathbf{A}_r$ are: $A_{r_x} = -x\omega_z^2 - y\dot{\omega}_z$; $A_{r_y} = -y\omega_z^2 + x\dot{\omega}_z$; and $A_{r_z} = 0$.

Coriolis acceleration

The Coriolis acceleration is determined from the familiar expression $\mathbf{A}_{cor} = 2\boldsymbol{\omega}_{base} \times \mathbf{V}_r$. $\boldsymbol{\omega}_{base}$ is the angular velocity of the base, that is $\boldsymbol{\omega}_{base} = \boldsymbol{\omega}_y = \omega_y\mathbf{j}$. $\mathbf{V}_r$ is the relative velocity of the point P, with $\mathbf{V}_r = \boldsymbol{\omega}_z \times \mathbf{r}_2$. The result is

$$\mathbf{A}_{cor} = 2\boldsymbol{\omega}_y \times \mathbf{V}_r = 2\boldsymbol{\omega}_y \times (\boldsymbol{\omega}_z \times \mathbf{r}_2) = 2(\boldsymbol{\omega}_y \cdot \mathbf{r}_2)\boldsymbol{\omega}_z - 2(\boldsymbol{\omega}_y \cdot \boldsymbol{\omega}_z)\mathbf{r}_2 =$$
$$2(y\omega_y)\omega_z\mathbf{k} - 2(0)\mathbf{r}_2 = 2y\omega_y\omega_z\mathbf{k}$$

The components of $\mathbf{A}_{cor}$ are then $A_{cor_x} = 0$, $A_{cor_y} = O$ and $A_{cor_z} = 2y\omega_y\omega_z$.

The total acceleration components of the point P are now

$$A_x = A_{C_x} + (z\dot{\omega}_y - y\dot{\omega}_z) - x(\omega_y^2 + \omega_z^2)$$
$$A_y = x\dot{\omega}_z - y\omega_z^2 \tag{8.1}$$
$$A_z = A_{C_z} - z\omega_y^2 - x\dot{\omega}_y + 2y\omega_y\omega_z$$

8.3 MOMENT EQUATIONS IN PLANE GYROSCOPIC MOTION BY NEWTON'S LAWS

Considering a particle of mass dm located at point P in Fig. 8.1, and taking $\mathbf{F}$ as the resultant force of all *external* and *internal* forces on the particle, *Newton's second law* now gives the equation

$$\mathbf{F} = (\mathrm{d}m)\mathbf{A} = A_x\mathrm{d}m\mathbf{i} + A_y\mathrm{d}m\mathbf{j} + A_z\mathrm{d}m\mathbf{k}$$

The moment $\mathbf{m}_C$ of the force $\mathbf{F}$ in the centre of mass is by definition $\mathbf{m}_C = \mathbf{r}_C \times \mathbf{F} = m_{C_x}\mathbf{i} + m_{C_y}\mathbf{j} + m_{C_z}\mathbf{k}$, where $\mathbf{r}_C = x\mathbf{i} + y\mathbf{j} + z\mathbf{k}$ is the position vector of the point P from the centre of mass as shown in Fig. 8.1.

We have now

$$\mathbf{m}_C = \mathbf{r}_C \times \mathbf{F} = \mathrm{d}m \begin{vmatrix} \mathbf{i} & \mathbf{j} & \mathbf{k} \\ x & y & z \\ A_x & A_y & A_z \end{vmatrix}$$

$$= (yA_z - zA_y)\mathrm{d}m\mathbf{i} - (xA_z - zA_x)\mathrm{d}m\mathbf{j} + (xA_y - yA_x)\mathrm{d}m\mathbf{k}$$

The x, y and z components of $\mathbf{m}_C$ are the *moments of the force* $\mathbf{F}$ *about the x-, y- and z-axes respectively*.

Summing up the moments for all the particles of the gyroscope, the contribution from the internal forces vanishes, and the result for all *the external forces* for the x-axis is

$$M_x = \int (yA_z - zA_y)dm = \int [yA_{Cz} - yz\omega_y^2 - yx\dot{\omega}_y + 2y^2\omega_y\omega_z - xz\dot{\omega}_z + yz\omega_z^2]dm$$

where the integration is extended over the total volume of the gyroscope.

With the x-, y- and z-axes located along principal axes at the centre of mass, we have $\int y \text{ dm} = 0$ and $\int xy\ dm = \int yz\ dm = \int xz\ dm = 0$. Substituting this in the expression for M_x leads to $M_x = 2\omega_y\omega_z\int y^2 \text{dm}$.

Now $I_x + I_y + I_z = 2\int (x^2 + y^2 + z^2)\text{d}m$, from which $2\int y^2\text{d}m = I_x + I_y + I_z - 2\int (x^2 + z^2)\text{d}m = I_x + I_z - I_y$; however; $I_x = I_y$, so that $2\int y^2\text{dm} = I_z$, and the final result is $M_x = I_z\omega_y\omega_z$. In a similar way we find $M_y = \int (zA_x - xA_z)\text{d}m = I_y\dot{\omega}_y$ and $M_z = \int (xA_y - yA_x)\text{d}m = I_z\dot{\omega}_z$.

The moment equations are now

$$\left.\begin{aligned} M_x &= I_z\omega_y\omega_z \\ M_y &= I_y\dot{\omega}_y \\ M_z &= I_z\dot{\omega}_z \end{aligned}\right\} \qquad (8.2)$$

8.4 MOMENT EQUATIONS BY THE MOMENT OF MOMENTUM

For a body in *general motion* we have for formula (7.18) for the moment of momentum $\mathbf{H}_C$ in the centre of mass:

$$\mathbf{H}_C = \int \mathbf{r}_C \times (\boldsymbol{\omega} \times \mathbf{r}_C)\text{d}m$$

For the gyroscope in Fig. 8.1 we have $\mathbf{r}_C = x\mathbf{i} + y\mathbf{j} + z\mathbf{k}$ and $\boldsymbol{\omega} = \omega_y\mathbf{j} + \omega_z\mathbf{k}$.

Using the formula $\mathbf{a} \times (\mathbf{b} \times \mathbf{c}) = (\mathbf{a}\cdot\mathbf{c})\mathbf{b} - (\mathbf{a}\cdot\mathbf{b})\mathbf{c}$, we obtain

$$\begin{aligned} &\mathbf{r}_C \times (\boldsymbol{\omega} \times \mathbf{r}_C) = (\mathbf{r}_C\cdot\mathbf{r}_C)\boldsymbol{\omega} - (\mathbf{r}_C\cdot\boldsymbol{\omega})\mathbf{r}_C = (x^2 + y^2 + z^2)(\omega_y\mathbf{j} + \omega_z\mathbf{k}) \\ &- (y\omega_y + z\omega_z)(x\mathbf{i} + y\mathbf{j} + z\mathbf{k}) = -(xy\omega_y + xz\omega_z)\mathbf{i} + \\ &[(x^2 + z^2)\omega_y - yz\omega_z]\mathbf{j} + [(x^2 + y^2)\omega_z - yz\omega_y]\mathbf{k} \end{aligned}$$

Multiplying by dm and summing up for all elements of the body, the x component of $\mathbf{H}_C$ is $H_{C_x} = \int (-xy\omega y - xz\omega_z)\text{d}m = -\omega_y\int xy\text{ d}m - \omega_z\int xz\text{d}m = -\omega_y I_{xy} - \omega_z I_{xz} = 0$, since xyz are principal axes with $I_{xy} = I_{xz} = I_{yz} = 0$.

$$
\begin{aligned}
H_{C_y} &= \int[(x^2+z^2)\omega_y - yz\omega_z]dm = \omega_y\int(x^2+z^2)dm - \omega_z\int yz\,dm = \\
&\omega_y I_y - \omega_z I_{yz} = \omega_y I_y \\
H_{C_z} &= \int[(x^2+y^2)\omega_z - yz\omega_y]dm = \omega_z\int(x^2+y^2)dm - \omega_y\int yz\,dm = \\
&\omega_z I_z - \omega_y I_{yz} = \omega_z I_z
\end{aligned}
$$

We have now $\mathbf{H}_C = \omega_y I_y \mathbf{j} + \omega_z I_z \mathbf{k}$, from which we find $\dot{\mathbf{H}}_C = \dot{\omega}_y I_y \mathbf{j} + \dot{\omega}_z I_y \mathbf{k} + \omega_y I_y \dot{\mathbf{j}} + \omega_z I_z \dot{\mathbf{k}}$; now $\dot{\mathbf{j}} = \boldsymbol{\omega} \times \mathbf{j} = \omega_y \mathbf{j} \times \mathbf{j} = \mathbf{o}$ and $\dot{\mathbf{k}} = \boldsymbol{\omega} \times \mathbf{k} = \omega_y \mathbf{j} \times \mathbf{k} = \omega_y \mathbf{i}$. Substituting gives $\dot{\mathbf{H}}_C = \omega_y\omega_z I_z \mathbf{i} + \dot{\omega}_y I_y \mathbf{j} + \dot{\omega}_z I_z \mathbf{k}$. Since $\mathbf{M}_C = \dot{\mathbf{H}}_C$ from equation (7.23) we have the moment equations

$$
M_x = I_z\omega_y\omega_z \qquad M_y = I_y\dot{\omega}_y \qquad \text{and} \qquad M_z = I_z\dot{\omega}_z
$$

as equations (8.2)

8.5 THE EQUATIONS OF MOTION FOR A GYROSCOPE IN PLANE MOTION

The acceleration of the centre of mass of the gyroscope in Fig. 8.1 is $\mathbf{A}_C = A_{C_x}\mathbf{i} + A_{C_z}\mathbf{k}$; this is in the plane of motion and $\mathbf{A}_C$ is the *absolute acceleration* of the centre of mass.

Applying Newton's second law to the centre of mass gives directly the equations

$$
\left.
\begin{aligned}
F_x &= MA_{C_x} \\
F_y &= 0 \\
F_z &= MA_{C_z}
\end{aligned}
\right\} \tag{8.3}
$$

where M is the total mass of the gyroscope, and F_x, F_y and F_z are the components in the xyz system of the *resultant* force $\mathbf{F}$ acting on the gyroscope. If we take the coordinates of the centre of mass (X, Z) in the fixed system, we may also take Newton's second law for the centre of mass in the form

$$
\left.
\begin{aligned}
F_X &= M\ddot{X} \\
F_Y &= 0 \\
F_Z &= M\ddot{Z}
\end{aligned}
\right\} \tag{8.4}
$$

The motion of the gyroscope is called a *precessional motion* about the y-axis with the *precessional angular velocity* ω_y, while the angular velocity ω_z is called the *spin velocity*. Introducing the notations $\omega_y = \omega_p, \dot{\omega}_y = \dot{\omega}_p, \omega_z = \omega_s, \dot{\omega}_z = \dot{\omega}_s, I_y = I_p$ and $I_z = I_s$, we have the following three moment equations for the gyroscope from equations (8.2):

$$\left.\begin{aligned} M_x &= I_s\omega_p\omega_s \\ M_y &= I_p\dot{\omega}_p \\ M_z &= I_s\dot{\omega}_s \end{aligned}\right\} \tag{8.5}$$

Equations (8.3) or (8.4) are the equations of motion for the centre of mass, while the last two of the equations (8.5) are the equations of rotational motion. The equations $F_y = F_Y = 0$ and $M_x = I_s\omega_p\omega_s$ are *additional* equations that are necessary to determine the forces that must act on the gyroscope to keep it in the plane of motion.

8.6 VECTOR EQUATION FOR THE GYROSCOPIC TORQUE

Because of its importance, the moment equation $M_x = I_s\omega_p\omega_s$ is often given a vector formulation.

M_x is *called the gyroscopic torque*, and denoting this by $\mathbf{G}_T$ we have $|\mathbf{G}_T| = M_x$. The equation $M_x = I_s\omega_p\omega_s$ may now be stated as a vector equation.

$$\mathbf{G}_T = I_s\boldsymbol{\omega}_p \times \boldsymbol{\omega}_s \tag{8.6}$$

The important expression (8.6) gives *both* the *magnitude* and the *direction* of the torque in one equation. From the vector cross-product, the magnitude of $\mathbf{G}_T$ is $|\mathbf{G}_T| = I_s\omega_p\omega_s \sin 90° = I_s\omega_p\omega_s$, as it should be. It may be seen from the vector cross-product that the direction of the torque is correct. In the case of plane motion the direction of $\mathbf{G}_T$ may also be determined by a rotation of the $\boldsymbol{\omega}_s$ vector of 90° in the plane of motion and in the rotational direction of $\boldsymbol{\omega}_p$.

It is noteworthy that the vectors $\boldsymbol{\omega}_p$ and $\boldsymbol{\omega}_s$ are in alphabetic order in equation (8.6).

It must be kept in mind that the torque vector $\mathbf{G}_T$ is perpendicular to the plane in which the torque acts and follows the right-hand screw rule. $\mathbf{G}_T$ is the torque that must act *on* the gyroscope to make it precess; the torque on the supports is $-\mathbf{G}_T$.

Once the $\boldsymbol{\omega}_s$ and $\boldsymbol{\omega}_p$ vectors have been located following the right-hand screw rule, the z-axis is taken along $\boldsymbol{\omega}_s$ and in its positive direction, while the y-axis is taken along $\boldsymbol{\omega}_p$ and in its positive direction; this determines the xyz system in each case.

8.7 EXAMPLES ON PLANE GYROSCOPIC MOTION

As a *first* example, consider the wheel in Fig. 8.2 which is rolling along a curve on a horizontal plane; the $\boldsymbol{\omega}_p$ and $\boldsymbol{\omega}_s$ vectors are introduced as shown, and the equation $\mathbf{G}_T = I_s\boldsymbol{\omega}_p \times \boldsymbol{\omega}_s$ gives the magnitude and direction of the gyroscopic torque that must act on the wheel from its bearing support. It may be seen that the equal and opposite torque acting on the support of the wheel, which may be the front wheel of a bicycle, tend to overturn the bicycle. The same situation occurs when a car is rounding a corner; the gyroscopic torque on the wheels has an overturning effect on the car, although the effect is generally small.

As a *second* example, consider the system in Fig. 8.3(a) which shows the rotor of a jet engine in an aircraft. The aircraft is flying in a curve of radius R in a horizontal

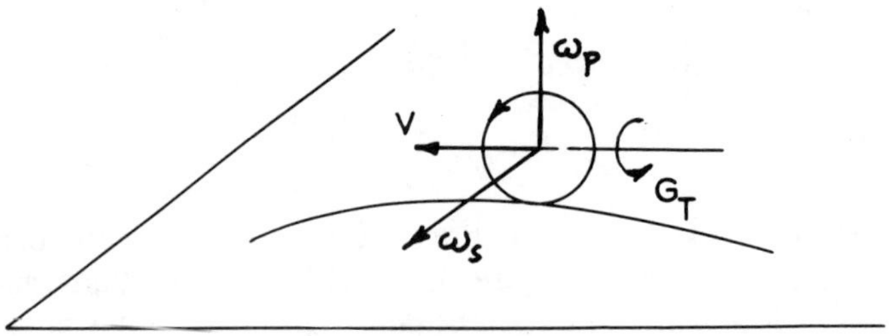

Fig. 8.2.

plane, with constant velocity V and constant angular velocity $\boldsymbol{\omega}_s$ of the rotor relative to the aircraft. The mass of the rotor is M and the moment of inertia is I_s about the axis of rotation. The spin-axis is forced to precess with angular velocity $\omega_p = V/R$ as shown. The gyroscopic torque as it acts on the rotor is shown in the figure and is $\mathbf{G}_T = I_s\boldsymbol{\omega}_p \times \boldsymbol{\omega}_s$.

Fig. 8.3(b) shows the forces acting on the rotor, assuming complete symmetry.

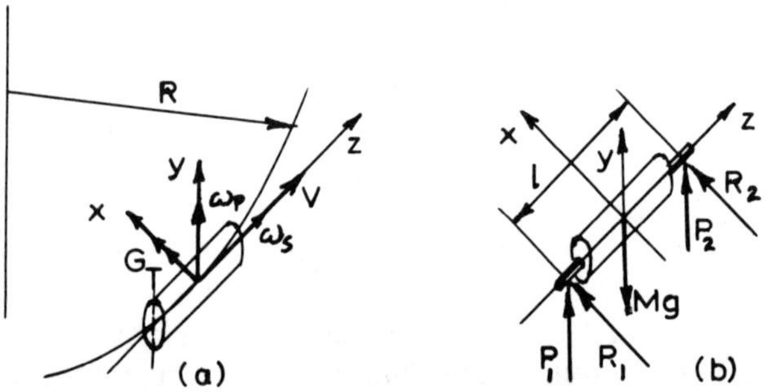

Fig. 8.3.

Equations (8.3) and (8.5) are

$$F_x = R_1 + R_2 = MR\omega_p^2$$

$$F_y = Mg - P_1 - P_2 = 0$$

$$M_x = P_1\frac{l}{2} - P_2\frac{l}{2} = G_T = I_s\omega_p\omega_s$$

$$M_y = R_1\frac{l}{2} - R_2\frac{l}{2} = 0$$

The solution is

$$R_1 = R_2 = \frac{M}{2} R\omega_p^2$$

$$P_1 = \frac{Mg}{2} + \frac{I_s}{l}\omega_p\omega_s$$

$$P_2 = \frac{Mg}{2} - \frac{I_s}{l}\omega_p\omega_s$$

The torque $-\mathbf{G}_T$ acts on the airframe with bearing pressures opposite to those shown *on* the rotor.

With the direction of $\boldsymbol{\omega}_s$ as shown, the gyroscopic effects result in an increased vertical bearing pressure on the rear bearing of the rotor, so that the nose of the aircraft would tend to rise.

As a numerical example, consider a velocity $V = 1200$ km/h, engine revolutions 15000 rpm, radius $R = 2500$ m and $I_s = 4$ kg m^2. We find

$$\omega_p = \frac{V}{R} = \frac{1200 \times 10^3}{3600 \times 2500} = 0.133 \text{ rad/s}$$

$$\omega_s = 500\pi = 1571 \text{ rad/s}$$

and

$$G_T = 4 \times 1571 \times 0.133 = 836 \text{ N m}$$

This torque is significant and must be taken into account in the design of the bearings of the rotor.

As a *third* example, consider the system shown in Fig. 8.4(a). The system consists of a right homogeneous circular disc of radius r and mass M. The disc is rigidly connected to a light, rigid, horizontal shaft of length R, which is located in a journal bearing at position A and a socket bearing at position B. A motor at C is able to rotate the entire system about the axis BC, while a small motor at B rotates the disc about the axis AB.

For simplicity we neglect wind resistance, friction and the mass of the shaft and the supports. The xyz system is introduced as shown at the centre of mass of the disc.

The disc is now given a *constant* angular velocity ω_s and the system is given a *constant* angular velocity ω_p, and the motors subsequently disconnected. The gravity force and the components of the bearing reactions are shown on the figure as they act on the gyroscope. The moments of inertia of the disc are $I_s = Mr^2/2$ and $I_p = Mr^2/4$. The centre of mass is moving in a horizontal circle of radius R and with constant angular velocity ω_p; the components of the acceleration of the centre of mass are then $A_{C_x} = 0$, $A_{C_y} = 0$ and $A_{C_z} = -R\omega_p^2$, directed towards point B.

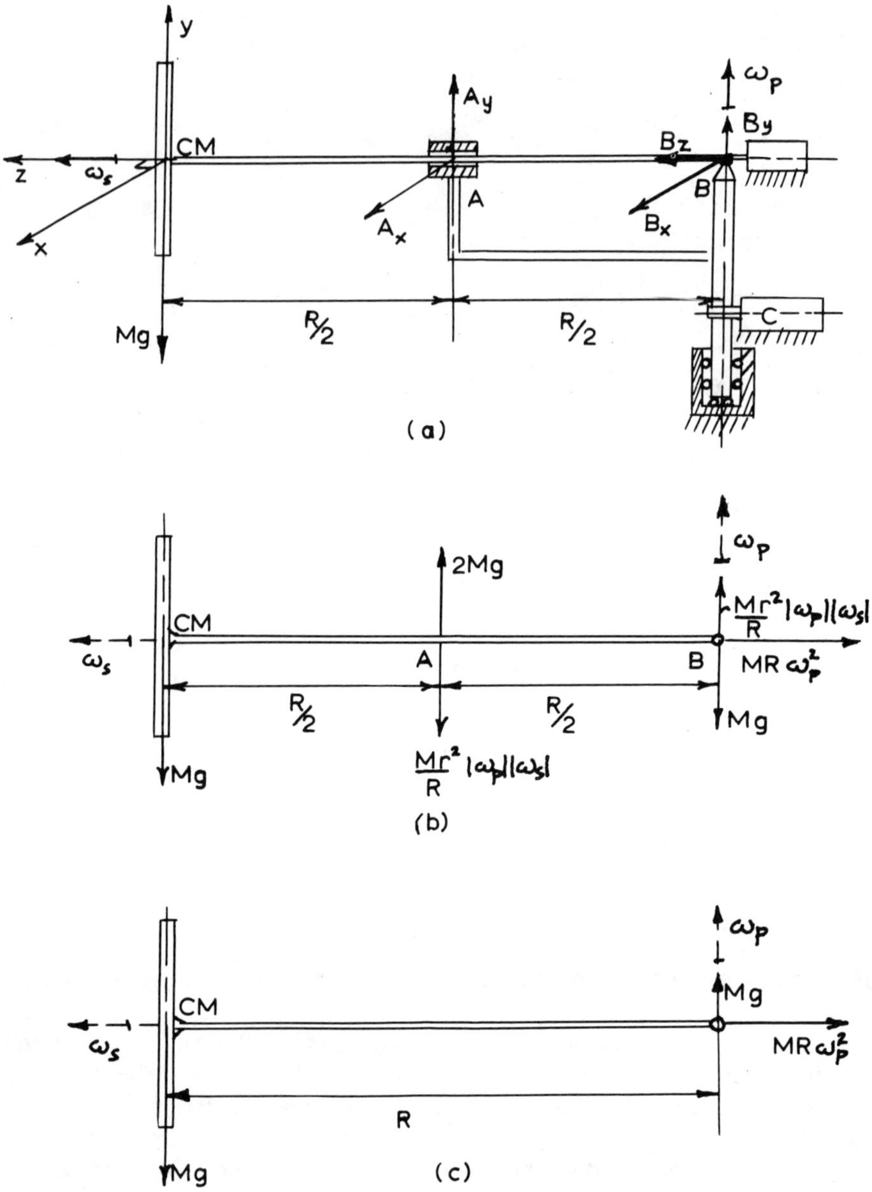

Fig. 8.4.

Applying the equations (8.3) and (8.5) to this system results in the following equations:

(a) $A_x + B_x = 0$

(b) $A_y + B_y - Mg = 0$

(c) $B_z = MA_{C_z} = -MR\omega_p^2$

(d) $M_x = A_y \frac{R}{2} + B_y R = \frac{Mr^2}{2} \omega_p \omega_s$

(e) $M_y = -A_x \frac{R}{2} - B_x R = I_p \dot{\omega}_p = 0$

(f) $M_z = I_s \dot{\omega}_s = 0$

The solution of these equations is straightforward; the equations (a) and (e) lead to

$$\begin{cases} A_x = 0 \\ B_x = 0 \end{cases}$$

Equations (b) and (d) may be solved directly to give

$$A_y = 2Mg - \frac{Mr^2}{R} \omega_p \omega_s$$

$$B_y = -Mg + \frac{Mr^2}{\mathrm{R}} \omega_p \omega_s$$

Fig. 8.4(b) shows a free-body diagram of the gyroscope with the acting forces. None of the forces does any work, so the kinetic energy of the system stays constant, with constant angular velocities ω_p and ω_s. It may be seen, by taking moments about the x-axis, that the acting forces give the necessary gyroscopic torque $I_s\omega_p\omega_s$ in the vertical plane, to produce the precessional motion in the horizontal plane.

The only force on the gyroscope from the journal bearing at A is A_y; we may set this force equal to zero with the result that $\omega_p = 2gR/r^2\omega_s$; for this relationship between ω_p and ω_s the bearing at A may be removed and the remaining forces on the gyroscope are shown in Fig. 8.4(c). In this special case the gyroscope precesses in the horizontal plane owing to the *gravity torque* $MgR = |\mathbf{G}_T|$.

The gyroscope seems to defy gravity. The precessional angular velocity ω_p becomes very small for large values of ω_s; for instance if the disc is of steel with r = 8 cm, thickness 2.5 cm and density 8×10^3 kg/cm^3, its mass will be 4.021 kg. If R = 20 cm and the speed of revolution is 20 000 rpm, we have $\omega_s = 2094.4$ rad/s and $\omega_p = (2 \times 9.81 \times 0.20 \times 10^4)/(64 \times 2094.4) = 0.2928$ rad/s, from which the precessional velocity is $N_p \simeq 2.80$ rpm $\simeq 0.047$ rps, this means that it takes about 21 s for one precessional revolution.

8.8 KINETIC ENERGY OF THE GYROSCOPE AND LAGRANGE'S EQUATIONS

The kinetic energy T of the gyroscope in Fig. 8.1 is

$$T = \frac{1}{2}\int V^2 \mathrm{d}m = \frac{1}{2}\int \mathbf{V}\cdot\mathbf{V}\mathrm{d}m$$

Taking the centre of mass as a pole, we have $\mathbf{V} = \mathbf{V}_C + \boldsymbol{\omega}\times\mathbf{r}_C$, and

$$\mathbf{V}\cdot\mathbf{V} = (\mathbf{V}_C + \boldsymbol{\omega}\times\mathbf{r}_C)\cdot(\mathbf{V}_C + \boldsymbol{\omega}\times\mathbf{r}_C) =$$
$$V_C^2 + 2\mathbf{V}_C\cdot(\boldsymbol{\omega}\times\mathbf{r}_C) + (\boldsymbol{\omega}\times\mathbf{r}_C)\cdot(\boldsymbol{\omega}\times\mathbf{r}_C)$$

The *first term* in the kinetic energy is now

$$\frac{1}{2}\int V_C^2 \mathrm{d}m = \frac{1}{2}V_C^2 \int \mathrm{d}m = \frac{1}{2}MV_C^2$$

where M is the total mass of the gyroscope and $\mathbf{V}_C$ the instantaneous velocity of the centre of mass.

The *second term* is

$$\int \mathbf{V}_C\cdot(\boldsymbol{\omega}\times\mathbf{r}_C)\mathrm{d}m = \mathbf{V}_C\cdot[\int\boldsymbol{\omega}\times\mathbf{r}_C\mathrm{d}m] = \mathbf{V}_C\cdot[\boldsymbol{\omega}\times\int\mathbf{r}_C\mathrm{d}m] = 0$$

since $\int \mathbf{r}_C \mathrm{d}m = \mathbf{0}$.

The *third term* is

$$\int \frac{1}{2}(\boldsymbol{\omega}\times\mathbf{r}_C)\cdot(\boldsymbol{\omega}\times\mathbf{r}_C)\mathrm{d}m$$

Using the vector formula $(\mathbf{a}\times\mathbf{b})\cdot\mathbf{c} = \mathbf{a}\cdot(\mathbf{b}\times\mathbf{c})$ we find that this term is $\frac{1}{2}\int\boldsymbol{\omega}\cdot[\mathbf{r}_C\times(\boldsymbol{\omega}\times\mathbf{r}_C)]\mathrm{d}m$

$$= \frac{1}{2}\boldsymbol{\omega}\cdot[\int\mathbf{r}_C\times(\boldsymbol{\omega}\times\mathbf{r}_C)\mathrm{d}m] = \frac{1}{2}\boldsymbol{\omega}\cdot\mathrm{H}_C$$

We have $\boldsymbol{\omega} = \omega_y\mathbf{j} + \omega_z\mathbf{k}$ and $\mathbf{H}_C = \omega_y I_y\mathbf{j} + \omega_z I_z\mathbf{K}$, so that $\frac{1}{2}\boldsymbol{\omega}\cdot\mathbf{H}_C = \frac{1}{2}I_y\omega_y^2 + \frac{1}{2}I_z\omega_z^2$.

The *total* kinetic energy is now

$$T = \frac{1}{2}MV_C^2 + \frac{1}{2}I_y\omega_y^2 + \frac{1}{2}I_z\omega_z^2 \tag{8.7}$$

The first two terms of (8.7) are the same as for a body in plane motion; the third term is due to the spinning motion of the gyroscope. The formula (8.7) holds only for a symmetrical gyroscope in plane motion.

The gyroscope in Fig. 8.1 has *four* degrees of freedom as a dynamics system. We may take the generalised coordinates as the coordinates (X, Z) of the centre of mass, and the angle of rotation ϕ about the y-axis and the angle of spin θ about the z-axis, where $\dot{\phi} = \omega_y = \omega_p$ and $\dot{\theta} = \omega_z = \omega_s$. The kinetic energy T of the gyroscope expressed in terms of the generalised coordinates is now from (8.7)

$$T = \frac{1}{2}M(\dot{X}^2 + \dot{Z}^2) + \frac{1}{2}I_y\dot{\phi}^2 + \frac{1}{2}I_z\dot{\theta}^2$$

The differentiations for Lagrange's equations are

$$\frac{\mathrm{d}}{\mathrm{d}t}\frac{\partial T}{\partial \dot{X}} = M\ddot{X} \qquad \frac{\partial T}{\partial X} = 0 \qquad \frac{\mathrm{d}}{\mathrm{d}t}\frac{\partial T}{\partial \dot{Z}} = M\ddot{Z} \qquad \frac{\partial T}{\partial Z} = 0$$

$$\frac{\mathrm{d}}{\mathrm{d}t}\frac{\partial T}{\partial \dot{\phi}} = I_y\ddot{\phi} \qquad \frac{\partial T}{\partial \phi} = 0 \qquad \frac{\mathrm{d}}{\mathrm{d}t}\frac{\partial T}{\partial \dot{\theta}} = I_z\ddot{\theta} \qquad \frac{\partial T}{\partial \theta} = 0$$

The acting forces on the gyroscope may be reduced to a resultant force $\mathbf{F}$ at the centre of mass, with components (F_X, F_Y, F_Z), and a couple $\mathbf{M}_C$ about the centre of mass with components $(M_x, M_y, \mathbf{M}_z)$. Fixing all the generalised coordinates except X, we find $\delta(\text{W.D.}) = Q_X\delta X = F_X\delta X$.

Similarly we find $Q_Z = F_Z$; $Q_\phi = M_y$ and $Q_\theta = M_z$.

We have now the Lagrange's *equations of motion* of the gyroscope:

$$\begin{aligned} F_X &= M\ddot{X} \\ F_Z &= M\ddot{Z} \\ M_y &= I_y\ddot{\phi} = I_p\dot{\omega}_p \\ M_z &= I_z\ddot{\theta} = I_s\dot{\omega}_s \end{aligned} \qquad (8.8)$$

The first two of equations (8.8) are the Newton equations (8.4) for the motion of the centre of mass in the plane of motion. The last two equations (8.8) are the moment equations (8.5) determined previously.

8.9 RATE GYROSCOPE

Gyroscopes are used in a number of instruments for direction indication, stabilisation and inertial guidance. These applications rely on the three-dimensional theory of the gyroscope; however, a rate gyroscope in plane motion may be considered here.

Fig. 8.5 shows a rate gyroscope. This is a single-degree-of-freedom gyro mounted as shown in a gimbal. The gyro is free to spin about its geometrical axis, while the gimbal is constrained to small angular rotations about the x-axis limited by the

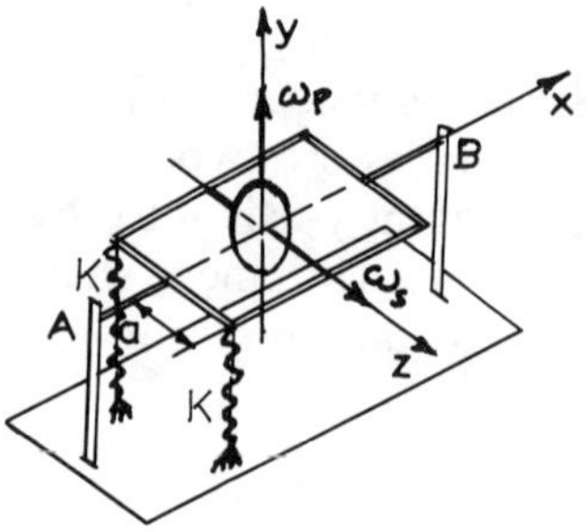

Fig. 8.5.

springs shown in the figure. The springs exert a moment about the axis AB, which is the x-axis as shown, of magnitude $M_x \simeq 2Ka^2\alpha$, where α is the *small* angle of rotation about the x-axis.

The gyro case is rigidly fastened to the vehicle for which the rate of turn is to be measured. If the angular velocity of precession is ω_p and the spin velocity is ω_s, and if these are *constant*, we have steady precession, and from equations (8.5) we have $M_x = I_s\omega_p\omega_s$, so that $2Ka\alpha^2 \simeq I_s\omega_p\omega_s$. The constant angle α may be determined from this expression for steady precession in the plane.

The rate gyro then gives a means of measuring the rate of rotation of a vehicle in the plane. If α is displayed on a dial, the pilot of a plane, for instance, can see the magnitude and direction of the turn of the aircraft.

Imperfections in the manufacture of the rate gyro may result in vibrations of the instrument; to reduce this to a minimum, the rate gyro is usually equipped with a damping mechanism or a dash-pot.

The rate gyro has many applications in aeronautics and in military fields — it is the controlling element in turn indicators for blind flying, gun sights and control gyros for ship stabilisers and for inertial navigation for stabilising platforms.

As an example of a rate gyro in plane motion (Fig. 8.5), consider a case of steady precession.

The given figures are $I_s = 4.08 \times 10^{-4}$ kg m^2, $\omega_s = 2 \times 10^4$ rpm $= 2094$ rad/s.

The vechicle is turning at a constant rate with one full turn in one minute, so that $\omega_p = 2\pi$ rad/min $= 0.1047$ rad/s. $Ka^2 = 0.25$ N cm.

From the expression $M_x \simeq 2Ka^2\alpha = I_s\omega_p\omega_s$, we find

$$\alpha \simeq \frac{I_s\omega_p\omega_s}{2Ka^2} = \frac{4.08 \times 10^{-4} \times 0.1047 \times 2094}{2 \times 0.25}$$

$$= 0.1789 \text{ rad} = 10^\circ.25$$

PROBLEMS

8.1 The armature of the motor of an electric car weighs 2670 N and rotates in a direction opposite to the rotation of the car wheels. The distance between its bearings is 0.610 m and its radius of gyration is 15.24 cm. The motor makes four revolutions to one revolution of the car wheels which have diameters of 0.84 m. If the car is moving forward around a curve of 30.5 m radius with a velocity of 6.10 m/s, find the total pressures on the bearings of the armature if the centre of the curve is to the right.

The centreline of the armature is parallel to the wheel centreline.

8.2 The low-pressure turbine rotor of a cargo steamer weighs 84.8 kN and has a radius of gyration of 0.381 m. The speed of rotation is 3300 rev/min in a clockwise sense when viewed from the stern of the ship. If the ship is pitching with an amplitude of 6° above and below the horizontal with a period of 10 s in simple harmonic motion, calculate the maximum torque which is transmitted to the turbine bed plates. State the direction of the torque in relation to the pitching motion of the ship.

The centreline of the turbine is on the longitudinal centreline of the ship.

8.3 An aircraft landing gear assembly is shown in Fig. 8.6. After taking off, the

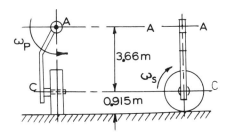

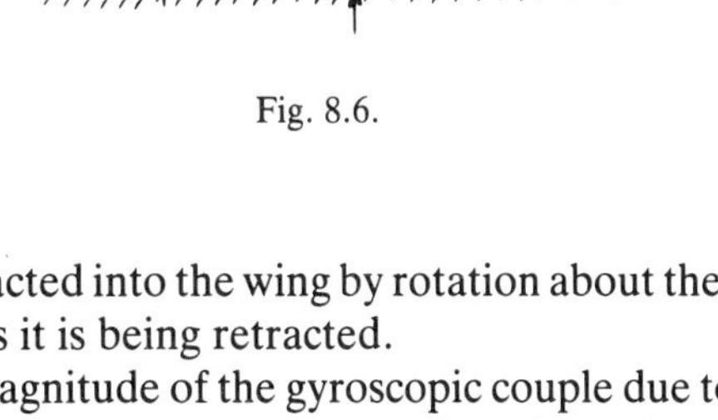

Fig. 8.6.

landing gear is retracted into the wing by rotation about the axis AA. The wheel continues to spin as it is being retracted.

Calculate the magnitude of the gyroscopic couple due to the spinning wheel if the weight of the wheel is 667 N, the radius of gyration about the axis CC is 0.549 m, the take-off speed is 290 km/h and the maximum speed of retraction ω_p is 3 rad/s. Treat as plane gyroscopic motion.

Indicate the direction of the gyroscopic couple and its action on the strut AC.

8.4 A right circular disc of mass $M = 7$ kg and radius $r = 8$ cm is attached to a shaft AB as shown in Fig. 8.7. The disc rotates at an angular velocity $\omega_2 = 100$ rad/s

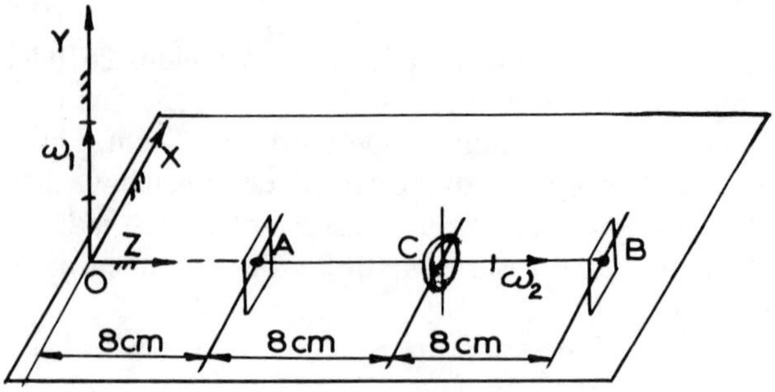

Fig. 8.7.

constant relative to the bearings. The bearings are mounted on a horizontal platform which rotates at a constant angular velocity $\omega_1 = 20$ rad/s about the axis OY. The mass of the shaft AB may be neglected, and it may be assumed that bearing A alone retains the system in the z direction. For the instant shown in Fig. 8.7, compute the components of the reactions on the shaft at A and B.

9

Vibrations with one degree of freedom

The case of a mass in linear motion under the action of a force proportional to the displacement was discussed in section 2.2.4. This was the case of a mass on a spring and was shown to be a simple harmonic vibratory motion.

To set up a more realistic mathematical model, consider the system shown in Fig. 9.1. The spring and the dash-pot are assumed light, and all the mass of the system is

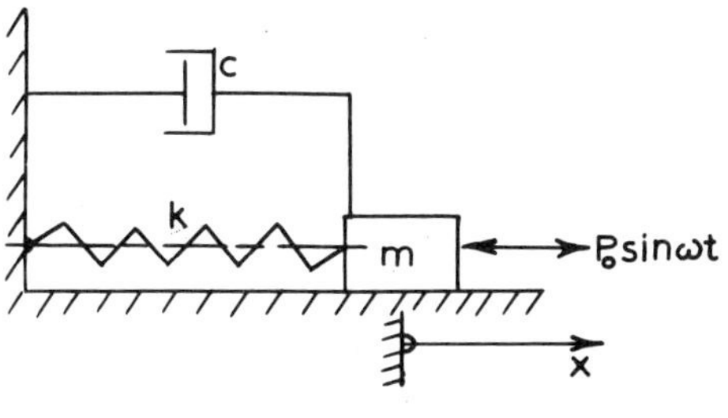

Fig. 9.1.

assumed concentrated in the body of mass m as shown. All damping in the system is assumed to be of the viscous friction type and concentrated in the dash-pot with damping coefficient c. The damping force is then proportional to the velocity and in the opposite direction, so that it may be expressed as $-c\dot{x}$.

An external exciting force $P = P_0 \sin \omega t$, of force amplitude P_0 and frequency ω, is acting on the body as shown. The displacement x is measured from the static equilibrium position and forces and motion are taken positive to the right.

Newton's second law gives directly the equations of motion:

$$m\ddot{x} = P_0 \sin \omega t - c\dot{x} - \mathrm{k}\mathrm{x} \quad \text{or} \quad \ddot{x} + \frac{c}{m}\dot{x} + \frac{\mathrm{k}}{\mathrm{m}}\mathrm{x} = \frac{\mathrm{P}_0}{m} \sin \omega t$$

Substituting the constants $\omega_0^2 = k/m$ and $n = c/2m$ leads to

$$\ddot{x} + 2n\dot{x} + \omega_0^2 x = \frac{P_0}{m} \sin \omega t \qquad (9.1)$$

The investigation of the motion is divided into four cases for convenience.

9.1 FREE VIBRATIONS WITHOUT DAMPING

If $P_0 = 0$ and $c = 0$, we have free vibrations without damping, and the equation of motion (9.1) takes the form

$$\ddot{x} + \omega_0^2 x = 0 \qquad (9.2)$$

The general solution of this equation is

$$x = A \cos \omega_0 t + B \sin \omega_0 t$$

With starting conditions $x = x_0$ and $\dot{x} = \dot{x}_0$ at $t = 0$, the solution is

$$x = \sqrt{[x_0^2 + (\dot{x}_0/\omega_0)^2]} \cos (\omega_0 t - \phi) \quad \tan \phi = \dot{x}_0/x_0\omega_0$$

as shown in Chapter 2. The motion is *simple harmonic motion*. The amplitude is $A_0 = \sqrt{[x_0^2 + (\dot{x}_0/\omega_0)^2]}$ and ϕ is the phase angle. The constant $\omega_0 = \sqrt{(k/m)}$ is the *natural circular frequency*; the *frequency is* $f = \omega_0/2\pi$ and the *period* is $\tau = 1/f$. If the static deflection of the spring is Δ_s under a load mg, we have $\Delta_s = mg/k$, and

$$\omega_0 = \sqrt{(kg/mg)} = \sqrt{(g/\Delta_s)}$$

If the spring is vertical, the frequency of the system may be determined by measuring the static deflection Δ_s occurring when the mass is attached to the spring.

Example 9.1
The vibratory system shown in Fig. 9.2 is in static equilibrium when the rigid light bar OB is horizontal. For small rotational motions of the bar about O, the spring may be assumed to stay vertical and the body of mass m to move in vertical rectilinear motion. Determine the frequency for small displacements of the system.

Solution
In the equilibrium position, the force in the spring is P. Taking moments about O gives the result $Pb = lmg$, or $P = (l/b)mg$.

If the mass is displaced a distance x, the point A is displaced $\Delta = xb/l$, and the total

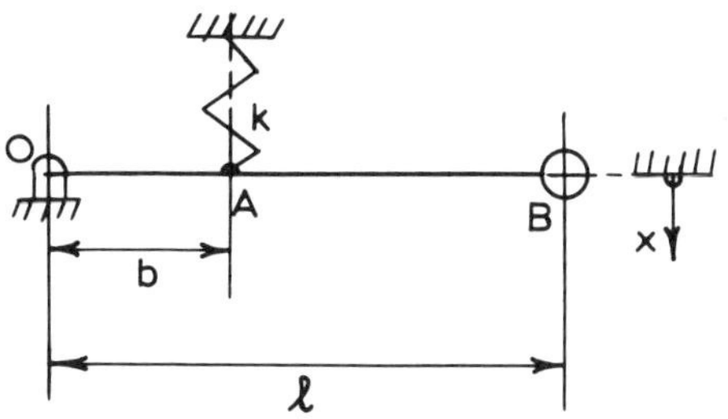

Fig. 9.2.

spring force is $P + k\Delta = (l/b)mg + kxb/l$. The force F acting on the mass from the bar is then found by taking moments about O:

$$Fl = \left(\frac{l}{b}\, mg + kx\,\frac{b}{l}\right) b \quad \text{or} \quad F = mg + kx(b/l)^2$$

The total restoring force on the mass is then

$$-(F - mg) = -k\left(\frac{b}{l}\right)^2 x$$

Newton's second law gives the equation of motion

$$m\ddot{x} = -k\left(\frac{b}{l}\right)^2 x \quad \text{or} \quad \ddot{x} + \frac{k}{m}\left(\frac{b}{l}\right)^2 x = 0$$

This is simple harmonic motion with circular frequency $\omega_0 = (b/l)\surd(k/m)$. If $b \to 0$, the frequency becomes very low; the maximum frequency is for $b = l$ when $\omega_0 = \surd(k/m)$.

The effective spring constant at the mass is evidently $k(b/l)^2$.

Example 9.2
In the system in Fig. 9.3 the body of mass m performs small vertical vibrations. The bar AB is rigid and massless, and C is the middle point of the bar. Determine the frequency of the vibrations.

Solution
In static equilibrium, the force in the middle spring is mg and that in the other two springs $mg/2$. If the mass is moved down a distance x from static equilibrium, the increase in the force in the middle spring is a force P, and the increase in the forces in the other two springs is $P/2$. If the displacements of points A and B are Δ_1 and Δ_2, the displacement of point C is $(\Delta_1 + \Delta_2/2$. We have $\Delta_1 = P/2k_1$ and $\Delta_2 = P/2k_2$, so that

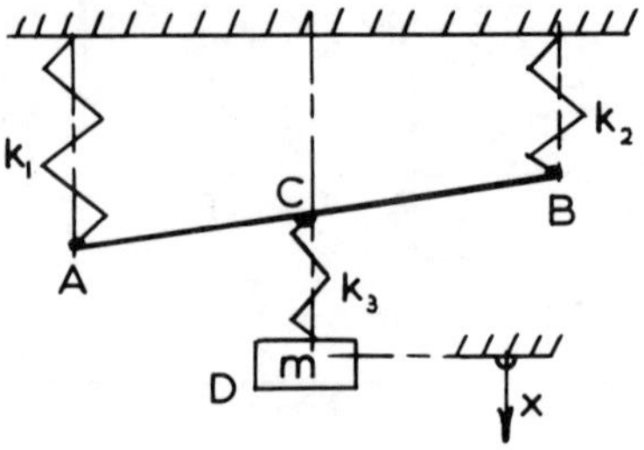

Fig. 9.3.

$$(\Delta_1+\Delta_2)\tfrac{1}{2} = \tfrac{1}{4}P\left(\frac{1}{k_1}+\frac{1}{k_2}\right)$$

The extension of the middle spring is $x - (\Delta_1+\Delta_2)$, and the force

$$P = k_3[x - \tfrac{1}{2}(\Delta_1+\Delta_2)] = k_3x - \tfrac{1}{4}Pk_3\left(\frac{1}{k_1}+\frac{1}{k_2}\right)$$

or

$$P = \frac{k_3x}{1+\frac{1}{4}k_3(1/k_1+1/k_2)}$$

When $x=1$, P is the total spring constant K, so that

$$K = \frac{4k_1k_2k_3}{4k_1k_2+k_3(k_1+k_2)}$$

The circular frequency is $\omega_0 = \sqrt{K/m}$.

9.2 FREE VIBRATIONS WITH VISCOUS DAMPING

In this case the equation of motion is obtained from (9.1), with $P_0 = 0$:

$$\ddot{x} + 2n\dot{x} + \omega_0^2 x = 0 \tag{9.3}$$

In solving this equation, we assume a solution of the form $x = \mathrm{e}^{st}$, where e is the base of the natural logarithms, e = 2.71828. . . ., t is the time and s is a constant that must be adjusted so that $x = \mathrm{e}^{st}$ satisfies equation (9.3).

Differentiating x gives $\dot{x} = s\mathrm{e}^{st}$ *and* $\ddot{x} = s^2\mathrm{e}^{st}$; substituting in (9.3), we find that s

must satisfy the equation $s^2 + 2ns + \omega_0^2 = 0$. This equation has the roots $s = -n \pm \sqrt{(n^2 - \omega_0^2)}$. Three possible cases occur for the roots s:

(a) $n^2 = \omega_0^2$
This means that the roots are $s_1 = s_2 = -n = -\omega_0$, and we have the solution $x = e^{-nt}$. It may be seen by direct substitution in (9.3), that a second solution is $x = te^{-nt}$. The solutions may be muliplied by an arbitrary constant, so a set of solutions is given by $x = C_1e^{-nt}$ and $x = C_2te^{-nt}$. The sum of two solutions is also a solution, so the *complete general solution* is

$$x = e^{-nt}(C_1 + C_2t)$$

The constants C_1 and C_2 may be determined in each particular case from the starting conditions $x = x_0$ and $\dot{x} = \dot{x}_0$ at $t = 0$; the result is

$$x = e^{-nt}[x_0 + (x_0\omega_0 + \dot{x}_0)t]$$

It is easy to show that for $t \to \infty$, $x \to 0$; the mass then creeps back to the equilibrium position after release.

If $\dot{x}_0$ is sufficiently large and directed towards equilibrium, it is possible for the mass to move through the equilibrium position; it will then creep back to equilibrium from the opposite side of the starting displacement. In no case will the curve $x = f(t)$ for the motion intercept the t-axis more than once.

The motion is thus an *aperiodic* motion and *not* a vibratory motion. The damping coefficient in this case is found from $n = c/2m = \omega_0$; therefore $\underline{c = 2m\omega_0}$. This value of the damping coefficient is called the *critical damping* $\underline{c_c = 2m\omega_0} = 2\sqrt{(km)}$, since this value of c is the *minimum value for aperiodic motion*. Introducing a *damping ratio* $d = c/c_c$, we have in this case $\underline{d = 1}$, and the system is called a *critically damped system*.

(b) $n^2 > \omega_0^2$
The roots of the characteristic equation are here

$$\left.\begin{matrix} s_1 \\ s_2 \end{matrix}\right\} = -n \pm \sqrt{(n^2 - \omega_0^2)}$$

These are *both real and negative*. A pair of solutions is $x = e^{s_1t}$ and $x = e^{s_2}t$. The general solution to equation (9.3) in this case is $x = C_1e^{s_1t} + C_2e^{s_2t}$, where C_1 and C_2 are arbitrary constants for the general solution; for any particular case, these constants may be found from the starting conditions $x = x_0$, $\dot{x} = \dot{x}_0$ at $t = 0$.

The motion is *aperiodic*, and the remarks concerning the motion for case (a) hold also in this case. The damping ratio is $d = c/c_c = c/2m\omega_0 = n/\omega_0$; since $n^2 > \omega_0^2$, we have in this $\underline{\text{case } d > 1}$. The system is called an *overdamped system*.

This type of motion is encountered in hydraulic door stops.

(c) $n^2 < \omega_0^2$
The roots of the characteristic equation are

$$\left.\begin{matrix} s_1 \\ s_2 \end{matrix}\right\} = -n \pm \sqrt{(n^2 - \omega_0^2)} = -n \pm i\sqrt{(\omega_0^2 - n^2)}$$

These are *conjugate complex roots*. Introducing the *real* constant $p_1 = \sqrt{(\omega_0^2 - n^2)}$, we have $s_1 = -n + p_1 i$ and $n_2 = -n - p_1 i$.

The general solution of equation (9.3) in this case is

$$x = C_1 e^{s_1 t} + C_2 e^{s_2 t} = e^{-nt}\,(C_1 e^{ip_1 t} + C_2 e^{-ip_1 t})$$

introducing Euler's formula, $e^{\pm i\theta} = \cos\theta \pm i \sin\theta$, we find

$$\begin{aligned} x &= e^{-nt}[C_1(\cos p_1 t + i \sin p_1 t) + C_2(\cos p_1 t - i \sin p_1 t)] \\ &= e^{-nt}[(C_1 + C_2)\cos p_1 t + (C_1 - C_2)\, i \sin p_1 t] \end{aligned}$$

Introducing new arbitrary constants $A_1 = C_1 + C_2$ and $A_2 = C_1 - C_2$, we may write this as

$$x = e^{-nt}(A_1 \cos p_1 t + A_2 i \sin p_1 t)$$

The solution is of the form $x = u + iv$; differentiating we find $\dot{x} = \dot{u} + i\dot{v}$ and $\ddot{x} = \ddot{u} + i\ddot{v}$. Substituting in equation (9.3), the result is

$$(\ddot{u} + 2n\dot{u} + \omega_0^2 u) + i(\ddot{v} + 2n\dot{v} + \omega_0^2 v) = 0$$

This expression is satisfied only if $\ddot{u} + 2n\dot{u} + \omega_0^2 u = 0$ and $\ddot{v} + 2n\dot{v} + \omega_0^2 v = 0$, which means that $x = u$ and $x = v$ satisfies equation (9.3), and a set of solutions is $x = e^{-nt}A_1 \cos p_1 t$ and $x = e^{-nt}A_2 \sin p_1 t$. *The general solution* of equation (9.3) is then in this case

$$x = e^{-nt}(A_1 \cos p_1 t + A_2 \sin p_1 t)$$

The constants A_1 and A_2 must be found from the starting conditions $x = x_0$, $\dot{x} = \dot{x}_0$ at $t = 0$; we find in this way the solution

$$x = e^{-nt}\left(x_0 \cos p_1 t + \frac{\dot{x}_0 + n x_0}{p_1} \sin p_1 t\right)$$

where $\omega_0 = \sqrt{(k/m)}$; $d = c/c_c = c/2m\omega_0 = n/\omega_0$, so that $\underline{d < 1}$ and

$$p_1 = \sqrt{(\omega_0^2 - n^2)} = \omega_0\sqrt{(1-d^2)}$$

This is called an *underdamped* case of motion. The solution may be simplified by writing

$$A \cos p_1 t + B \sin p_1 t = C \cos (p_1 t - \phi) = C \cos p_1 t \cos \phi + C \sin p_1 t \sin \phi$$

so that $C \cos \phi = A$, $C \sin \phi = B$, and $C = (A^2 + B^2)$ with $\tan \phi = B/A$. We find

$$C = \sqrt{\left[x_0^2 + \left(\frac{\dot{x}_0 + nx_0}{p_1}\right)^2\right]} \qquad \tan \phi = \frac{\dot{x}_0 + nx_0}{p_1 x_0}$$

$$x = e^{-nt} C \cos (p_1 t - \phi) \tag{9.4}$$

Fig. 9.4 shows a graph of this function. The curve is called a *damped sine curve*, and

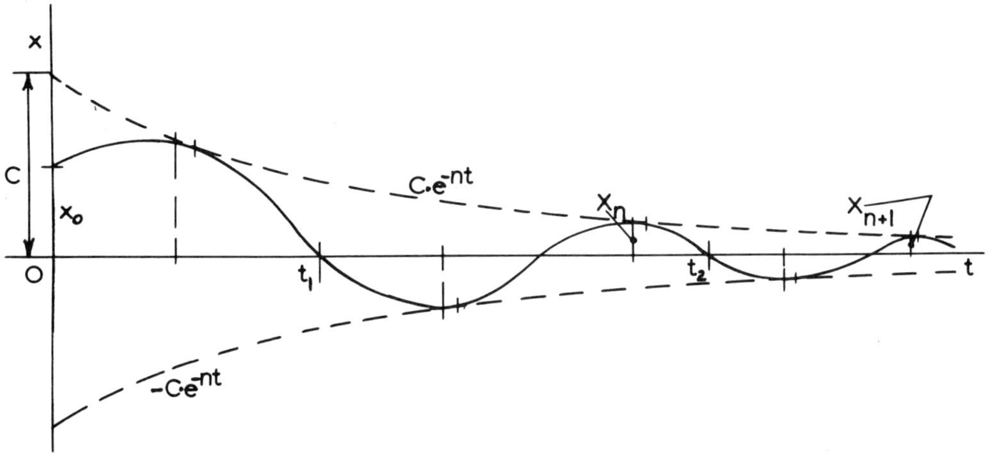

Fig. 9.4.

shows the motion of damped free vibrations. When $\cos (p_1 t - \phi) = \pm 1$, the curve touches the curves $x = \pm Ce^{-nt}$, which are exponentially decreasing as shown. The maximum displacements occur slightly before the tangential points. The points of intersection with the t-axis are found for $x = 0$, and at these points $\cos (p_1 t - \phi) = 0$.

The curve is not periodic in the usual sense of the word, since the displacements are diminishing; however, it is customary to talk about a 'period' as the time from $t = t_1$ to $t = t_2$ as shown. We find

$$p_1 t_1 - \phi = \frac{\pi}{2} \quad \text{and} \quad p_1 t_2 - \phi = \frac{5\pi}{2}$$

and the period is then $\tau = t_2 - t_1 = 2\pi/p_1$. The constant $p_1 = \omega_0\sqrt{(1-d^2)}$ is called the *damped natural circular frequency*. The *damped natural frequency* is

$$f = \frac{1}{\tau} = \frac{p_1}{2\pi} = \frac{\omega_0}{2\pi}\sqrt{(1-d^2)}$$

It may be seen that damping slows down the motion — the frequency has a smaller value than the undampened case, where $d = 0$.

The *rate of decay* of the motion may be defined as the ratio X_n/X_{n+1}. The maximum amplitudes occur approximately at those points where $\cos(p_1 t - \phi) = 1$, so that

$$X_n = Ce^{-nt_n} \quad X_{n+1} = Ce^{-n(t_n+\tau)} \quad X_n/X_{n+1} = e^{n\tau} = e^{\delta}$$

The constant δ defining the rate of decay is determined by

$$\delta = \ln\left(\frac{X_n}{X_{n+1}}\right) = n\tau = d\omega_0\tau$$

δ is called the *logarithmic decrement*. Since

$$\tau = \frac{2\pi}{p_1} = \frac{2\pi}{\omega_0\sqrt{(1-d^2)}}$$

we find

$$\delta = d\omega_0\tau = \frac{2\pi d}{\sqrt{(1-d^2)}} \tag{9.5}$$

If the damping is negligible, $d = c/c_c \to 0$ and $\delta \to 0$; for increasing damping, δ increases. For *small damping*, $d^2 \ll 1$, and we have *approximately* $\delta = 2\pi d$.

Example 9.3

The system in Fig. 9.1 is executing free damped vibrations ($P_0 = 0$) and $m = 7.66$ kg, $k = 52.5$ N/cm and $c = 0.438$ N s/cm. Determine c_c, δ, X_n/X_{n+1} and the damped natural frequency.

Given that the system is released from rest with $x_0 = 1$ cm, determine the approximate displacement after 10 cycles and the time elapsed.

Solution

$$\omega_0 = \sqrt{(k/m)} = \sqrt{(52.5 \times 100/7.66)} = 26.2 \text{ rad/s}$$

$$c_c = 2m\omega_0 = 2 \times 7.66 \times 26.2 = 401 \text{ N s/m} = 4.01 \text{ N s/cm}$$

$$d = c/c_c = 0.438/4.01 = 0.109.$$

The damping is about 11% of critical damping

$$\delta = \frac{2\pi d}{\sqrt{(1-d^2)}} = \frac{2\pi \times 0.109}{\sqrt{(1-0.109^2)}} = 0.690$$

$X_n/X_{n+1} = e^{\delta} = 1.994$; we have approximately $X_{n+1} = \frac{1}{2}X_n$. The damped natural frequency is

$$f_1 = \frac{p_1}{2\pi} = \frac{\omega_0}{2\pi}\sqrt{(1-d^2)} = \frac{26.2}{2\pi}\sqrt{(1-0.109^2)} = 4.14 \text{ cycles/second}$$

The undamped natural frequency is

$$f_0 = \frac{\omega_0}{2\pi} = 4.17 \text{ cycles/second}$$

Damping has lowered the frequency by less than 1% in this case.

If the system is released with $\dot{x}_0 = 0$ and $x_0 = 1$ cm at $t = 0$, we find that the maximum displacement after 10 cycles is approximately $(\frac{1}{2})^{10} = \frac{1}{1024} < \frac{1}{1000}$ cm. The time elapsed is $10/4.14 = 2.41$ s. The vibrations have clearly been damped out, for all practical purposes, after, say, 5 s.

Example 9.4

Fig. 9.5 shows a damped vibratory system. The system constants are $k = 87.5$ N/cm,

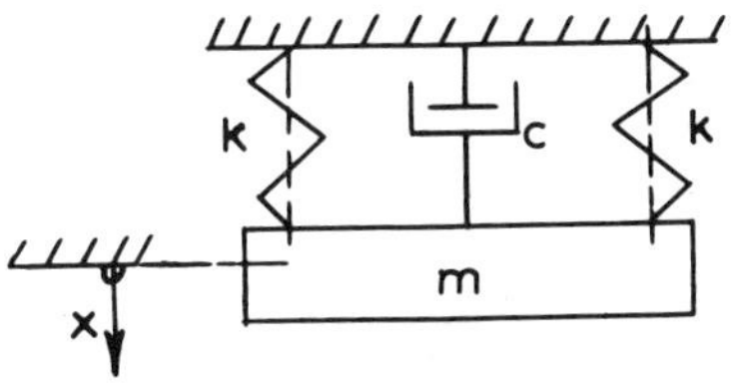

Fig. 9.5.

$m = 22.7$ kg and $c = 3.50$ N s/cm. The system is initially at rest, when the body of mass m is given an impact which starts it moving in the positive x direction with initial velocity $\dot{x}_0 = 12.7$ cm/s. Determine the frequency of the ensuring damped vibrations, the logarithmic decrement and the maximum displacement from the equilibrium position.

$$\omega_0 = \sqrt{(k/m)} = \sqrt{\left(\frac{2 \times 87.5 \times 100}{22.7}\right)} = 27.8 \text{ rad/s}$$

$$c_c = 2m\omega_0 = 2 \times 22.7 \times 27.8 = 1261 \text{ N s/m} = 12.61 \text{ N s/cm}$$

$$d = \frac{c}{c_c} = \frac{3.50}{12.61} = 0.277$$

$$f_1 = \frac{27.8}{2\pi} \sqrt{(1 - 0.277^2)} = 4.25 \text{ cycles/s}$$

$$\delta = \frac{2\pi d}{\sqrt{(1 - d^2)}} = 1.81$$

From the solution (9.4), $x = e^{-nt} C \cos(p_1 t - \phi)$, the maximum displacement occurs approximately when $\cos(p_1 t - \phi) = 1$ the first time; this happens at a time $t = t_1$, and $p_1 t_1 - \phi = 0$, or $t_1 = \phi/p_1$ and $x_{max} = e^{-nt_1} C$.

$$p_1 = \omega_0 \sqrt{(1 - d^2)} = 26.7 \text{ rad/s}$$

From (9.4)

$$\tan \phi = (\dot{x}_0 + nx_0)/p_1 x_0 \rightarrow \infty \quad \phi = \pi/2 \quad n = c/2m = 350/(2 \times 22.7) = 7.71 \text{ s}^{-1}$$

$$C = \dot{x}_0/p_1 = 12.70/26.7 = 0.476 \text{ cm} \quad nt_1 = n\phi/p_1 = 0.453$$

We have now $x_{max} = e^{-0.453} \times 0.476 = 0.303$ cm.

9.3 FORCED VIBRATIONS WITHOUT DAMPING

If we assume that there is no damping in the system of Fig. 9.1, we have $c = 0$, and the equation of motion is, from (9.1),

$$\ddot{x} + \omega_0^2 x = \frac{P_0}{m} \sin \omega t \tag{9.6}$$

The general solution to this non-homogenous equation is known to be the sum of the general solution to the *homogeneous equation* $\ddot{x} + \omega_0^2 x = 0$, and *a particular solution* to the *total equation* (9.6). The general solution to the homogeneous equation is

$x = A \cos \omega_0 t + B \sin \omega_0 t$, with $\omega_0 = \sqrt{(k/m)}$. The constants A and B are arbitrary. This solution has been discussed before under free undamped vibrations.

To find a particular solution to (9.6), we substitute a function of the same form as the right-hand function $(P_0/m) \sin \omega t$. Taking $x = x_0 \sin \omega t$, we have $\ddot{x} = -x_0\omega^2 \sin \omega t$; substituting in (9.6) and cancelling $\sin \omega t$, we find that $x = x_0 \sin \omega t$ is a solution to (9.6), if $x_0(\omega_0^2 - \omega^2) = P_0/m$, or

$$x_0 = \frac{P_0}{m(\omega_0^2 - \omega^2)} = \frac{P_0}{k - m\omega^2} = \frac{P_0}{k} \frac{1}{1 - (\omega/\omega_0)^2}$$

Introducing $x_s = P_0/k$, where x_s is *the static deflection* for a force P_0, we have

$$x = x_0 \sin \omega t = x_s \frac{1}{1 - (\omega/\omega_o)^2} \sin \omega t$$

The general solution to (9.6) is thus

$$x = (A \cos \omega_0 t + B \sin \omega_0 t) + \frac{x_s}{1 - (\omega/\omega_0)^2} \sin \omega t \tag{9.7}$$

The constants A and B are found from the starting conditions $x = x_0$, $\dot{x} = \dot{x}_0$ at $t = 0$.

The first two terms in (9.7) are the *free vibrations* of the system. In any practical case, some damping due to air resistance, dry friction and internal material damping is always present; the free vibration part of the solution is therefore soon damped out. This part of the solution is called the *transient state*. We are mainly interested in the last term of the solution (9.7), which is a vibratory motion with frequency ω equal to the forcing frequency, and a constant amplitude.

$$x_0 = x_s \frac{1}{1 - (\omega/\omega_0)^2}$$

This solution is called the *steady-state solution*. The factor

$$\left| \frac{x_0}{x_s} \right| = \frac{1}{|1 - (\omega/\omega_o)^2|}$$

is called *the magnification factor*, since it is the factor by which the static deflection x_s must be multiplied to give the maximum deflection x_0. We are usually only interested in the magnitude of the vibrations, any phase difference being of minor importance; only the numerical value of the magnification factor is therefore of interest.

The variation of the magnification factor with the ratio (ω/ω_0) is shown in Fig. 9.6 for $d = 0$. It will be seen that for $\omega/\omega_0 \to 0$, $x_0/x_s \to 1$. For $\omega/\omega_0 \to 1$, $|x_0/x_s| \to \infty$, and *this condition is called a resonance condition*. We say that the force is in resonance

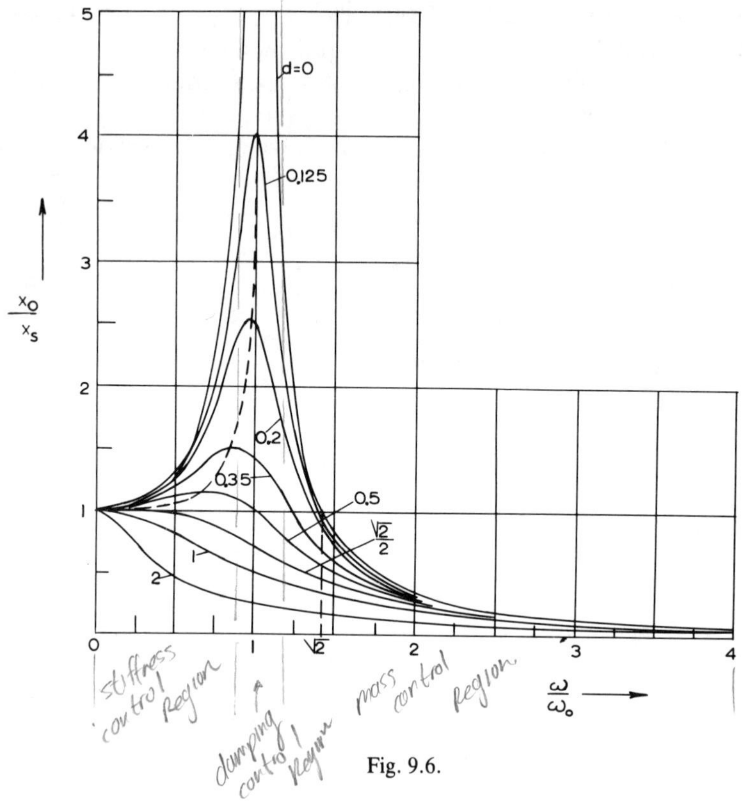

Fig. 9.6.

with the natural frequency of the system. For large values of (ω/ω_0), we find that $x_0/x_s \rightarrow 0$, which means that for large values of the forcing frequency compared to the natural frequency, the body is practically motionless.

Example 9.5

For the system in Fig. 9.5, the system constants are $k = 87.5$ N/cm, $m = 22.7$ kg and $c = 0$. The system is vibrated by an external force in the x direction: $P = 201 \sin(13t)$ N.

Determine the magnification factor and the steady-state motion of the mass.

Solution

$$\omega_0 = \sqrt{(k/m)} = \sqrt{(2 \times 87.5 \times 100/22.7)} = 27.8 \text{ rad/s}$$

The magnification factor is $1/[1 - (13/27.8)^2] = 1.28$.

$$x_s = \frac{P_0}{k} = \frac{201}{175} = 1.15 \text{ cm} \quad x_0 = 1.15 \times 1.28 = 1.47 \text{ cm}$$

The steady-state motion is $x = x_0 \sin \omega t = 1.47 \sin 13t$ cm.

9.4 FORCED VIBRATIONS WITH VISCOUS DAMPING

Turning finally to the complete system in Fig. 9.1, the equation of motion is, from (9.1),

$$\ddot{x} + 2n\dot{x} + \omega_0^2 x = \frac{P_0}{m} \sin \omega t$$

with $\omega_0^2 = k/m$ and $n = c/2m$.

The equation is a second-order non-homogenous ordinary differential equation with constant coefficients. The corresponding homogeneous equation is $\ddot{x} + 2n\dot{x} + \omega_0^2 x = 0$, with the solution obtained before:

$$x = e^{-nt} (A_1 \cos p_1 t + A_2 \sin p_1 t)$$

where $p_1 = \omega_0 \sqrt{(1 - d^2)}$, and A_1 and A_2 are arbitrary constants. This solution is for $c < c_c$; if $c > c_c$, the solution for these cases must be substituted instead.

It is known that the general solution to equation (9.1) is the sum of the general solution to the corresponding homogeneous equation and *a particular solution* to equation (9.1).

The forcing function is a sine function; to find a particular solution we assume a function of the same type as a solution. Taking $x = A \sin \omega t + B \cos \omega t$, where A and B are *constants* to be determined so that this is a solution to (9.1), we find by substitution in (9.1):

$$[A(\omega_0^2 - \omega^2) - 2n\omega B - P_0/m] \sin \omega t + [2n\omega A + (\omega_0^2 - \omega^2)B] \cos \omega t = 0$$

To satisfy this equation for all values of time t, the square brackets must vanish. This gives the equations

$$A(\omega_0^2 - \omega^2) - 2n\omega B = P_0/m$$
$$2n\omega A + (\omega_0^2 - \omega^2) B = 0$$

Substituting $\omega_0^2 = k/m$ and $n = c/2m$, we obtain the solution

$$A = \frac{P_0(k - m\omega^2)}{(k - m\omega^2)^2 + c^2\omega^2} \qquad B = \frac{-P_0 c\omega}{(k - m\omega^2)^2 + c^2\omega^2}$$

The general solution to (9.1) is thus (for $c < c_c$)

$$x = e^{-nt}(A_1 \cos p_1 t + A_2 \sin p_1 t) + (A \sin \omega t + B \cos \omega t)$$

The constants A_1 and A_2 must be determined *by the starting conditions* $x = x_0$, $\dot{x} = \dot{x}_0$ at $t = 0$, while A and B have the above values.

The solution thus consists of two parts, the first part being that of the free damped vibrations, the *transient state*, superimposed on the forced vibrations with frequency ω equal to the forcing frequency, the *steady state*.

As was shown before, the transient state soon vanishes because of the damping, and we are mainly interested in the steady-state vibrations

$$x = A \sin \omega t + B \cos \omega t$$

Writing this as

$$x = x_0 \sin (\omega t - \phi) = (x_0 \cos \phi) \sin \omega t - (x_0 \sin \phi) \cos \omega t$$

we find $x_0 \cos \phi = A$ and $x_0 \sin \phi = -B$, so that $x_0 = \sqrt{(A^2 + B^2)}$ and $\tan \phi = -B/A$. Substituting the expressions for A and B found before, we obtain

$$x_0 = \frac{P_0}{\sqrt{[(k - m\omega^2)^2 + c^2\omega^2]}} \quad \text{and} \quad \tan \phi = \frac{c\omega}{k - m\omega^2}$$

The steady-state motion is thus $x = x_0 \sin (\omega t - \phi)$, which is simple harmonic motion with frequency ω equal to the forcing frequency. The amplitude is x_0 and the phase angle ϕ determined by the above expression.

To obtain the formulae in a *dimensionless form*, more convenient for analysis, we substitute $\omega_0^2 = k/m$, $d = c/c_c$, $c_c = 2m\omega_0$ and $x_s = P_0/k$; the result is

$$\frac{x_0}{x_s} = \frac{1}{\sqrt{\{[1 - (\omega/\omega_0)^2]^2 + (2d\omega/\omega_0)^2\}}} \tag{9.8}$$

x_0/x_s is thus a function of (ω/ω_0) and d. Fig. 9.6 shows this relationship for various values of the damping ratio d. It may be seen that for small values of (ω/ω_0), $x_0 \simeq x_s$. When $(\omega/\omega_0) \to 1$, x_0/x_s increases to a value close to the maximum value, and this is approximately the *resonance condition*. When ω is large compared to ω_0, the mass is practically stationary; this is the case for ω/ω_0 greater than about 3.

The phase angle was determined by the expression $\tan \phi = c\omega/(k - m\omega^2)$; introducing $\omega_0^2 = k/m$, $d = c/c_c$ and $c_c = 2m\omega_0$, this may be given the *dimensionless form*

$$\tan \phi = \frac{2d\omega/\omega_0}{1 - (\omega/\omega_0)^2} \tag{9.9}$$

Fig. 9.7 shows the phase angle ϕ as a function of ω/ω_0 for various values of d. For $d = 0$, the phase angle is $\phi = 0$ for $(\omega/\omega_0) < 1$ and $\phi = 180°$ for $\omega/\omega_0 > 1$. There is a *discontinuous jump* at $\omega/\omega_0 = 1$. For all values of d the phase angle is $\phi = 90°$ at the condition $\omega/\omega_0 = 1$, which is the resonance condition for *small* damping.

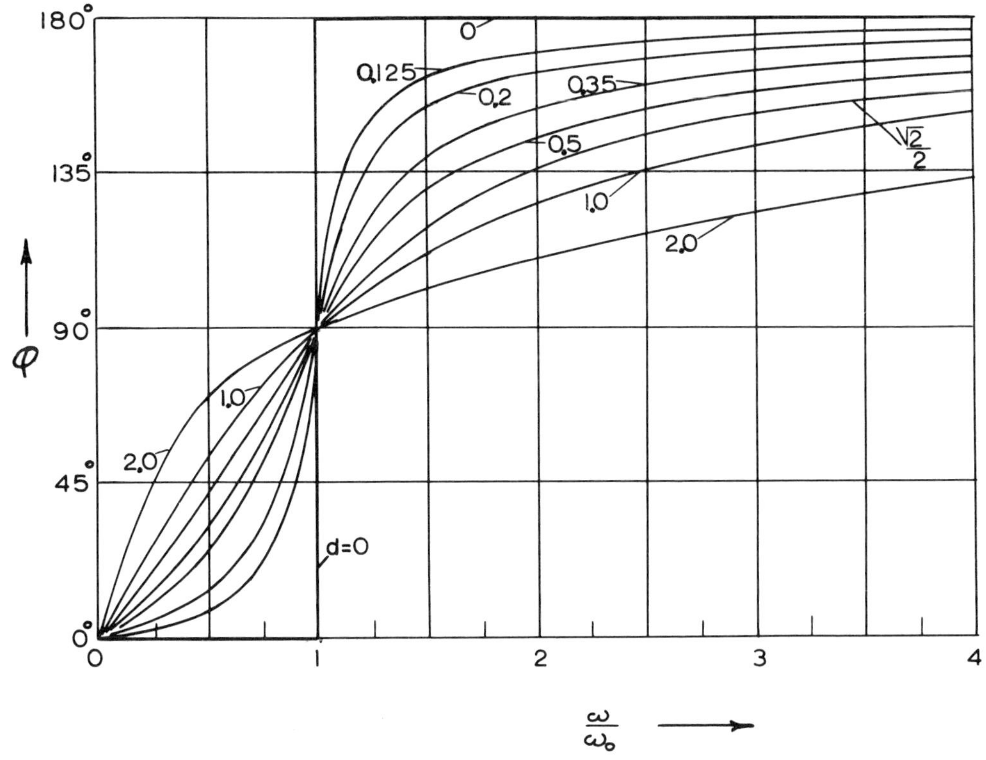

Fig. 9.7.

Example 9.6

The system in Fig. 9.1 has the system constants $k = 43.8$ N/cm, $m = 18.2$ kg, $c = 1.49$ N s/cm, $P_0 = 44.5$ N and $\omega = 15$ rad/s. Determine the steady-state motion of the mass.

Solution

$$x_s = \frac{P_0}{k} = \frac{44.5}{43.8} = 1.016 \text{ cm} \quad \omega_0 = \sqrt{\frac{4380}{18.2}} = 15.5 \text{ rad/s}$$

$$c_c = 2m\omega_0 = 2 \times 18.2 \times 15.5 = 564 \text{ N s/m} = 5.64 \text{ N s/cm}$$

$$d = \frac{c}{c_c} = \frac{1.49}{5.64} = 0.265 \quad (\omega/\omega_0) = \frac{15}{15.5} = 0.967$$

$$1 - (\omega/\omega_0)^2 = 0.065 \times 2d(\omega/\omega_0) = 0.513$$

From (9.8)

$$x_0 = \frac{1.016}{\sqrt{(0.065^2 + 0.513^2)}} = 1.835 \text{ cm}$$

From (9.9)

$$\tan\phi = \frac{0.513}{0.065} = 7.90 \quad \phi = 82.7° = 1.45 \text{ rad}$$

The steady-state motion is

$$x = x_0 \sin(\omega t - \phi) = 1.835 \sin(15t - 1.45) \text{ cm}$$

9.5 VIBRATION ISOLATION

The forces transmitted to the ground by a vibrating system are transmitted through the spring and damper, since these are the only connections to the ground.

The steady-state motion is $x = x_0 \sin(\omega t - \phi)$, with $\dot{x} = x_0\omega \cos(\omega t - \phi)$. The transmitted force P_T is

$$P_T = kx + c\dot{x} = kx_0 \sin(\omega t - \phi) + cx_0\omega \cos(\omega t - \phi)$$

The maximum value P_{T_m} of the transmitted force is

$$P_{T_m} = \sqrt{(k^2x_0^2 + c^2x_0^2\omega^2)} = x_0\sqrt{(k^2 + c^2\omega^2)}$$
$$= kx_0 \sqrt{\left[1 + \left(\frac{c\omega}{k}\right)^2\right]}$$

which is superimposed on the weight force. Substituting $c = dc_c = 2m\omega_0 d$ gives

$$P_{T_m} = kx_0 \sqrt{\left[1 + \left(\frac{2m\omega_0\omega d}{k}\right)^2\right]} = kx_0 \sqrt{\left[1 + \left(2d\frac{\omega}{\omega_0}\right)^2\right]}$$

If the springs were completely rigid, the maximum transmitted force would be P_0, so the ratio P_{T_m}/P_0 is a measure of the quality of the vibration isolation of the system. This dimensionless ratio is called the *transmissibility* of the system and is denoted t_r.

Introducing equation (9.8) for x_0, we have

$$t_r = \frac{P_{T_m}}{P_0} = \sqrt{\left[\frac{1 + (2d\omega/\omega_0)^2}{[1 - (\omega/\omega_0{}^2]^2 + (2d\omega/\omega_0)^2}\right]} \qquad (9.10)$$

In an ideal case, t_r should be zero; in practical cases it is made as small as possible.

If there is no damping, we have $d=0$ and $t_r = 1/|1-(\omega/\omega_0)^2)|$; for $\omega/\omega_0 \to 1$ we obtain $t_r \to \infty$. For $\omega/\omega_0 = \sqrt{2}$ the maximum force is transmitted in full for all values of d. The effect of damping in lowering the transmitted force may be seen for $\omega/\omega_0 = 1$ and $d = 0.2$, which gives $t_r = 2.693$.

The mounting springs should be designed with as small a value of k as possible so that $\omega_0 = \sqrt{k/m}$ is as small as possible, and ω/ω_0 large, in order to minimise the transmitted force.

Actually damping is detrimental for $\omega/\omega_0 > \sqrt{2}$, since a greater damping ratio gives a greater transmitted force in this region; for instance, if $\omega/\omega_0 = 3$, we find $t_r = 0.608$ for $d = 1$ and $t_r = 0.1931$ for $d = 0.2$, so the greater damping *in this region* leads to a greater transmitted force. However, some damping is necessary to avoid dangerous force amplitudes at the approximate resonance condition $\omega/\omega_0 = 1$, so that a *compromise* is necessary.

Example 9.7
A refrigerator unit weighs 290 N and it to be supported on three strings of spring constant k each.

Given that the unit operates at 580 rev/min., determine the value of the spring constant k if only 10% of the shaking force of the unit is to be transmitted to the support. Determine also the static deflection due to the weight of the unit when the unit is placed on the srings. Damping may be neglected.

Solution
From equation (9.10) we find, for $d = 0$,

$$t_r = \frac{1}{-[1-(\omega/\omega_0)^2]} = 0.10$$

from which $(\omega/\omega_0)^2 = 11$, and $\omega/\omega_0 = 3.32$. $\omega = \pi \times 580/30 = 60.7$ rad/s, so that $\omega_0 = \omega/3.32 = 18.3$ rad/s. Thus $K = m\omega_0^2 = (290/9.81) \times 18.3^2 = 9880$ N/m = 98.8 N/cm. Since $K = 3k$, we have $k = 98.8/3 = 32.9$ N/cm. The static deflection is

$$\Delta_s = W/K = 290/98.8 = 2.93 \text{ cm}$$

9.6 VIBRATING SUPPORT

In many practical cases, instruments have to be mounted on supports that are vibrating, as in the case of an aircraft instrument panel. The vibrations may cause malfunction of the instruments, and it becomes necessary to isolate the instruments as far as possible from the supports.

To investigate this situation, consider the system shown in Fig. 9.8. The displacement of the vibrating support is given by the function $y = y_0 \sin \omega t$, and the velocity is $\dot{y} = y_0\omega \cos \omega t$. The absolute motion of the body of mass m is given by the

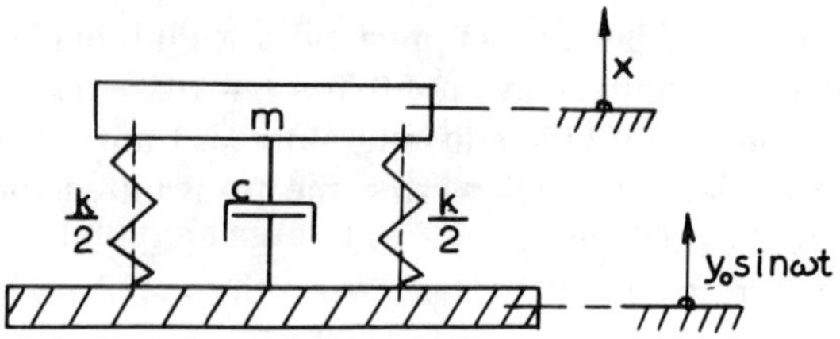

Fig. 9.8.

coordinate x as shown. The velocity of the damper piston relative to the damper cylinder is $\dot{x} - \dot{y}$, while the extension of the spring is $x - y$. The equation of motion of the body of mass m is thus

$$m\ddot{x} + c(\dot{x} - \dot{y}) + k(x - y) = 0$$

Substituting the expressions for y and $\dot{y}$, we obtain

$$\begin{aligned} m\ddot{x} + c\dot{x} + kx &= cy_0\omega \cos \omega t + ky \sin \omega t \\ &= \sqrt{(c^2y_0^2\omega^2 + k^2y_0^2)} \sin (\omega t - \beta) \\ &= y_0\sqrt{(k^2 + c^2\omega^2)} \sin (\omega t - \beta) = P_0 \sin (\omega t - \beta) \end{aligned}$$

where $P_0 = y_0\sqrt{(k^2 + c^2\omega^2)}$.

For a given frequency ω, the vibrating support then acts as a sinusoidal exciting force of frequency ω and force amplitude P_0. The phase angle β is of no particular interest in this case.

The equation of motion is of the same type as equation (9.1), and the same solution therefore applies. We find

$$\begin{aligned} P_0 &= y_0\sqrt{(k^2 + c^2\omega^2)} = y_0k \sqrt{\left[1 + \left(\frac{c\omega}{k}\right)^2\right]} \\ &= y_0k \sqrt{\left[1 + \left(\frac{2m\omega_0\omega d}{k}\right)^2\right]} = y_0k \sqrt{\left[1 + \left(\frac{2d\omega}{\omega_0}\right)^2\right]} \end{aligned}$$

From the expression (9.8) for the steady-state amplitude x_0, we find the amplitude ratio:

$$\left(\frac{x_0}{y_0}\right) = \sqrt{\left(\frac{1 + (2d\omega/\omega_0)^2}{[1 - (\omega/\omega_0)^2]^2 + (2d\omega/\omega_0)^2}\right)} \tag{9.11}$$

This is exactly the same as expression (9.10) for the transmissibility t_r.

The ratio x_0/y_0 shows the effectiveness of the vibration mounting in reducing the amplitude of the body of mass m. For any value of $\omega/\omega_0 > \sqrt{2}$, the amplitude of the body will be less than that of the vibrating support.

Damping is again detrimental in the region $\omega/\omega_0 > \sqrt{2}$, but a compromise must be made, since *damping is essential* to prevent large amplitudes in passing through the *approximate resonance condition* $\omega/\omega_0 = 1$.

Example 9.8

An aircraft instrument board, including the instruments, weighs 178 N, and is supported on four rubber mounts with spring constant 224 N/cm each. The total damping ratio is $d = 0.10$. The rubber supports are mounted on a vibrating surface which has a frequency equal to the engine revolutions.

Determine the percentage of the vibrating motion of the supporting surface that is transmitted to the instrument panel at an engine revolution of (a) 2200 rev/min, (b) 1500 rev/min. (c) Determine the percentage transmissibility for (a) and (b) if $d = 0$.

Solution

The percentage of the motion transmitted is, from equation (9.11),

$$\frac{x_0}{y_0} \times 100\% = 100\sqrt{\left(\frac{1 + (2d\omega/\omega_0)^2}{[1 - (\omega/\omega_0)^2]^2 + (2d\omega/\omega_0)^2}\right)}$$

(a) $d = 0.10$, $\omega = (\pi/30) \times 2200 = 230$ rad/s.

$$\omega_0 = \sqrt{(K/m)},\ K = 4k = 4 \times 224 = 896 \text{ N/cm} = 89\ 600 \text{ N/m}$$

$$m = 178/9.81 \text{ kg} \quad \omega_0 = \sqrt{(89600 \times 9.81/178)} = 70.2 \text{ rad/s}$$

$$\omega/\omega_0 = 230/70.2 = 3.28 \quad 1 - (\omega/\omega_0)^2 = -9.74$$

$$2d\omega/\omega_0 = 0.656$$

$$(x_0/y_0) \times 100\% = 100\sqrt{[1.430/(9.74^2 + 0.430)]} = \underline{12.3\%}$$

(b) $\omega = \pi/30 \times 1500 = 157$ rad/s $\quad \omega_0 = 70.2$ rad/s

$$\omega/\omega_0 = 2.24 \quad 2d\omega/\omega_0 = 0.448$$

$$(x_0/y_0) \times 100\% = 100\sqrt{[1.20/(4.01^2 + 0.20)]} = \underline{27.3\%}$$

(c) $d = 0$; *for part* (*a*),

$$(x_0/y_0) \times 100\% = 100/|1 - (\omega/\omega_0)^2| = 100/9.74 = \underline{10.3\%}$$

For question (*b*):

$$(x_0/y_0) \times 100\% = 100/4.01 = \underline{24.9\%}$$

In both cases the transmitted motion is somewhat *smaller* for $d=0$, since $\omega/\omega_0 > \sqrt{2}$ in both cases.

9.7 FORCED VIBRATIONS DUE TO A ROTATING UNBALANCE

Fig. 9.9 shows a body of total mass m which is supported on springs with total spring

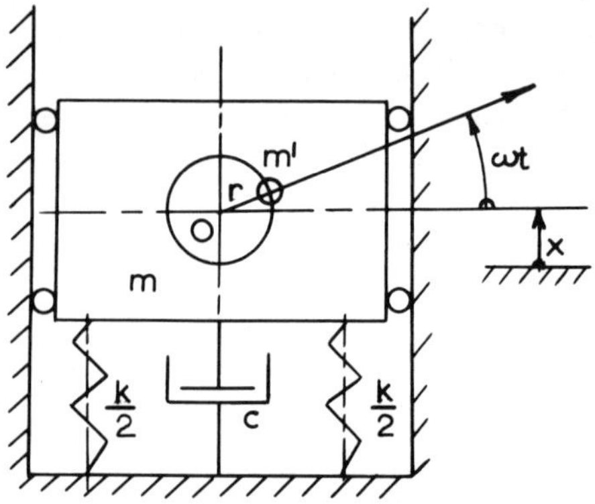

Fig. 9.9.

constant k. The motion is damped by a dash-pot of damping coefficient c. The body is guided to move in the vertical direction only. An unbalance of magnitude m^1r is rotating at a constant angular velocity ω as shown.

Assuming that the vertical motion is small, the changes in the unbalanced force due to the motion of the centre O may be ignored. The force on the unbalance is $m^1r\omega^2$ directed towards O and its acts therefore as a rotating exciting force of constant force amplitude $m^1r\omega^2$ on the body as shown. Resolving the force in a vertical and horizontal component, the vertical exciting force is $m^1r\omega^2 \sin \omega t$.

The equation of motion of the body is thus

$$m\ddot{x} + c\dot{x} + kx = (m^1r\omega^2) \sin \omega t$$

This is of the same form as equation (9.1) if we take $P_0 = m^1r\omega^2$, and the solutions to equation (9.1) may be applied directly to the case of a rotating unbalance.

Considering the *steady-state vibrations* only, the solution is

$$x = x_0 \sin (\omega t - \phi)$$

with

$$x_0 = \frac{P_0}{\sqrt{[(k - m\omega^2)^2 + c^2\omega^2]}}$$

as found before.

as found before.

Substituting $P_0 = m^1 r\omega^2$, $\omega_0^2 = k/m$, $d = c/c_c$ and $c_c = 2m\omega_0$, we obtain the solution in *dimensionless form*:

$$\left(\frac{mx_0}{m'r}\right) = \frac{(\omega/\omega_0)^2}{\surd[1-(\omega/\omega_0)^2]^2 + (2d\omega/\omega_0)^2} \tag{9.12}$$

If there is no damping, $d = 0$ and

$$\left(\frac{mx_0}{m'r}\right) = \frac{(\omega/\omega_0)^2}{|1-(\omega/\omega_0)^2|}$$

for the resonance condition $\omega/\omega_0 \to 1$ we obtain $mx_0/m'r \to \infty$. If $\omega/\omega_0 = 1$ and $d = 0.2$, we find from (9.12) that $mx_0/m'r = 2.50$, which shows the effect of damping in lowering the amplitude x_0 of the mass when the system goes through resonance. If $\omega \gg \omega_0$, or $\omega/\omega_0 \to \infty$, we find $mx_0/m'r \to 1$, so that $x_0 \to m^1 r/m$ for all values of damping, with x_0 a small fraction of r.

For larger values of ω/ω_0, say $\omega/\omega_0 > 5$, the effect of damping is negligible; however, damping is essential for passing through resonance safely.

The expression for the phase angle ϕ is the same as for forced vibrations:

$$\tan\phi = \frac{c\omega}{k - m\omega^2} = \frac{2d\omega/\omega_0}{1-(\omega/\omega_0)^2}$$

For a rotating unbalance system, the transmissibility

$$t_r = \frac{P_{T_m}}{P_0} = \frac{kx_0\surd[1+(2d\omega/\omega_0)^2]}{m'r\omega^2}$$

Substituting x_0 from equation (9.12) gives the result

$$t_r = \sqrt{\left[\frac{1+(2d\omega/\omega_0)^2}{[1-(\omega/\omega_0)^2]^2 + (2d\omega/\omega_0)^2}\right]}$$

This expression is exactly the same as (9.10).

Example 9.9

Four equal vibration mounts are used to support a machine which has a total weight of 1985 N. The machine has a rotating unbalance of 3.46 kg cm, and runs at 600 rev/min. The total damping is 30% of the critical damping.

Given that 20% of the exciting force is transmitted to the foundation, determine

(a) the spring constant of each mount, (b) the deflection of the springs when the machine is placed on them statically, (c) the maximum transmitted vibratory force, (d) the maximum amplitude of vibration.

The problem may be treated as a system with one degree of freedom, and only vertical translational motion considered.

Solution

(a) The transmissibility is

$$t_r = \sqrt{\left[\frac{1+(2d\omega/\omega_0)^2}{[1-(\omega/\omega_0)^2]^2+(2d\omega/\omega_0)^2}\right]} = 0.20$$

Introducing $x = (\omega/\omega_0)^2$ and $d = c/c_c = 0.30$, we obtain

$$\frac{1+0.36x}{1-2x+x^2+0.36x} = 0.04 \quad \text{or} \quad x^2 - 10.65 - 24 = 0$$

Hence

$$x = 5.32 + \sqrt{(28.3+24)} = 12.56$$

$$\omega = (\pi/30) \times 600 = 62.8 \text{ rad/s} \quad \omega_0^2 = \omega^2/12.56 = 314$$

$$K = m\omega^2 = (1985/9.81) \times 314 = 63\ 500 \text{ N/m} = 635 \text{ N/cm}$$

For each mount, $k = K/4 = 158.6$ N/cm.

(b) The static deflection $\Delta_s = W/K = 1985/635 = 3.12$ cm.

(c) The maximum transmitted vibratory force is found from $t_r = P_{T_m}/P_0 = 0.20$.

$$P_0 = m^1 r\omega^2 = \frac{3.46 \times 1}{100} \times 62.8^2 = 136.7 \text{ N}$$

$P_{T_m} = 0.20\, P_0 = 27.34$ N, which is superimposed on the weight force 1985 N.

(d) The maximum vibration amplitude x_0 is determined from $P_{T_m} = kx_0\sqrt{[1 + (2d\omega/\omega_0)^2]}$.

$$(2d\omega/\omega_0)^2 = 0.36 \times 12.56 = 4.5216$$

$$x_0 = 27.34/635\ \sqrt{5.5216} = \underline{0.018 \text{ cm}}$$

x_0 may also be determined directly from (9.12), which gives

$$\frac{mx_0}{m^1r}=\frac{x}{\sqrt{(1-x)^2+4.5216)}}=1.0686$$

so that $x_0 = 1.0686 \times 3.46 \times 9.81/1985 = 0.018$ cm.

PROBLEMS

9.1 A spherical ball of radius R floats half submerged in water. Supposing that the ball is depressed slightly and released, write the differential equation of motion and determine the period of vibration. If $R = 0.61$ m, what is the numerical value of the period?

9.2 Determine the equation for the period of small oscillations of the system shown in Fig. 9.10.

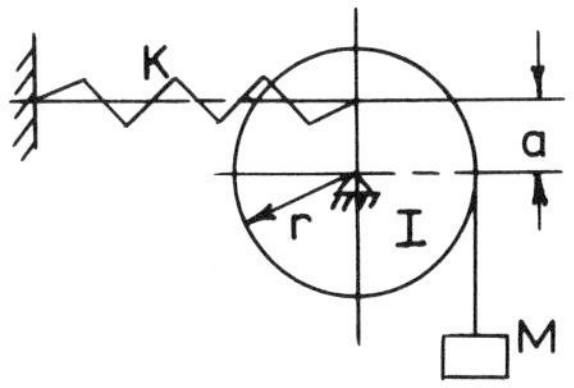

Fig. 9.10.

9.3 Find the frequency of small vibrations of the inverted pendulum in Fig. 9.11.

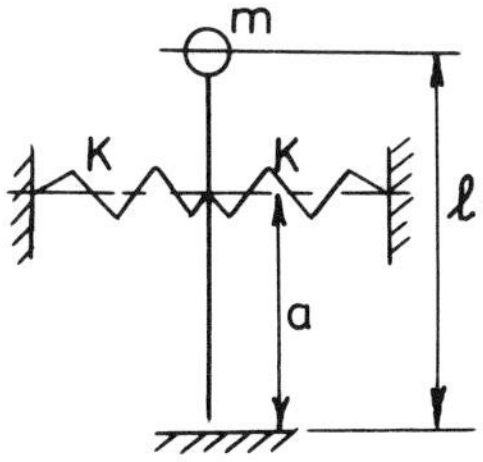

Fig. 9.11.

Assume that K is sufficiently large, so that the pendulum is stable.

9.4 Determine the damped natural frequency and the critical damping coefficient for the system shown in Fig. 9.12.

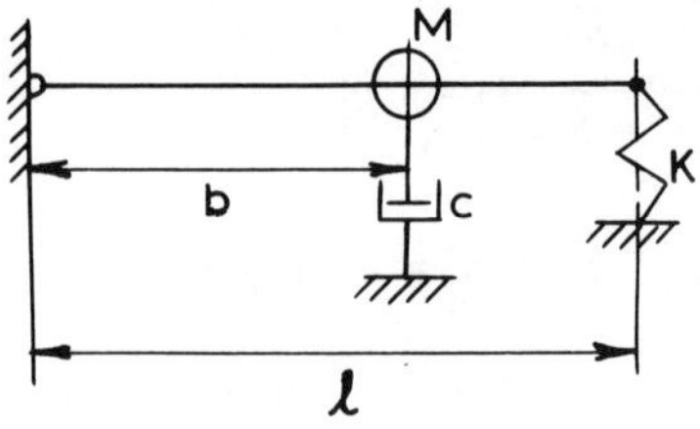

Fig. 9.12.

9.5 An undamped spring–mass system, which under gravity has a static deflection of 2.54 cm, is acted upon by a sinusoidal exciting force which has a frequency of 4 cycles/second. Find the damping ratio if the amplitude of the steady-state forced vibrations is reduced to half of the amplitude of the undamped system.

9.6 A mass of 0.980 kg is attached to the end of a spring with a stiffness of 7 N/cm. Determine the critical damping coefficient.

Given that the damping coefficient $c = 0.252$ N s/cm, determine the damped natural frequency and compare it with the natural frequency of the undamped system.

9.7 An instrument of mass 9.07 kg is to be spring mounted on a vibrating surface which has a sinusoidal motion of amplitude 0.397 mm and frequency 60 cycles/second. If the instrument is mounted rigidly on the surface, what is the maximum force to which it is subjected? Find the spring constant for the support which will limit the maximum acceleration of the instrument to $g/2$. Assume that negligible damping forces have caused the transient vibrations to die out.

9.8 A machine having a total weight of 89 000 N has an unbalance such that it is subjected to a force of amplitude 22 200 N at a frequency of 600 cycles/minute. Find the spring constant for the supporting springs if the maximum force transmitted to the foundation is 2220 N. Assume that damping may be neglected.

Answers to problems

CHAPTER 1

1.1 $x = gt/n + g/2n^2$.

1.2 $t = \dfrac{s}{V} + \dfrac{V}{2}\left(\dfrac{1}{a_1} + \dfrac{1}{a_2}\right)$, $t_{\min} = \dfrac{s}{V} + \dfrac{V}{a}$.

1.3 (a) $\mathbf{a} = 2\mathbf{i} + 6\mathbf{j}$, 6.32 m/s²; (b) $y^2 = x^3$.

1.4 $\mathbf{a} = \mathbf{a}_n + \mathbf{a}_t = 8\downarrow + 6\leftarrow$; 10 m/s².

1.5 $\mathbf{V} = 2\mathbf{r}_1 + 6\boldsymbol{\phi}_1$; 6.32 m/s; $\mathbf{a} = -18\mathbf{r}_1 + 14\boldsymbol{\phi}_1$; 22.8 m/s².

1.6 $\omega = 1.305$ rad/s↻, $\dot{\omega} = 1.430$ rad/s²↻.

1.7 3 m/s; 11.46 m, $-\pi/2$ m/s².

CHAPTER 2

2.1 $S = \dfrac{M_1 g(\sin\alpha + \mu\cos\alpha + 2)}{1 + 4M_1/M_2}$; 932 N.

2.2 8/5; 6/5; −38/5.

2.3 $f = \dfrac{1}{2\pi}\sqrt{\left(\dfrac{T(a+b)}{m\,ab}\right)}$.

2.4 $R = 15.95$ N.

2.5 75.8 m; 10 230 N/m.

2.6 2.38 km/s; 11.2 km/s.

2.7 $\Delta = 3$ cm; 2.8 9 blows.

CHAPTER 3

3.1 $m\ddot{x} - mx\dot{\theta}^2 + mg(1 - \cos\theta) + K(x - x_0) = 0$;
$mx^2\ddot{\theta} + 2mx\dot{\theta}\dot{x} + mgx \sin\theta = 0$.

3.2 $\sin\phi\ddot{\theta} + 2\dot{\theta}\dot{\phi}\cos\phi = 0$; $\ddot{\phi} - \dot{\theta}^2 \sin\phi \cos\phi + (g/l)\sin\phi = 0$.
$\ddot{\phi} + (g/l)\sin\phi = 0$, simple pendulum; $\omega^2 = g/(l\cos\alpha)$ for $\phi = \alpha$, conical pendulum.

3.3 $m\ddot{x} + (AE/l^3)x^3 = 0$; $\ddot{x} + (2S/ml)x = 0$.

3.4 (a) $T = \frac{1}{2}m(l_0 + x)^2\dot{\theta}^2 + \frac{1}{2}m\dot{x}^2$;
$V = (\frac{1}{2}Kx^2 + mgx) - mg[(l_0 + x)\cos\theta - l_0]$;
(b) $\ddot{x} + (c/m)\dot{x} + (K/m)x - (l_0 + x)\dot{\theta}^2 + g(1 - \cos\theta) = 0$;
$(l_0 + x)\ddot{\theta} + 2\dot{x}\dot{\theta} + g\sin\theta = 0$.

3.5 $c\dot{x}_1 + (K_1 + K_2)x_1 - K_2x_2 = 0$
$m\ddot{x}_2 + K_2x_2 - K_2x_1 = F_0\cos\omega t$.

3.6 $\ddot{\theta} + \frac{1}{2}\ddot{\phi}\cos(\phi - \theta) - \frac{1}{2}\dot{\phi}\sin(\phi - \theta) + (g/l)\sin\theta = 0$;
$\ddot{\phi} + \ddot{\theta}\cos(\phi - \theta) + \dot{\theta}^2\sin(\phi - \theta) + (g/l)\sin\phi = 0$.

CHAPTER 4

4.1 0.5 m/s ↓ ; 2.52 rad/s ↺ .

4.2 $-5.41\mathbf{i} - 9.39\mathbf{j}$; 10.8 m/s; $4.78\mathbf{i} - 8.28\mathbf{j}$; 9.56 m/s.

4.3 0.238 m/s ↘ ; 2.34 rad/s ↺ .

4.4 2.05 m/s at 188°, 12.8 m/s² at 333.8°; 8.75 rad/s) ↺ , 59 rad/s² ↻ .

4.5 3.60 m/s at 9°; 70 m/s² at 120°; 11.4 rad/s ↺ ; 632 rad/s² ↺ .

4.6 $\mathbf{V} = 3\mathbf{i} + 18.86\mathbf{j}$; 19.1 m/s; $\mathbf{A} = -18.86\mathbf{i} + 6\mathbf{j}$; 19.8 m/s^2.

4.7 $13.28\mathbf{j}$ m/s; $1064\mathbf{i}$ m/s^2.

4.8 $3000\mathbf{i} + 9.43\mathbf{j}$ m/s; $-5.92\mathbf{i} + 3770\mathbf{j}$ m/s^2.

4.9 0.28 m/s at 342°; 4.7 m/s^2 at 309°; 0.40 m/s at 297.5°;
3.43 m/s^2 at 302.5°; 4.65 rad/s ↻; 6.90 rad/s^2 ↺.

CHAPTER 5

5.1 11.83 mm from flange outer face; (1) 8.83×10^4 mm^4.
(b) 2.27×10^4 mm^4.

5.2 $-\frac{1}{72}a^2b^2$; $\frac{1}{24}a^2b^2$.

5.3 $\phi = 18.5°$; $I_{x_1} = 1211 \times 10^4$ mm^4; $I_{y_1} = 85 \times 10^4$ mm^4.

5.4 $\frac{5}{3}b^2M$; $\frac{3}{2}r^2M$.

5.5 $I_x = \frac{1}{3}M(b^2 + d^2)$.

CHAPTER 6

6.1 $\ddot{\theta} + (g/l)\sin\alpha\sin\theta = 0$; $\dfrac{1}{2\pi}\sqrt{\left(\dfrac{g}{l}\sin\alpha\right)}$.

6.2 2206 N m.

6.3 $\dfrac{r_1}{r_2}\dfrac{I_2\omega_1}{\mu P}$; $\dfrac{I_1I_2r_1\omega_1}{\mu P(I_1r_2^2 + I_2r_1^2)}$; $\dfrac{I_1\omega_1r_2^2}{I_1r_2^2 + I_2r_1^2}$.

6.4 (a) $A_x = -B_x = (Ml^2/24L)\dot{\omega}\sin 2\beta$;
$\left.\begin{matrix}A_Y\\B_Y\end{matrix}\right\} = \frac{1}{2}M(g \pm (l^2\omega^2/12L)\sin 2\beta)$; $A_Z = B_Z = 0$,
(b) $T = \frac{1}{12}Ml^2\dot{\omega}\sin^2\beta$.

6.5 73.2 rad/s^2; $A_X = -B_X = -123.8$ N; $A_Y = -33.1$ N,
$B_Y = -78$ N, $A_Z = B_Z = 0$.

6.6 (a) $A_Z + B_Z = 0$, $A_X + B_X = 0$, $A_Y + B_Y - Mg = 0$.
(b) $I_{XZ} = 0$, $I_{YZ} = \frac{1}{2}Ma^2$, $I_Z = \frac{1}{3}Ma^2$.
(c) $\dot{\omega} = 3T/Ma^2$;

$$A_X = -B_X = -\frac{T}{2a}, \left.\begin{matrix} A_Y \\ B_Y \end{matrix}\right\} = \frac{M}{2}\left(g + \frac{-a\omega^2}{3}\right); \; A_Z = B_Z = 0.$$

6.7 $M_L = 154$ g, $M_R = 52.5$ g, both in 4th quadrant, angles to horizontal 83.1° and 50.5°.

CHAPTER 7

7.1 $8M_1g/(8M_1 + 3M_2)$; $3M_1M_2/(8M_1 + 3M_2)g$.

7.2 (a) $\dfrac{P(\cos\alpha - r/R)}{M[1 + (r_C/R)^2]}$; (b) α limited to 60°, motion to right;
(c) 1.612 m/s²; (d) 0.355.

7.3 29.6°.

7.4 1.37 s; 1.46 s.

7.5 $\frac{1}{4}V(1 + \cos\alpha)^2$.

7.6 16 rad/s, 14.2 m/s at 338.7°.

7.7 $\ddot{x}_C = -\ddot{x} = \dfrac{g\sin\alpha\cos\alpha}{3 - \cos^2\alpha}$; $\ddot{y}_C \downarrow = 2g/(3 + 2\cot^2\alpha)$.

7.8 (a) $(M + m)\ddot{x} + m(R - r)\dot{\theta}^2\cos\theta + m(R - r)\ddot{\theta}\sin\theta + Kx = 0$,
$\frac{3}{2}(R - r)\ddot{\theta} + \ddot{x}\sin\theta + g\sin\theta = 0$;
(c) $\sqrt{[\frac{2}{3}g/(R - r)]}$; (d) $\sqrt{[K/(M + m)]}$.

7.9 $(M + m)\ddot{x} - ml\dot{\theta}^2\sin\theta + ml\cos\theta\,\ddot{\theta} + Kx = P(t)$;
$ml^2\ddot{\theta} + ml\cos\theta\,\ddot{x} + mgl\sin\theta = 0$; S.H.M.

CHAPTER 8

8.1 Left bearing 1215 N, right bearing 1455 N, thrust bearing 332 N.
8.2 28 600 N m in a horizontal plane, anti-clockwise seen from above for bow moving down.

8.3 5410 N m twisting AC clockwise seen from above.

8.4 $A_X = B_X = 0$; $A_Y = -448$ N, $B_Y = 0$; $A_Z = -245.7$ N, $B_Z = 314.4$ N.

CHAPTER 9

9.1 1.28 S.

9.2 $2\pi \sqrt{\left(\frac{I + Mr^2}{Ka^2}\right)}$.

9.3 $\frac{1}{2\pi l} \sqrt{\left(\frac{2Ka^2}{m} - gl\right)}$.

9.4 $\frac{1}{2\pi} \sqrt{\left[\frac{K}{M}\left(\frac{l}{b}\right)^2 - \left(\frac{c}{2M}\right)^2\right]}$; $2\frac{l}{b}\sqrt{(KM)}$.

9.5 0.428.

9.6 0.505 Ns/cm; 3.82 cycles/s, 4.41 cycles/s.

9.7 512 N; 1030 N/cm.

9.8 39 800 N/cm.

Appendix
SI units in mechanics

DEFINITIONS

Basic SI units

Metre (m); unit of length. The metre is defined as the distance light travels in a vacuum in $(2.997925\times10^8)^{-1}$ seconds.
Kilogram (kg); unit of mass. The kilogram is represented by the mass of the international prototype kilogram.
The international prototype is in the custody of the Bureau International des Poids et Mésures (BIPM), Sèvres, near Paris.
Second(s); unit of time interval. The second is defined as the time interval occupied by 9192 631 770 cycles of vibration of the caesium — 133 atom.

Derived SI units having special names

Hertz (Hz); unit of frequency. The number of repetitions of a regular occurrence in one second.
Newton (N); unit of force. That force which, applied to a mass of 1 kilogram, gives it an acceleration of 1 metre per second per second.
Pascal (Pa); unit of pressure. The pressure produced by a force of 1 newton applied, uniformly distributed, over an area of 1 square metre.
Joule (J); unit of energy, including work and quantity of heat. The work done when the point of application of a force of 1 newton is displaced through a distance of 1 metre in the direction of the force.
Watt (W); unit of power. The watt is equal to 1 joule per second.
Sttandard gravity or standard acceleration. Denoted by g=9·80665 m/s^2; 9·81 for practical work.

LIST OF SI UNITS

Plane angle	rad (radian)
solid angle	sr (steradian)

length	m (metre)
area	m^2
volume	m^3
time	s (second)
angular velocity	rad/s
velocity	m/s
frequency	Hz (hertz)
rotational frequency	l/s
mass	kg (kilogram)
density (mass density)	kg/m^3
momentum	kg m/s
momentum of momentum, angular momentum	kg m^2/s
moment of inertia	kg m^2
force	N (newton)
moment of force	N m
pressure and stress	N/m^2
viscosity (dynamic)	N s/m^2
surface tension	N/m
energy, work	J (joule)
power	W (watt)

Index